Principles of Biostatistics

Principles of Biostatistics

Marcello Pagano

Harvard School of Public Health

Kimberlee Gauvreau

Harvard School of Public Health

Duxbury Press
An Imprint of Wadsworth Publishing Company
Belmont, California

Sponsoring Editors: *Michael J. Sugarman* and *Alex Kugushev*
Marketing Representative: *John Moroney*
Editorial Assistant: *Carol Ann Benedict*
Production Coordinator: *Marlene Thom*
Manuscript Editor: *Stacey C. Sawyer*
Interior Design: *Katherine Minerva*
Cover Design: *Susan Haberkorn*
Art Coordinator: *Lisa Torri*
Interior Illustration: *Gloria Langer*
Typesetting: *SuperScript Typography*
Printing and Binding: *R. R. Donnelley & Sons*

This book is printed on acid-free recycled paper.

1 2 3 4 5 6 7 8 9 10—97 96 95 94 93

Library of Congress Cataloging in Publication Data

Pagano, Marcello, [date–]
Principles of biostatistics / Marcello Pagano, Kimberlee Gauvreau.
p. cm.
Includes bibliographical references and index.
ISBN 0-534-14064-5
1. Biometry I. Gauvreau, Kimberlee, [date–]. II. Title.
QH323.5.P34 1992
574'.01'5195–dc20 92-32841
CIP

This book is dedicated with love to
Phyllis, John-Paul, Marisa, Loris, Alice, and Lilian.
David, Lois, Karen, and Tami.

Preface

This book was written for students of the health sciences and serves as an introduction to the study of biostatistics—the use of numerical techniques to extract information from data and facts. Because numbers are more precise than words, they are particularly well suited for communicating scientific results.

However, just as one can lie with words, one can also lie with numbers. Indeed, numbers and lies have been linked for quite some time; there is even a book entitled *How to Lie with Statistics.* This association may owe its origin, or its affirmation at the very least, to the British prime minister Benjamin Disraeli. Disraeli is credited by Mark Twain as having said "There are three kinds of lies: lies, damned lies, and statistics." One has only to observe any modern political campaign to be convinced of the abuse of statistics. But enough about lies; this book adopts the position of Frederick Mosteller, who said "It is easy to lie with statistics, but it is easier to lie without them."

Principles of Biostatistics is aimed at the student who wishes to learn modern research methods. It is based on a required course offered at the Harvard School of Public Health. A large number of health professionals in the rest of the Harvard medical area attend as well. The course is as old as the School itself, which attests to its importance. It spans 16 weeks of lectures and laboratory sessions. Each week includes two 50-minute lectures and one 2-hour lab. The entire class is together for the lectures but divides into smaller groups, headed by teaching assistants, for the lab sessions. These labs reinforce the material covered in the lectures, review the homework assignments, and introduce the computer into the course. We have included the lab materials—except those dealing with the homework assignments and specific computer commands—in the sections called Further Applications. These sections present additional examples or a different perspective of the material already covered. The labs are designed to provoke discussion, although they are sufficiently complete for the reader who is not using the book as a course text to benefit from reading them.

This book has evolved over the years to include topics that we believe can be covered at some depth in one American semester. Clearly, some choices had to be made; we hope that we have chosen well. In our course, we have sufficient time to cover most of the material in the first 20 chapters. However, there is enough material to allow the instructor some flexibility in the choice of topics. For example, some instructors may choose to omit the sections covering grouped data (Section 3.3), Chebychev's inequality (Section 3.4), and the Poisson distribution (Section 7.3) or the chapter on the analysis of variance (Chapter 12), if they consider these concepts to be less important than others.

Some say that statistics is the study of variability and uncertainty. We believe there is some truth to this adage and have used it as a guide by dividing the book into three parts. The first five chapters deal with collections of numbers and ways in which to summarize, explore, and explain them; these methods are easily comprehensible. The next two chapters consider probability and serve as an introduction to the tools needed for the subsequent investigation of uncertainty. It is only in the eighth chapter and thereafter that we distinguish between populations and samples and begin to investigate the inherent variability introduced by sampling, thus progressing to inference. We think that this modular introduction to the quantification of uncertainty is justified by the success achieved by our students. Postponing the slightly more difficult parts until a solid foundation has been established makes it easier for the reader to comprehend them.

Throughout the text we have used data drawn from published studies to exemplify biostatistical concepts. Not only is real data more meaningful, it is usually more interesting as well. Of course, we must take care not to use examples in which the subject matter is too esoteric or too complex. To this end, we have been guided by the backgrounds and interests of our students–mostly public health and clinical research–to choose examples that best illustrate the concepts at hand.

There is some risk involved in using published data. We cannot guarantee that all the examples are honest and that the data were properly collected; for this we must rely on the reputations of our sources. We do not belittle the importance of this consideration. The value of our inference depends critically on the worth of the data, and we strongly recommend that a good deal of effort be expended on evaluating their quality. We assume that this is understood by the reader.

More than once we have used examples in which the population of the United States is broken down along racial lines. In reporting these official statistics we follow the lead of the government agencies that release them. We do not wish to reify this racial categorization, since in fact the observed differences may well be due to socioeconomic factors rather than the implied racial ones. One option would be to ignore these statistics; however, that would hide inequities that exist in our health system–inequities that need to be eliminated. We focus attention on the problem in the hope of stimulating interest in promoting solutions.

We have minimized the use of mathematical notation because of its well-deserved reputation of being the ultimate jargon. On one hand, it can be over used and intimidate even the most ardent scholar. On the other hand, we should not eliminate its use entirely; it has been developed over the ages to be helpful in communicating results. We hope that in this respect we have written a succinct and understandable text.

Over and above their precision, there is something more to numbers—maybe a little magic—that makes them fun to study. The fun is in the conceptualization more so than the calculations, and we are fortunate to have the computer to do the drudge work. This allows students to concentrate on the ideas. In other words, the computer allows the instructor to teach the poetry of statistics and not the plumbing.

To take advantage of the computer, one needs a good statistical package. We use Stata, which is available from the Computing Resource Center in Santa Monica, California. We find this statistical package to be one of the best on the market today; it is user friendly, accurate, powerful, reasonably priced, and works on a number of different platforms, including DOS, Unix, and Macintosh. Furthermore, the output from this package is acceptable to the Federal Drug Administration in New Drug Approval submissions. Other packages are available, and this book can be supplemented by any one of them. We strongly recommend that some statistical package be used.

Some of the exercises in the text require the use of the computer. To help the reader, we have included the data sets used in the review exercises both in Appendix B and on the disk that accompanies the book. The disk contains each data set in two different formats: an ASCII file (the "raw" suffix) and a Stata file (the "dta" suffix). There are also many exercises that do not require the computer. Answers to selected exercises are supplied in Appendix C. As always, active learning yields better results than passive observation does. To this end, we cannot stress enough the importance of the exercises and urge the reader to attempt as many as time permits.

A debt of gratitude is owed a number of people: President Derek Bok for providing the support that got this book off the ground, Dr. Michael K. Martin for calculating all the statistical tables in Appendix A, and John-Paul Pagano for assisting in the editing. We thank the individuals who reviewed the manuscript: Stewart Anderson, University of Pittsburgh; Ronald Cody, Robert Wood Johnson Medical School; Charles Davis, University of Iowa; Janice Derr, Pennsylvania State University; Christiana Drake, University of California, Davis; David Gjertson, University of California, Los Angeles; James Godbold, Mt. Sinai School of Medicine; Maureen Lahiff, University of California, Berkeley; Nicholas Lange, Brown University; John Lefante, Tulane Medical Center; and Donald Slymen, San Diego State University. Our thanks also go to Karen U. Abbett and Judith Nahas for assistance in typing an early version of the text and to Stacey Sawyer, whose editing suggestions have resulted in an improved book. The teaching assistants who have helped us teach the course and who have made many valuable suggestions, we thank. Probably the most deserving of thanks are the students who have taken the course over the years and who have tolerated us as we learned how to teach it. We are still learning.

Marcello Pagano
Kimberlee Gauvreau

Boston, Massachusetts
August 1992

Contents

CHAPTER 1

Introduction

H. G. Wells is reputed to have said that statistical reasoning will one day be as important to good citizenship as the ability to read and write. Indeed, statistics play an important role in many decision-making processes. Before a new drug can be marketed, for instance, the U.S. Food and Drug Administration requires that it be subjected to a clinical trial—an experimental study involving human subjects. The data from this study must be compiled and analyzed to determine whether the drug is effective. In addition, the U. S. government's decisions regarding Social Security and public health programs rely in part on predictions about the longevity of the nation's population; as a result, it must be able to predict the number of years that a particular individual will live. There are many other questions that we would like to answer as well. Where should a government invest its resources if it wishes to reduce infant mortality? Does the use of a seat belt decrease the chance of death in a motor vehicle accident? Should a mastectomy always be recommended to a patient with breast cancer? What factors increase the risk that an individual will develop coronary heart disease? To address these issues and others, we rely on the methods of biostatistics.

The study of statistics investigates the collection, organization, analysis, and interpretation of numerical data. The concepts of statistics may be applied to a number of fields that include business, psychology, and agriculture. When focus is on the biological and health sciences, we use the term *biostatistics*.

Historically, statistics have been used to tell a story with numbers. Numbers often communicate ideas more succinctly than do words. For example, for the following data, the picture is quite clear. In 1979, 48 persons in Japan, 34 in Switzerland, 52 in Canada, 58 in Israel, 21 in Sweden, 42 in Germany, 8 in England, and 10,728 in the United States were killed by handguns [1]. The power of these numbers is obvious; the message would remain clear even if we were to correct for differences in population size.

As a second example, consider the following quotation taken from an editorial in *The Boston Globe* [2]:

> Lack of contraception is linked to an exceptionally high abortion rate in the Soviet Union—120 abortions for every 100 births, compared with 20 per 100 births in Great Britain, where access to contraception is guaranteed. Inadequate support for family planning in the United States has resulted in 40 abortions for every 100 births—a lower rate than the Soviet Union, but twice as high as most industrialized nations.

In this case, a great deal of information is contained in only three numbers: 120, 20, and 40. The statistics provide some insight into the consequences of differing attitudes toward family planning.

In both these examples, the numbers provide a concise summary of certain aspects of the situation being studied. Surely the numerical explanation of the handgun data is more illuminating than if we had been told that some people got killed in Japan, fewer in Switzerland, more in Canada, still more in Israel, but far fewer in Sweden, and so forth. Both examples deal with very complex situations, yet the numbers convey the essential information. Of course, no matter how powerful, no statistic will convince everyone that a given conclusion is true. The handgun data are often brushed away with the aphorism, "Guns don't kill people, people do." This should not be surprising; after all, there are still members in the Flat Earth Society. The aim of a biostatistical study is to provide the numbers that contain information about a certain situation and to present them in such a way that valid interpretations are possible.

If we wish to study the effects of a new diet, we might begin by measuring the changes in body mass for all individuals who have been placed on the diet. Similarly, if we want to investigate the success of a certain therapy for treating breast cancer, we would look at the lengths of time that women treated with this therapy survive beyond diagnosis of the disease. However, the collections of numbers are not very informative until we start combining them in some way. Descriptive statistics are methods for organizing and summarizing data that help us to describe the attributes of a group or population. In Chapter 2, we examine both tabular and graphical descriptive techniques. The graphical capabilities of computers have made this type of summarization more feasible than in the past, and a whole new mode of presentation is now available for even the most modest analyses.

Chapter 3 goes beyond the graphical techniques presented in Chapter 2 and introduces numerical summary measures. By definition, a summary captures only a particular aspect of the data being studied; consequently, it is also important to have an idea of how well the summary represents the data as a whole. For example, we might wish to know how long AIDS patients survive after diagnosis with one of the characteristic opportunistic infections. If we calculate an average survival time, is this average representative of all patients? Furthermore, how useful would the measure be for planning future health service needs? Chapter 3 investigates descriptive techniques that help us to answer questions such as these.

Data that take on only two values require special attention. The most common example of this type of data is probably the categorization of being either alive or dead. If we denote the former state by 0 and the latter by 1, we are able to classify a group of individuals using these numbers and to average the results. In this way, we can summarize the mortality associated with the group. Chapter 4 deals exclusively with measurements that can take on only two values. The notion of dividing a group into subgroups or classes based on some characteristic such as age or gender is introduced as well. For example, we might wish to study the mortality of females separately from that of males. Finally, this chapter examines techniques that allow us to make valid comparisons among groups that may differ in composition.

Chapter 5 introduces the life table, one of the most important techniques available for study in the health sciences. Life tables are used by public health professionals to characterize the well-being of a population and by insurance companies to predict how long a particular individual will live. In this chapter, the study of mortality begun in Chapter 4 is extended to incorporate the actual time to death for each individual; this results in a more refined analysis.

In summary, the first five chapters of the text demonstrate that the extraction of important information from a collection of numbers is not precluded by their variability. Despite the variability, the data often exhibit a certain regularity as well. For example, if we look at the annual mortality rates of teenagers in the United States for each of the last ten years, we do not see much discrepancy in the numbers. Is this just a coincidence, or is it indicative of a natural underlying stability in the mortality rate? To answer questions such as this, we need to study the principles of probability.

Probability theory resides within what is known as an axiomatic system; we start with some basic truths and then build up a logical system around them. In its pure form, the system has no practical value. Its practicality comes from knowing how to use the theory to yield useful approximations. An analogy can be drawn with geometry, a subject that most students are exposed to early in their schooling. Although it is impossible for an ideal straight line to exist other than in our imaginations, that has not stopped us from constructing some wonderful buildings based on geometric calculations. The same is true of probability theory; although it is not practical in its pure form, its basic principles—which we investigate in Chapter 6—can be applied to provide a means of quantifying uncertainty.

One important application of probability theory occurs in diagnostic testing. Uncertainty arises because, despite their manufacturers' claims, no available tests are perfect. Consequently, there are a number of important questions that must be answered. For instance, can we conclude that every blood sample that tests positive for HIV actually harbors the virus? Furthermore, all the units in the Red Cross blood supply have tested negative for the virus; does this mean that there are no contaminated samples? If there are some contaminated samples, how many might there be? To answer these questions, we must rely on the average or long-term behavior of the tests; probability theory allows us to quantify this behavior.

Chapter 7 extends the notions of probability and introduces some common probability distributions. These mathematical models are useful in the remainder of the text.

The early chapters of this book cover the variability that exists in a collection of numbers. Subsequent chapters move on to another form of variability—the variability that arises when we draw a sample of observations from a population. Suppose we would like to know whether a putative vaccine will help protect the world from AIDS. Since the population being studied is very large, it is extremely unlikely that we would have the time or the resources necessary to examine each and every individual. In other situations, the population may include future patients; we might want to know how both patients who currently have a disease and individuals who will subsequently develop it will react to a certain treatment. To answer these types of questions, it is common to select a sample from the population of interest and, on the basis of this sample, infer what would happen to the group as a whole. Consequently, it is important to know whether the sample is in fact representative of the population.

If we choose two different samples, it is unlikely that we would end up with precisely the same sets of numbers. Similarly, if we study a group of patients in Boston, we will get different results than if we study a group in Rome. Despite this difference, we would like to be able to use one or both of the samples to draw some conclusion about the entire population of patients. The remainder of the text is concerned with the topic of statistical inference.

Chapter 8 investigates the properties of the sample mean when repeated samples are drawn from a population and introduces the central limit theorem. This theorem provides a foundation for quantifying the uncertainty associated with the conclusions that are drawn.

For any study to be of practical value, we must be able to extrapolate its findings to a larger group or population. To this end, the methods of confidence intervals and hypothesis testing are introduced in Chapters 9 and 10. These techniques provide a means of reaching a conclusion about the population we have sampled, while at the same time having some knowledge about the chance that the conclusion is incorrect. The ideas are first applied to the mean of a single population. In Chapter 11, the theory is extended to the comparison of two population means; it is further generalized to the comparison of three or more means in Chapter 12. Chapter 13 continues the development of the same concepts, but introduces techniques that allow the relaxation of some of the assumptions necessary to carry out the tests. Chapters 14 through 16 develop inferential methods that can be applied to enumerated data or counts rather than continuous measurements.

Inference can also be used to explore the relationships among a number of attributes. If a full-term baby whose gestational age is 39 weeks is born weighing 4 kilograms, or 8.8 pounds, no one would be surprised. If the baby's gestational age is only 22 weeks, however, then his or her weight would be cause for alarm. Why? We know that birth weight tends to increase with gestational age, and, although it is extremely rare to find a baby weighing 4 kilograms at 22 weeks, it is not uncommon at 39 weeks. The study of the degree to which two factors are related is known as correlation analysis; this is the topic of Chapter 17. If we wish to predict the outcome of one factor based on the value of another, then regression is the appropriate technique. Simple linear regression is investigated in Chapter 18 and is extended to the multiple regression setting in Chapter 19. If the outcome of interest can take on only two values, then an alternative technique must be applied; logistic regression is explored in Chapter 20.

In Chapter 21, the inferential methods corresponding to life tables are introduced. These techniques enable us to draw conclusions about the mortality of a population based on a sample of individuals drawn from it.

Finally, Chapter 22 examines an issue that is fundamental in inference–the concept of the representativeness of a sample. Several different methods for selecting representative samples are described. The concept of bias and various problems that can arise when choosing a sample are discussed as well. Common sense plays an important role in sampling, as it does throughout the rest of the book.

1.1 Review Exercises

1. Design a study aimed at investigating an issue that you believe might influence the health of the world. Briefly describe the data that you will require, how you will obtain them, how you intend to analyze them, and the method you will use to present your results. Keep this study design and reread it after you have completed the text.

2. Consider the following quotation regarding rapid population growth [3]:

> 512 million people were malnourished in 1986–1987, up from 460 million in 1979–1981.

   a. Suppose that you sympathize with the point being made. Justify the use of these numbers.
   b. Are you sure that the numbers are correct? Do you think it is possible that 513 million people were malnourished in 1986–1987 rather than 512 million?

3. In addition to stating that "the Chinese have eaten pasta since 1100 B.C.," the label on a box of pasta shells claims that "Americans eat 11 pounds of pasta per year," whereas "Italians eat 60 pounds per year." Do you believe that these statistics are accurate? Would you use these numbers as the basis for a nutritional study?

References

[1] McGervey, J. D., *Probabilities in Everyday Life*, Chicago: Nelson-Hall, 1986.
[2] "The Pill's Eastern Europe Debut," *The Boston Globe*, January 19, 1990, 10.
[3] United Nations Population Fund, "Family Planning: Saving Children, Improving Lives," New York: Jones & Janello.

CHAPTER 2

Data Presentation

Every study or experiment yields a set of data. Its size can range from a few measurements to many thousands of observations. A complete set of data, however, will not necessarily provide an investigator with information that can be easily interpreted. For example, Table 2.1 lists by row the first 2,560 cases of AIDS reported to the Centers for Disease Control. Each individual was classified as either suffering from Kaposi's sarcoma, designated by a 1, or not suffering from the disease, represented by a 0. Although Table 2.1 displays the entire set of outcomes, it is extremely difficult to characterize the data. We cannot even identify the relative proportions of 0s and 1s. Between the raw data and the reported results of the study lies some intelligent and imaginative manipulation of the numbers that was done using the methods of descriptive statistics.

Descriptive statistics are a means of organizing and summarizing observations. They provide us with an overview of the general features of a set of data. Descriptive statistics can assume a number of different forms; among these are tables, graphs, and numerical summary measures. In this chapter, we discuss the various methods of displaying a set of data. Before we decide which technique is the most appropriate in a given circumstance, however, we must first determine what kind of data we have.

2.1 Types of Numerical Data

2.1.1 Nominal Data

In the study of biostatistics we encounter many different types of numerical data. The different types have varying degrees of structure in the relationship between possible values. One of the simplest types of data is *nominal data,* in which values fall into unordered categories. As in Table 2.1, numbers are often used to represent the categories. As another example, males might be assigned the value 1 and females the value 0. Although the attributes are labeled with numbers rather than words, both the order and the magnitude of the numbers are unimportant. We could just as easily let 1 represent females and 0 designate males. Numbers are used largely for the sake of convenience; numerical values allow us to use computers to perform complex analyses of the data.

Nominal data that take on one of two distinct values–such as male and female–are said to be either *dichotomous* or *binary*, depending on whether the

Table 2.1 Outcomes indicating whether an individual had Kaposi's sarcoma for the first 2,560 AIDS patients reported to the Centers for Disease Control in Atlanta, Georgia

00000000	00010100	00000010	00001000	00000001	00000000	10000000	00000000
00101000	00000000	00000000	00011000	00100001	01001100	00000000	00000010
00000001	00000000	00000010	01100000	00000000	00000100	00000000	00000000
00100010	00100000	00000101	00000000	00000000	00000001	00001001	00000000
00000000	00010000	00010000	00010000	00000000	00000000	00000000	00000000
00000000	00000000	00000000	00001000	00000000	00010000	10000000	00000000
00100000	00000000	00001000	00000010	00000000	00000100	00000000	00010000
00000000	00000000	00000100	00001000	00001000	00000101	00000000	01000000
00010000	00000000	00010000	01000000	00000000	00000000	00000101	00100000
00000000	00000000	00000100	00000000	01000100	00000000	00000001	10100000
00000100	00000000	00010000	00000000	00001000	00000000	00000010	00100000
00000000	00000000	00000000	10001000	00001000	00000000	01000000	00000000
00000000	00001100	00000000	00000000	10000011	00000001	11000000	00001000
00000000	00000000	00000000	00000000	01000000	00000001	00010001	00000000
10000000	00000000	01000000	00000000	00000000	01010100	00000000	00010100
00000000	00000000	00000000	00001010	00000101	00000000	00000000	00010000
00000000	00000000	00000000	00000001	00000100	00000000	00000000	00001000
11000000	00000100	00000000	00000000	00000000	00000000	00000000	00001000
11000000	00010010	00000000	00001000	00000000	00111000	00000001	01001100
00000000	01100000	00100010	10000000	00000000	00000010	00000001	00000000
01000010	01000100	00000000	00010000	00000000	01000000	00000001	00000000
01000000	00000001	00000000	10000000	01000000	00000000	00000000	00000100
00000000	00000000	01000010	00000000	00000000	00000000	00000000	00000000
00000000	00000010	00001010	00001001	10000000	00000000	00000010	00000000
00000000	01000000	00000000	00001000	00000000	01000000	00010000	00000000
00001000	01000010	01001111	00100000	00000000	00100000	00000000	10000001
00000001	00000000	01000000	00000000	00000000	00000000	00000000	01000000
00000000	00000000	00100000	01000000	00100000	00000000	00000011	00000000
01000000	00000100	10000001	00000001	00001000	00000100	00001000	00001000
00100000	00000000	00000000	00000000	00000010	01000001	00010011	00000000
00000000	10000000	10000000	00000000	00000000	00001000	01000000	00000000
00001000	00000000	01000010	00011000	00000001	00001001	00000000	00000001
01000010	01001000	01000000	00000010	00000000	10000000	00000100	00000000
00000010	00000000	00000000	00000010	00000000	00100100	00000000	10110100
00001100	00000100	00001010	00000000	00000000	00000000	00000000	00000000
00000010	00000000	00000000	00000000	00100000	10100000	00001000	00000000
01000000	00000000	00000000	00100000	00000000	01000001	00010010	00010001
00000000	00100000	00110000	00000000	00010000	00000000	00000100	00000000
00010100	00000000	00001001	00000001	00000000	00000000	00000000	00000000
00000010	00000100	01010100	10000001	00001000	00000000	00010010	00010000

Greek or the Latin root for "two" is preferred. However, not all nominal data need be dichotomous. Often there are three or more possible categories into which the observations can fall. For example, persons may be grouped according to their blood type, where 1 represents type O, 2 is type A, 3 is type B, and 4 is type AB. Again, the sequence of these values is not important. The numbers simply serve as labels for the different blood types, just as the letters do. We must keep this in mind when we perform arithmetic operations on the data. For example, an average blood type of 1.8 for a given population is meaningless. One arithmetic operation that can be interpreted, however, is the proportion of individuals that fall into each group. An analysis of the data in Table 2.1 shows that 9.6% of the AIDS patients suffered from Kaposi's sarcoma and 90.4% did not.

2.1.2 Ordinal Data

When the order among categories becomes important, the data are referred to as *ordinal data*. For example, injuries may be classified according to their level of severity, where 1 represents a fatal injury, 2 is severe, 3 is moderate, and 4 is minor. Here a natural order exists among the groupings; a smaller number represents a more serious injury. We are still not concerned with the magnitude of these numbers, however. We could have let 4 represent a fatal injury and 1 a minor one. Furthermore, the difference between a fatal injury and a severe injury is not necessarily the same as the difference between a moderate injury and a minor one, even though both pairs of outcomes are one unit apart. As a result, many arithmetic operations still do not make sense when applied to ordinal data. Table 2.2 provides a second example of ordinal data; the scale displayed is used by oncologists to classify the performance status of patients enrolled in a clinical trial [1].

2.1.3 Ranked Data

In some situations we have a group of observations that are first arranged from highest to lowest according to magnitude and then assigned numbers that

Table 2.2 Eastern Cooperative Oncology Group's classification of patient performance status

Status	Definition
0	Patient fully active, able to carry on all predisease performance without restriction
1	Patient restricted in physically strenuous activity but ambulatory and able to carry out work of a light or sedentary nature
2	Patient ambulatory and capable of all self-care but unable to carry out any work activities; up and about more than 50% of waking hours
3	Patient capable of only limited self-care; confined to bed or chair more than 50% of waking hours
4	Patient completely disabled; not capable of any self-care; totally confined to bed or chair

correspond to each observation's place in the sequence. This type of data is known as *ranked data*. As an example, consider all possible causes of death in the United States. We could make a list of all of these causes, along with the number of lives that each one claimed in 1988. If the causes are ordered from the one that resulted in the greatest number of deaths to the one that caused the fewest and then assigned consecutive integers, the data are said to have been ranked. Table 2.3 lists the ten leading causes of death in the United States in 1988 [2]. Note that cerebrovascular diseases would be ranked third whether they caused 480,000 deaths or 98,000. In assigning the ranks, we disregard the magnitudes of the observations and consider only their relative positions. Even with this imprecision, it is amazing how much information the ranks contain. In fact, it is sometimes better to work with ranks than with the original data; this point is explored further in Chapter 13.

2.1.4 Discrete Data

For *discrete data*, both ordering and magnitude are important. In this case, the numbers represent actual measurable quantities rather than mere labels. In addition, discrete data are restricted to taking on only specified values—often integers or counts—that differ by fixed amounts; no intermediate values are possible. Examples of discrete data include the number of motor vehicle accidents in Massachusetts in 1990, the number of times that a woman has given birth, the number of new cases of AIDS reported in the United States during a one-year period, and the number of beds available in a particular hospital.

Note that a natural order exists among the possible data values. If we are interested in the number of times that a woman has given birth, for instance, a larger number indicates that a woman has had more children. Furthermore, the difference between one and two births is the same as the difference between four and five births. Finally, the number of births is restricted to the nonnegative

Table 2.3 Ten leading causes of death in the United States, 1988

Rank	Cause of Death	Total Deaths
1	Diseases of the heart	765,156
2	Malignant neoplasms	485,048
3	Cerebrovascular diseases	150,517
4	Accidents and adverse effects	97,100
5	Chronic obstructive pulmonary diseases	82,853
6	Pneumonia and influenza	77,662
7	Diabetes mellitus	40,368
8	Suicide	30,407
9	Chronic liver disease and cirrhosis	26,409
10	Nephritis, nephrotic syndrome, and nephrosis	22,392

integers; a woman cannot give birth 3.4 times. Because it is meaningful to measure the distance between possible data values for discrete observations, arithmetic rules can be applied. However, the outcome of an arithmetic operation performed on two discrete values is not necessarily discrete itself. Suppose, for instance, that one woman has given birth three times, whereas another has given birth twice. The average number of births for these two women is 2.5.

2.1.5 Continuous Data

Data that represent measurable quantities but are not restricted to taking on certain specified values are known as *continuous data*. In this case, the difference between any two possible data values can be arbitrarily small. Examples of continuous data include time, the serum cholesterol level of a patient, the concentration of a pollutant, and temperature. In all instances, fractional values are possible. Since we are able to measure the distance between two observations in a meaningful way, arithmetic operations can be applied. The only limiting factor for a continuous observation is the degree of accuracy with which it can be measured; consequently, we often see time rounded off to the nearest second and weight rounded off to the nearest pound or gram. The more accurate our measuring instruments, however, the greater the amount of detail that can be achieved in our recorded data.

At times we might require a lesser degree of detail than that afforded by continuous data; hence we occasionally transform continuous observations into either discrete or ordinal ones. In a study of the effects of maternal smoking on newborns, for example, we might first record the birth weights of a large number of infants and then categorize the infants into three groups: those who weigh less than 1,500 grams, those who weigh between 1,500 and 2,500 grams, and those who weigh more than 2,500 grams. Although we have the actual measures of birth weight, we are not concerned whether a particular child weighs 1,560 grams or 1,580 grams; we are only interested in the number of infants who fall into each category. Ordinal data are often easier to handle than continuous data and thus simplify the analysis. In our example, however, there is a consequential loss of detail in the information about the infants. In general, the degree of precision required in a given set of data depends on the questions that are being studied.

Section 2.1 described a gradation of numerical data that ranges from nominal to continuous. As we progressed, the nature of the relationship between possible data values became increasingly complex. Distinctions must be made among the various types of data because different techniques are used to analyze them. As previously mentioned, it does not make sense to speak of an average blood type of 1.8; it does make sense, however, to refer to an average temperature of 24.55° C.

2.2 Tables

Now that we are able to differentiate among the various types of data, we must learn how to identify the statistical techniques thatare most appropriate for describing each kind. Although a certain amount of information is lost when data are summarized, a great deal can also be gained. A *table* is perhaps the simplest means of summarizing a set of observations and can be used for all types of data.

2.2.1 Frequency Distributions

One type of table that is commonly used to evaluate data is known as a *frequency distribution*. For nominal and ordinal data, a frequency distribution consists of a set of classes or categories along with the numerical counts that correspond to each one. Table 2.4, which specifies the numbers of cigarettes smoked per adult in the United States (numerical count) in various years (classes or categories), presents the data in a frequency distribution format [3].

To display discrete or continuous data in the form of a frequency distribution, we must break down the range of values of the observations into a series of distinct, nonoverlapping intervals. If there are too many intervals, the summary is not much of an improvement over the raw data. If there are too few, then a great deal of information is lost. Although it is not necessary, intervals are often constructed so that they all have equal widths; this facilitates comparisons among classes. Once the upper and lower limits for each interval have been selected, the number of observations whose values fall within each pair of limits is counted, and the results are arranged as a table. As part of a National Health Examination Survey, for example, the serum cholesterol levels of 1,067 25- to 34-year-old males were recorded to the nearest milligram per 100 milliliters [4]. The observations were then subdivided into intervals of equal width; the frequencies corresponding to each interval are presented in Table 2.5.

Table 2.5 gives us an overall picture of what the data look like; it shows how the values of serum cholesterol level are distributed across the intervals. Note that the observations range from 80 to 399 mg/100 ml, with relatively few measurements at the ends of the range and a large proportion of the values falling between 120 and 279 mg/100 ml. The interval 160–199 mg/100 ml contains the greatest

Table 2.4 Cigarette consumption per person 18 years of age or older, United States, 1900–1990

Year	Number of Cigarettes
1900	54
1910	151
1920	665
1930	1,485
1940	1,976
1950	3,522
1960	4,171
1970	3,985
1980	3,851
1990	2,828

Table 2.5 Absolute frequencies of serum cholesterol levels for 1,067 U.S. males, aged 25 to 34 years, 1976–1980

Cholesterol Level (mg/100 ml)	Number of Men
80–119	13
120–159	150
160–199	442
200–239	299
240–279	115
280–319	34
320–359	9
360–399	5
Total	1,067

number of observations. Table 2.5 provides us with a much better understanding of the data than would a list of 1,067 cholesterol level readings. Although we have lost some information–given the table, we can no longer recreate the raw data values–we have also extracted important information that helps us to understand the distribution of serum cholesterol levels for this group of males.

The fact that one kind of information is gained while another is lost holds true even for the simple dichotomous data in Table 2.1. We might feel that we do not lose anything by summarizing these data, but in fact we do. For example, if there is some type of time trend in the observations–perhaps the proportion of AIDS patients with Kaposi's sarcoma is either increasing or decreasing as the epidemic matures–this information is lost in the summary.

Tables are most informative when they are not overly complex. As a general rule, tables and the columns within should always be clearly labeled. If units of measurement are involved, such as mg/100 ml for the serum cholesterol levels in Table 2.5, then they should be specified.

2.2.2 Relative Frequency

It is sometimes useful to know the proportion of values that fall into a given interval in a frequency distribution rather than the absolute number. The *relative frequency* for an interval is the percentage of the total number of observations that appear in that interval. The relative frequency is computed by dividing the number of values within an interval by the total number of values in the table and then multiplying by 100. In Table 2.5, for example, the relative frequency in the 80–119 mg/100 ml class is $(13/1,067) \times 100 = 1.2\%$; similarly, the relative frequency in the 120–159 mg/100 ml class is $(150/1,067) \times 100 = 14.1\%$. The relative frequencies for all intervals in a table sum to 100%.

Relative frequencies are useful for comparing sets of data that contain unequal numbers of observations. Table 2.6 displays the absolute and relative

Table 2.6 Absolute and relative frequencies of serum cholesterol levels for 2,294 U.S. males, 1976–1980

Cholesterol Level (mg/100 ml)	Ages 25–34		Ages 55–64	
	Number of Men	Relative Frequency (%)	Number of Men	Relative Frequency (%)
80–119	13	1.2	5	0.4
120–159	150	14.1	48	3.9
160–199	442	41.4	265	21.6
200–239	299	28.0	458	37.3
240–279	115	10.8	281	22.9
280–319	34	3.2	128	10.4
320–359	9	0.8	35	2.9
360–399	5	0.5	7	0.6
Total	1,067	100.0	1,227	100.0

frequencies of serum cholesterol level readings for the 1,067 25- to 34-year-olds depicted in Table 2.5 as well as a group of 1,227 55- to 64-year-olds. Because there are more men in the older age group, it is inappropriate to compare the columns of absolute frequencies for the two groups. Comparing the relative frequencies, however, is meaningful. We can see that, in general, the older men have higher serum cholesterol levels than the younger men; the younger men have a greater proportion of observations below 200 mg/100 ml, whereas the older men have a greater proportion above this value.

The *cumulative relative frequency* for an interval is the percentage of the total number of observations that have a value less than or equal to the upper limit of the interval. The cumulative relative frequency is calculated by summing the relative frequencies for the specified interval and all previous ones. Thus, for the group of 25- to 34-year-olds in Table 2.6, the cumulative relative frequency of the second interval is $1.2 + 14.1 = 15.3\%$; similarly, the cumulative relative frequency of the third interval is $1.2 + 14.1 + 41.4 = 56.7\%$. Like relative frequencies, cumulative relative frequencies are useful for comparing sets of data that contain unequal numbers of observations. Table 2.7 lists the cumulative relative frequencies for the serum cholesterol levels of the two groups of males in Table 2.6.

According to the data in Table 2.5, older men tend to have higher serum cholesterol levels than younger men do. This is a sort of generalization we hear quite often; for instance, it might also be said that men are taller than women or that women live longer than men. The generalization about serum cholesterol does not mean that every 55- to 64-year-old male has a higher cholesterol level than every 25- to 34-year-old male; neither does it mean that the serum cholesterol level of every man increases with age. What the statement does indicate is that for a given cholesterol level, the proportion of younger men with a reading less than or equal to this value is greater than the proportion of older men with a reading less than or equal to the value. This pattern is more obvious in Table 2.7 than it is in Table 2.6. For example, 56.7% of the 25- to 34-year-olds have a serum cholesterol level less than or equal to 199 mg/100 ml, whereas only 25.9% of the

Table 2.7 Relative and cumulative relative frequencies of serum cholesterol levels for 2,294 U.S. males, 1976–1980

	Ages 25–34		Ages 55–64	
Cholesterol Level (mg/100 ml)	Relative Frequency (%)	Cumulative Relative Frequency (%)	Relative Frequency (%)	Cumulative Relative Frequency (%)
80–119	1.2	1.2	0.4	0.4
120–159	14.1	15.3	3.9	4.3
160–199	41.4	56.7	21.6	25.9
200–239	28.0	84.7	37.3	63.2
240–279	10.8	95.5	22.9	86.1
280–319	3.2	98.7	10.4	96.5
320–359	0.8	99.5	2.9	99.4
360–399	0.5	100.0	0.6	100.0

55- to 64-year-olds fall into this category. Because the relative proportions for the two groups follow this trend in every interval in the table, the two distributions are said to be *stochastically ordered.* For any specified level, a larger proportion of the older men have serum cholesterol readings above this value than do the younger men; therefore, the distribution of levels for the older men is stochastically larger than the distribution for the younger men.

2.3 Graphs

A second way to summarize and display data is through the use of *graphs,* or pictorial representations of numerical data. Graphs should be designed so that they convey at a single glance the general patterns in a set of data. Although they are easier to read than tables, graphs often supply a lesser degree of detail. Once again, however, the loss of detail may be accompanied by a gain in understanding of the data. The most informative graphs are relatively simple and self-explanatory. Like tables, they should be clearly labeled, and units of measurement should be indicated.

2.3.1 Bar Charts

Bar charts are a popular type of graph used to display a frequency distribution for nominal or ordinal data. In a *bar chart,* the various categories into which the observations fall are presented along a horizontal axis. A vertical bar is drawn above each category such that the height of the bar represents either the frequency or the relative frequency of observations within that class. The bars should be of equal width and should be separated from one another so as not to imply continuity. For example, Figure 2.1 is a bar chart that displays the data relating

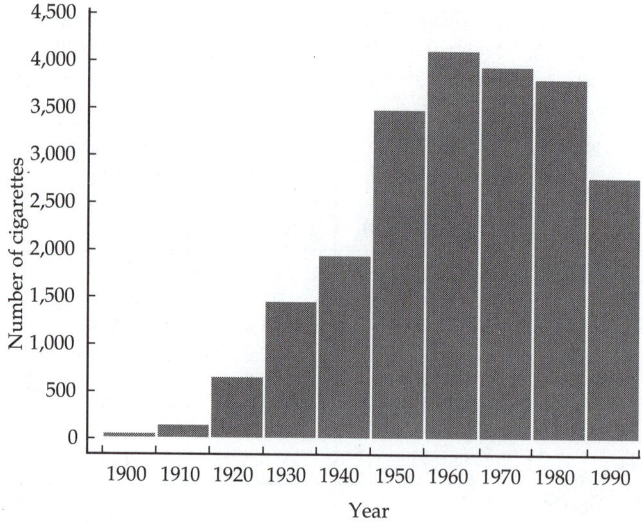

Figure 2.1. Bar chart: Cigarette consumption per person 18 years of age or older, United States, 1900–1990

to cigarette consumption in the United States presented in Table 2.4. Note that, when represented in the form of a graph, the trend in cigarette consumption over the years is more apparent than it is in the table.

2.3.2 Histograms

Perhaps the most commonly used type of graph is the histogram. Whereas a bar chart is a pictorial representation of a frequency distribution for either nominal or ordinal data, a *histogram* depicts a frequency distribution for discrete or continuous data. The horizontal axis displays the true limits of the various intervals. The true limits of an interval are the points that separate it from the intervals on either side. For example, the boundary between the first two classes of serum cholesterol level in Table 2.5 is 119.5 mg/100 ml; it is the true upper limit of the interval 80–119 and the true lower limit of 120–159. The vertical axis of a histogram depicts either the frequency or relative frequency of observations within each interval.

The first step in constructing a histogram is to lay out the scales of the axes. The vertical scale should begin at zero; if it does not, visual comparisons among the intervals may be distorted. Once the axes have been drawn, a vertical bar centered at the midpoint is placed over each interval. The height of the bar marks off the frequency associated with that interval. As an example, Figure 2.2 displays a histogram constructed from the serum cholesterol level data in Table 2.5.

In reality, the frequency associated with each interval in a histogram is represented not by the height of the bar above it but by the bar's area. Thus, in Figure 2.2, 1.2% of the total area corresponds to the 13 observations that lie between 79.5 and 119.5 mg/100 ml, and 14.1% of the area corresponds to the 150

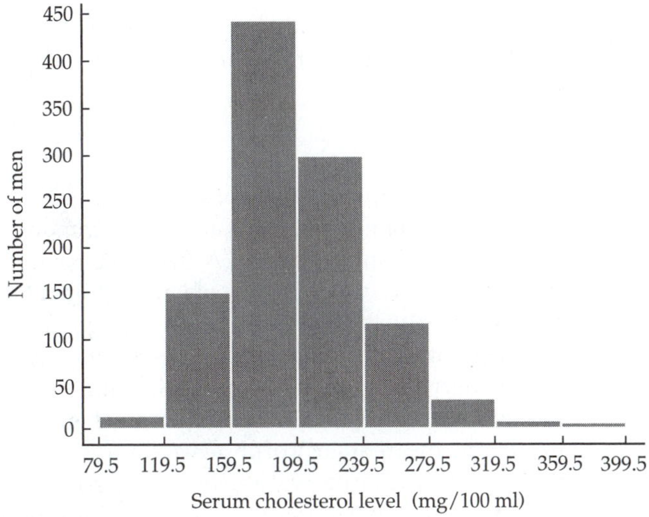

Figure 2.2. Histogram: Absolute frequencies of serum cholesterol levels for 1,067 U.S. males, aged 25 to 34 years, 1976–1980

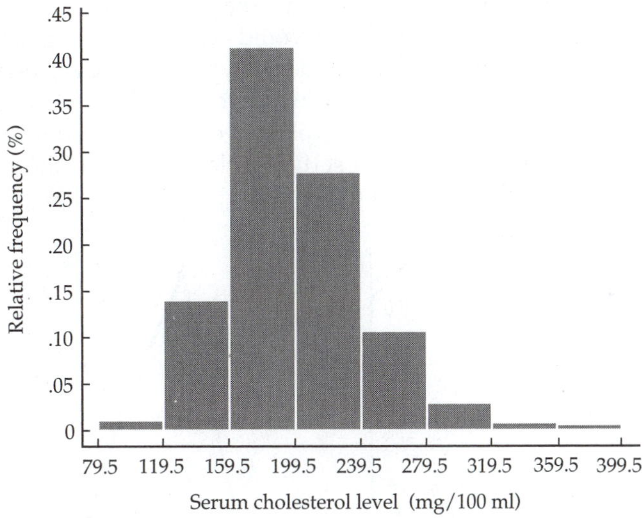

Figure 2.3. Histogram: Relative frequencies of serum cholesterol levels for 1,067 U.S. males, aged 25 to 34 years, 1976–1980

observations between 119.5 and 159.5 mg/100 ml. The area of the entire histogram sums to 100%, or 1. Note that the proportion of the total area corresponding to an interval is equal to the relative frequency of that interval. As a result, a histogram displaying relative frequencies—such as Figure 2.3—will have the same shape as a histogram which displays absolute frequencies. Because it is the area of each bar that represents the relative proportion of observations in an interval, care must be taken when constructing a histogram with unequal interval widths; the height must vary along with the width so that the area of each bar remains in proper proportion.

2.3.3 Frequency Polygons

The frequency polygon, another commonly used graph, is similar to the histogram in many respects. A *frequency polygon* uses the same two axes as a histogram. It is constructed by placing a point at the center of each interval such that the height of the point is equal to the frequency or relative frequency associated with that interval. Points are also placed on the horizontal axis at the midpoints of the intervals immediately preceding and immediately following the intervals that contain observations. The points are then connected by straight lines. As in a histogram, the frequency of observations for a particular interval is represented by the area within the interval and beneath the line segment. For example, Figure 2.4 is a frequency polygon of the serum cholesterol data in Table 2.5. Compare it with the histogram in Figure 2.2, which displays the same set of data. If the total number of observations in the data set were to increase steadily, we could decrease the widths of the intervals in the histogram accordingly and still have an adequate number of measurements in each class; in this case, the histogram and the frequency polygon would become indistinguishable.

Because they can be easily superimposed, frequency polygons are superior to histograms for comparing sets of data. Figure 2.5 displays the frequency polygons of the serum cholesterol level data presented in Table 2.6. Since the older men tend to have higher serum cholesterol levels, their polygon lies to the right of the polygon for the younger men.

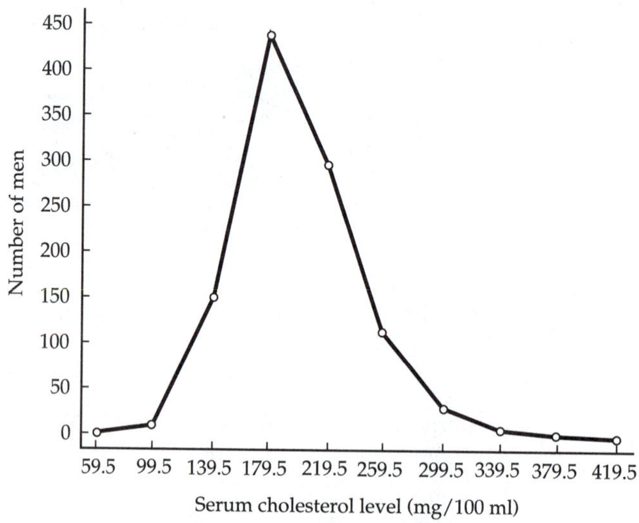

Figure 2.4. Frequency polygon: Absolute frequencies of serum cholesterol levels for 1,067 U.S. males, aged 25 to 34 years, 1976–1980

Although its horizontal axis is the same as that for a standard frequency polygon, the vertical axis of a *cumulative frequency polygon* displays cumulative relative frequencies. A point is placed at the true upper limit of each interval; the height of the point represents the cumulative relative frequency associated with that interval. The points are then connected by straight lines. Like frequency polygons, cumulative frequency polygons may be used to compare sets of data. This is illustrated in Figure 2.6. By noting that the cumulative frequency polygon for 55- to 64-year-old males lies to the right of the polygon for 25- to 34-year-old males for each value of serum cholesterol level, we can see that the distribution for older men is stochastically larger than the distribution for younger men.

Cumulative frequency polygons can also be used to obtain the *percentiles* of a set of data. Roughly, the 95th percentile is the value that is greater than or equal to 95% of the observations and less than or equal to the remaining 5%. This definition is only approximate because taking 95% of an integer does not typically result in another integer; consequently, there is often some rounding or interpolation involved. In Figure 2.6, the 50th percentile of the serum cholesterol levels for the group of 25- to 34-year-olds—the value that is greater than or equal to half of the observations and less than or equal to the other half—is approximately 193 mg/100 ml; the 50th percentile for the 55- to 64-year-olds is about 226 mg/100 ml.

Percentiles are useful for describing the shape of a distribution. For example, if the 40th and 60th percentiles of a set of data lie an equal distance away from the midpoint, and the same is true of the 30th and 70th percentiles, the 20th and 80th percentiles, and all other pairs of percentiles that sum to 100, then the data are *symmetric*; that is, the distribution of values has the same shape on each side of

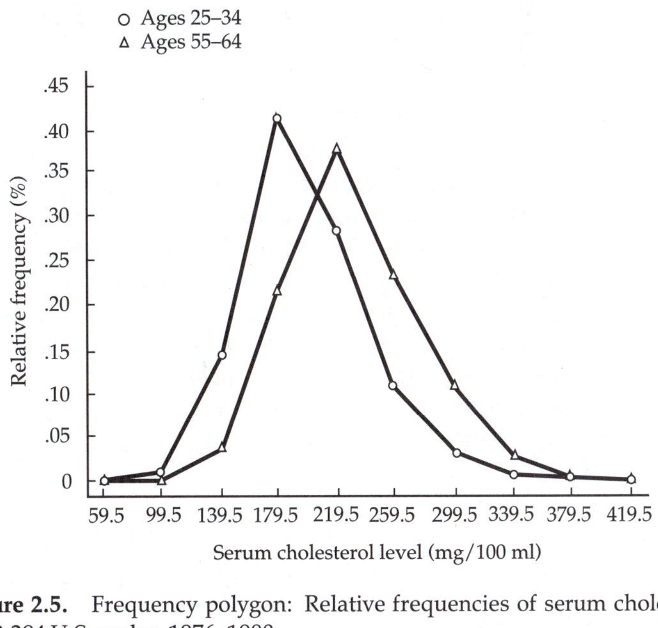

Figure 2.5. Frequency polygon: Relative frequencies of serum cholesterol levels for 2,294 U.S. males, 1976–1980

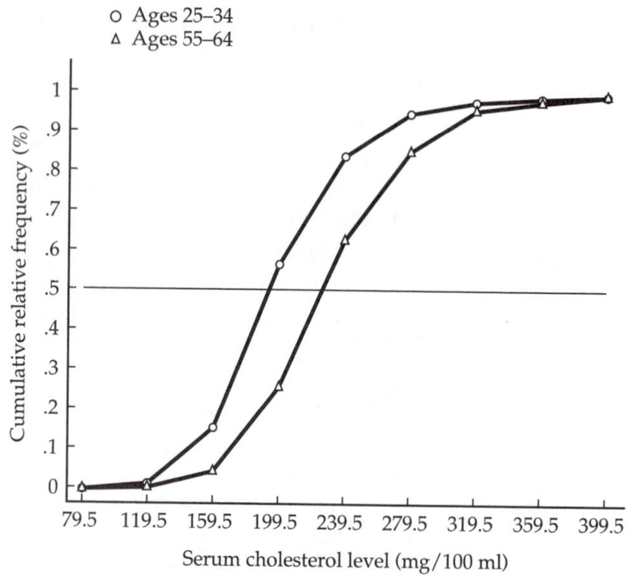

Figure 2.6. Cumulative frequency polygon: Cumulative relative frequencies of serum cholesterol levels for 2,294 U.S. males, 1976–1980

the 50th percentile. Alternatively, if there are a number of outlying observations on only one side of the midpoint, then the data are said to be *skewed*. If these observations are smaller than the rest of the values, the data are skewed to the left; if they are larger than the other measurements, then the data are skewed to the right. The various shapes that a distribution of data may assume are discussed further in Chapter 3.

2.3.4 One-Way Scatter Plots

Another type of graph that can be used to summarize a set of discrete or continuous observations is the one-way scatter plot. A *one-way scatter plot* uses a single horizontal axis to display the relative position of each data point. As an example, Figure 2.7 depicts the crude death rates for all 50 states and the District of Columbia, from a low of 393.9 per 100,000 population in Alaska to a high of 1,242.1 per 100,000 population in Washington, D.C. [2]. An advantage of the one-way scatter plot is that, since each observation is represented individually, no information is lost; a disadvantage is that the graph may be difficult to read if many data points lie close together.

Figure 2.7. One-way scatter plot: Crude death rates for the United States, 1988

2.3.5 Box Plots

Box plots are similar to one-way scatter plots in that they require a single axis; instead of plotting each and every observation, however, they display only a summary of the data [5]. Figure 2.8 is a box plot of the crude death rate data displayed in Figure 2.7. The central box extends from the 25th percentile, 804.0 per 100,000, to the 75th percentile, 953.9 per 100,000. The 25th and 75th percentiles of a data set are called the *quartiles* of the data. The line running between the quartiles at 886.3 deaths per 100,000 population marks the 50th percentile of the data set; half the observations are less than or equal to 886.3 per 100,000, whereas the other half are greater than or equal to this value. If the 50th percentile lies approximately halfway between the two quartiles, this suggests that the observations in the center of the data set are roughly symmetric. The lines projecting out from the box extend to the adjacent values. The *adjacent values* are the most extreme observations that are not more than 1.5 times the height of the box beyond either quartile. In Figure 2.8, the adjacent values are 648.8 per 100,000 and 1062.4 per 100,000. All points outside this range are represented by circles; these observations are considered to be *outliers*, which are not typical of the rest of the data.

It should be noted that the preceding explanation is merely one way to define a box plot; other definitions exist and exhibit varying degrees of complexity [6]. In addition, although a box plot conveys a fair amount of information about the distribution of a set of data points, an even greater amount of information can be displayed by combining the one-way scatter plot and the box plot, as in Figure 2.9.

2.3.6 Two-Way Scatter Plots

Unlike the other graphs that we have discussed, a *two-way scatter plot* is used to depict the relationship between two different continuous measurements. Each point on the graph represents a pair of values; the scale for one quantity is marked on the horizontal axis, or x-axis, and the scale for the other on the

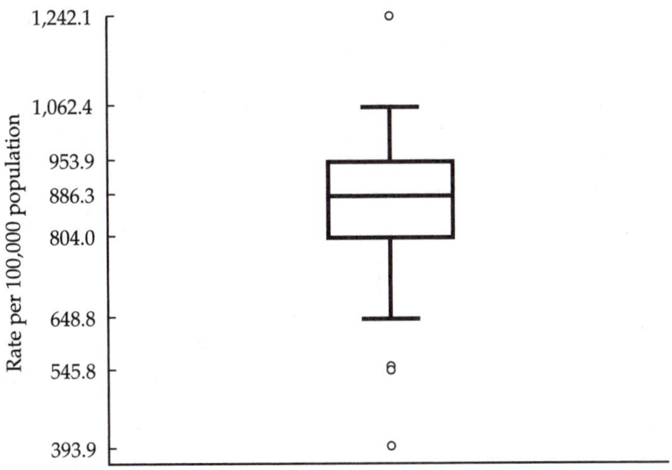

Figure 2.8. Box plot: Crude death rates for the United States, 1988

Figure 2.9. One-way scatter plot and box plot: Crude death rates for the United States, 1988

vertical axis, or y-axis. For example, Figure 2.10 plots two simple measures of lung function–forced vital capacity (FVC) and forced expiratory volume in one second (FEV_1)–for 19 asthmatic subjects who participated in a study investigating the physical effects of sulphur dioxide [7]. Note that the individual represented by the point that is farthest to the left had an FEV_1 measurement of 2.0 liters and an FVC measurement of 2.8 liters. As might be expected, there is a strong relationship between these two quantities; FVC increases in magnitude as FEV_1 increases.

2.3.7 Line Graphs

A *line graph* is similar to a two-way scatter plot in that it can be used to illustrate the relationship between continuous quantities. Once again, each point on the graph represents a pair of values. In this case, however, each value on the x-axis has a single corresponding measurement on the y-axis; adjacent points are connected by straight lines. Most commonly, the scale along the horizontal axis represents time. Consequently, we are able to trace the change in the quantity on the vertical axis over a specified period. As an example, Figure 2.11 displays the trends in the reported rates of malaria–including changes that are the result of identifiable sources–that occurred in the United States between 1940 and 1989 [8].

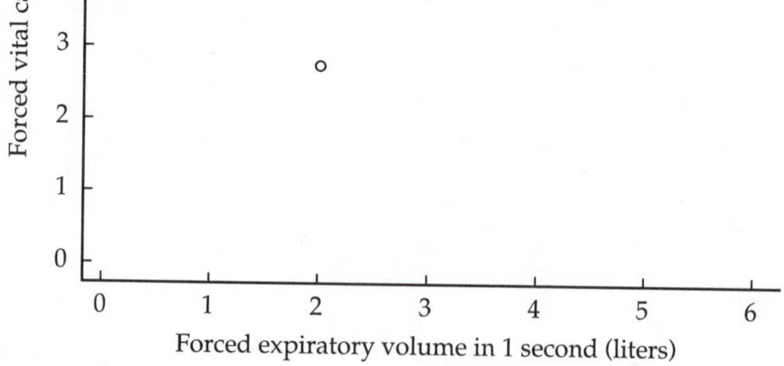

Figure 2.10. Two-way scatter plot: Forced vital capacity versus forced expiratory volume in one second for nineteen asthmatic subjects

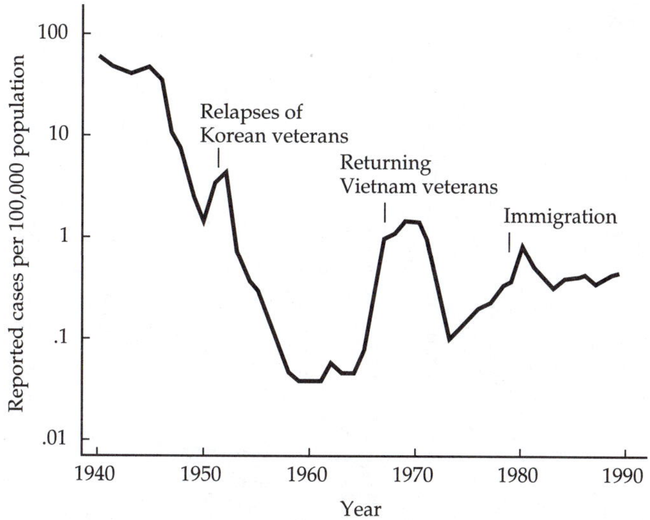

Figure 2.11. Line graph: Reported rates of malaria by year, United States, 1940–1989

Note the log scale on the vertical axis; this scale allows us to depict a large range of observations while still showing variation among the smaller values.

To compare two or more groups with respect to a given quantity, it is possible to plot more than one measurement along the y-axis. Suppose that we are concerned with the rising costs of health care. To investigate this problem, we might wish to compare the variations in cost that have occurred under two different health care systems in recent years. Figure 2.12 (on the following page) depicts the trends in health care expenditures in both the United States and Canada between 1970 and 1989 [9].

In this section we have not attempted to examine all possible types of graphs. Instead, we have included only a selection of the more common ones. It should be noted that many other imaginative displays exist [10]. As a general rule, however, too much information should not be squeezed into a single graph. A relatively simple illustration is often the most effective.

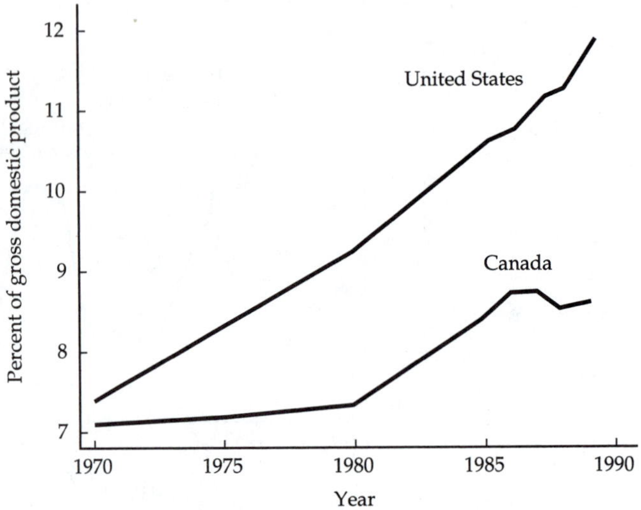

Figure 2.12. Line graph: Health care expenditures as a percentage of gross domestic product for the United States and Canada, 1970–1989

2.4 Further Applications

Suppose that we wish to reduce the number of childhood deaths caused by injury. We first need to understand the nature of the problem. Displayed below is a set of data indicating the causes of death for 100 children between the ages of 5 and 9 who died as a result of injury [11]. The data are nominal; 1 represents a motor vehicle accident, 2 represents drowning, 3 is a fire in the home, 4 is homicide, and 5 designates other causes, including suffocation, falls, and poisoning. Given these data, what are we able to conclude about childhood injury deaths?

1	5	3	1	2	4	1	3	1	5
2	1	1	5	3	1	2	1	4	1
4	1	3	1	5	1	2	1	1	2
5	1	1	5	1	5	3	1	2	1
2	3	1	1	2	1	5	1	5	1
1	2	5	1	1	1	3	4	1	1
1	1	2	1	1	2	1	1	2	3
3	3	1	5	2	3	5	1	3	4
1	1	2	4	5	4	1	5	1	5
5	1	1	5	1	1	5	1	1	5

Although the entire set of outcomes is available, it is extremely difficult to make any type of statement about these data. If we wished to summarize the observations, however, we could begin by constructing a frequency distribution. For nominal and ordinal data, a frequency distribution is a table made up of a list of categories or classes along with the numerical counts that correspond to each one. To construct a frequency distribution for the set of data shown above, we would begin by listing the various causes of death; we would then count the number of children who died as a result of each of these causes. These observations are displayed in frequency distribution format in Table 2.8. Using this table, we are able to see that 48% of the injury deaths of children in this age group are the result of motor vehicle accidents, 14% are caused by drowning, 12% by house fires, 7% by homicide, and 19% by other causes.

Table 2.8 Injury deaths of 100 children between the ages of 5 and 9 years, United States, 1980–1985

Cause	Number of Deaths
Motor vehicle	48
Drowning	14
House fire	12
Homicide	7
Other	19
Total	100

Table 2.9 Absolute frequencies of birth weights for 3,751,275 infants born in the United States, 1986

Birth Weight (grams)	Number of Infants
0–499	4,843
500–999	17,487
1,000–1,499	23,139
1,500–1,999	49,112
2,000–2,499	160,919
2,500–2,999	597,738
3,000–3,499	1,376,008
3,500–3,999	1,106,634
4,000–4,499	344,390
4,500–4,999	62,769
5,000–5,500	8,236
Total	3,751,275

Discrete and continuous data can also be displayed in the form of a frequency distribution. To begin, the range of values of the outcomes must be subdivided into a series of nonoverlapping intervals. The numbers of observations that fall within each pair of limits are then counted and arranged in a table. Suppose that we are interested in studying the consequences of low birth weight among newborns. To put the magnitude of the problem into context, we first examine the distribution of birth weights for all infants born in the United States in 1986 [12]. We separate these observations into intervals of equal width; the corresponding frequencies are displayed in Table 2.9. This table provides us with more information about the distribution of birth weights than would a list of 3,751,275 measurements. We can see that most of the observations lie between 2,000 and 4,499 grams; relatively few measurements fall outside this range. The intervals 3,000–3,499 and 3,500–3,999 grams contain the largest numbers of values.

We might also be interested in finding the relative frequency associated with each interval in the table. The relative frequency is the percentage of the total number of observations that lie within an interval. The relative frequencies for the birth weights displayed in Table 2.9—which we compute by dividing the number of values in the interval by the total number of measurements in the table and multiplying by 100—are shown in Table 2.10. The table indicates that $36.7 + 29.5 = 66.2\%$ of the birth weights are between 3,000 and 3,999 grams, and $4.3 + 15.9 + 36.7 + 29.5 + 9.2 = 95.6\%$ are between 2,000 and 4,499 grams.

In addition to tables, we can also use graphs to summarize and display data. For example, we could illustrate the nominal data in Table 2.8 using the bar chart in Figure 2.13. The categories into which the observations fall are placed along the horizontal axis; the vertical bars represent the frequency of observations in each class. The graph emphasizes that a large proportion of childhood injury deaths are the result of motor vehicle accidents.

Table 2.10 Relative frequencies of birth weights for 3,751,275 infants born in the United States, 1986

Birth Weight (grams)	Relative Frequency (%)
0–499	0.1
500–999	0.5
1,000–1,499	0.6
1,500–1,999	1.3
2,000–2,499	4.3
2,500–2,999	15.9
3,000–3,499	36.7
3,500–3,999	29.5
4,000–4,499	9.2
4,500–4,999	1.7
5,000–5,500	0.2
Total	100.0

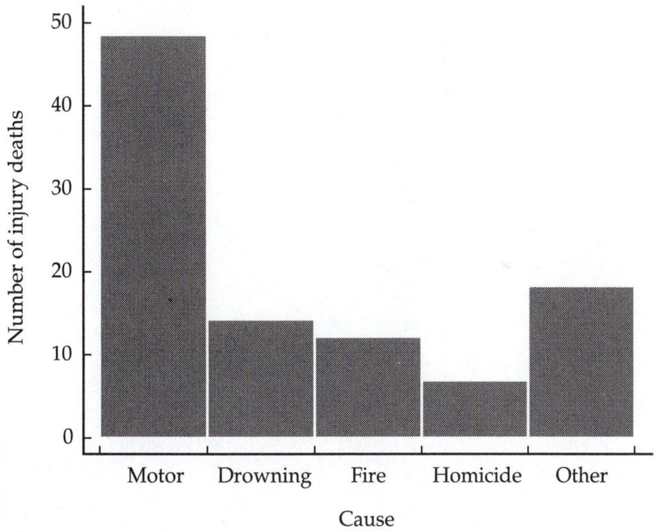

Figure 2.13. Injury deaths of 100 children between the ages of 5 and 9 years, United States, 1980–1985

A *stacked bar chart* can be used to convey a greater amount of information in a single picture. In this type of graph, bars that represent the frequency of observations in two or more different subgroups are placed on top of one another. As an example, Figure 2.14 displays the mortality rates per 1,000 births in France for four categories of infants—those who were stillborn, those who died less than one week after they were born, those who died between 7 and 27 days after they were born, and those who survived for more than 27 days but less than one year [13]. Since each of these rates is decreasing over time, the graph is able to make a powerful statement about the overall mortality.

Of the various graphical displays that can be applied to discrete or continuous data, the histogram is perhaps the most commonly used. Like a bar chart, it is a pictorial representation of a frequency distribution. The horizontal axis displays the true limits of the intervals into which the observations fall; the vertical axis depicts the frequency or relative frequency of observations within each interval. As an example, Figure 2.15 is a histogram of the birth weight data summarized in Table 2.9. Looking at the graph, we can see that the data are skewed to the left.

A line graph is one type of display that can be used to illustrate the relationship between continuous quantities. Each point on the line represents a pair of values; the line itself allows us to trace the change in the quantity on the y-axis which corresponds to a change along the x-axis. Figure 2.16, like Figure 2.1, depicts data relating to cigarette consumption in the United States. Note that the line graph shows more detail than the corresponding bar chart.

Over the years, the use of computers in statistics has increased dramatically. As a result, many formerly time-consuming calculations can now be performed much more efficiently using a statistical package. A *statistical package* is a series of programs that have been designed to analyze numerical data. A variety of packages are available; in general, they differ with respect to the commands they use and the format of the output they produce.

One statistical package that is both powerful and relatively easy to use is called *Stata*. Stata is an interactive program that helps us manage, display, and

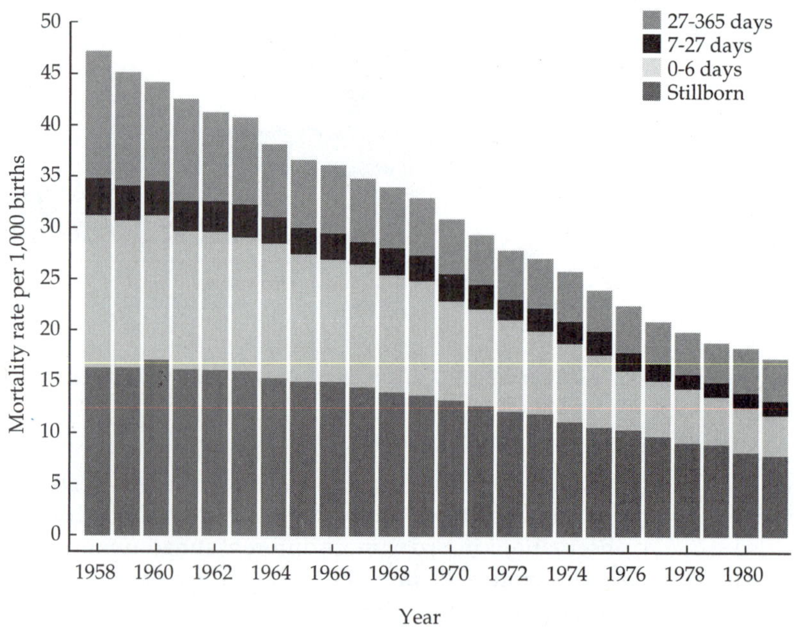

Figure 2.14. Infant and perinatal mortality in France, 1978–1981

analyze data. Observations or measurements are saved in columns; each column is assigned a variable name. We then use these names to execute specific analytical procedures or commands. When it is appropriate, we will reproduce output from Stata to illustrate what the computer is able to do.

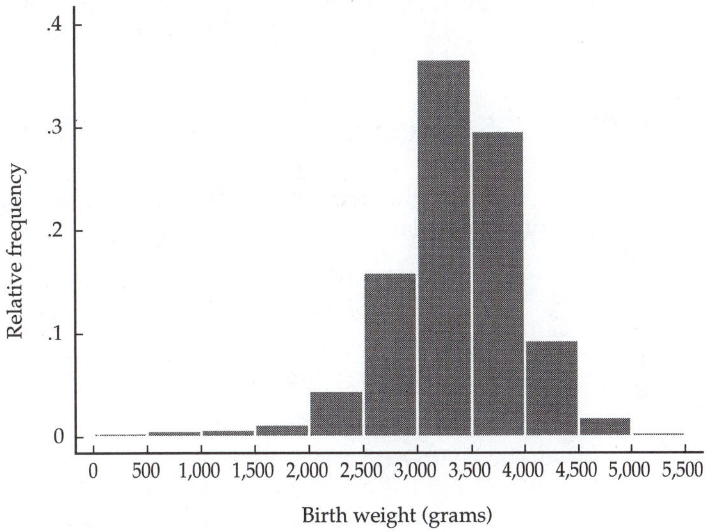

Figure 2.15. Relative frequencies of birth weights for 3,751,275 infants born in the United States, 1986

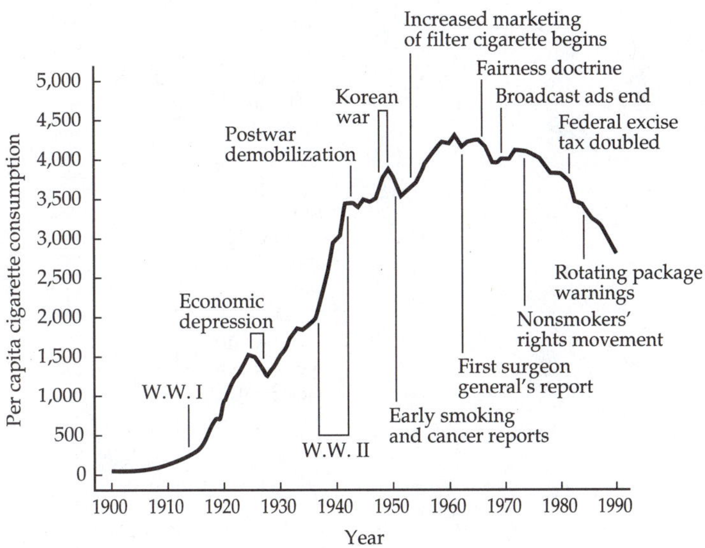

Figure 2.16. Cigarette consumption per person 18 years of age or older, United States, 1900–1990

Computers are particularly useful in constructing graphs. For example, Figures 2.13 through 2.16 were all generated in Stata. To create Figure 2.16 we saved the years between 1900 and 1990 under the variable name `year` and saved the values of per capita cigarette consumption under the name `cigarette`. Stata plotted the points representing each pair of values; the points were connected and labels were added by means of the appropriate command options.

2.5 Review Exercises

1. State whether each of the following observations is an example of discrete or continuous data.

   a. The number of suicides in the United States in 1990
   b. The concentration of lead in a sample of water
   c. The length of time that a cancer patient survives after diagnosis
   d. The number of previous miscarriages an expectant mother has had

2. Below is a list of 23 countries and the per capita health care expenditure associated with each one in 1989 [9].

Nation	Per Capita Expenditure (U.S.$)
Australia	1,032
Austria	1,093
Belgium	980
Britain	836
Canada	1,683
Denmark	912
Finland	1,067
France	1,274
Germany	1,232
Greece	371
Iceland	1,353
Ireland	658
Italy	1,050
Japan	1,035
Luxembourg	1,193
Netherlands	1,135
New Zealand	820
Norway	1,234
Portugal	464
Spain	644
Sweden	1,361
Switzerland	1,376
United States	2,354

   a. Rank these countries according to per capita health care expenditure.

b. Construct a histogram of these data.
c. Describe the shape of the histogram.

3. The table below categorizes 10,614,000 office visits to cardiovascular disease specialists in the United States by the duration of each visit [14]. A duration of 0 minutes implies that the patient did not have face-to-face contact with the specialist.

Duration (minutes)	Number of Visits (thousands)
0	390
1–5	227
6–10	1,023
11–15	3,390
16–30	4,431
31–60	968
61+	185
Total	10,614

The statement is made that office visits to cardiovascular disease specialists are most often between 16 and 30 minutes long. Do you agree with this statement? Why or why not?

4. The frequency distribution below displays the numbers of cases of pediatric AIDS reported in the United States between 1983 and 1989 [8].

Year	Number of Cases
1983	122
1984	250
1985	455
1986	848
1987	1,412
1988	2,811
1989	3,098

Construct a bar chart to illustrate these data.

5. The relative frequencies of blood lead concentrations for two groups of workers in Canada—one examined in 1979 and the other in 1987—are displayed below [15].

Blood Lead (μg/dl)	1979 (%)	1987 (%)
<20	11.5	37.8
20–29	12.1	14.7
30–39	13.9	13.1
40–49	15.4	15.3
50–59	16.5	10.5
60–69	12.8	6.8
70–79	8.4	1.4
≥80	9.4	0.4

a. In which of the two years do the workers tend to have lower blood lead levels?
b. Compute the cumulative relative frequencies for each group of workers and use these data to construct a pair of cumulative frequency polygons.
c. For which group of workers is the distribution of blood lead levels stochastically larger?

6. A frequency distribution for the serum zinc levels of 462 males between the ages of 15 and 17 is displayed below [16]. The observations are stored on the accompanying disk in a data set called `serzinc`. The 462 serum zinc measurements, which were recorded in micrograms per deciliter, are saved under the variable name `zinc`.

United States Males, Ages 15–17	
Serum Zinc Level (μg/dl)	**Number of Males**
50–59	6
60–69	35
70–79	110
80–89	116
90–99	91
100–109	63
110–119	30
120–129	5
130–139	2
140–149	2
150–159	2

a. Compute the relative frequency associated with each interval. What can you conclude about this distribution of serum zinc levels?
b. Produce a histogram of the data. The observations should be broken down into the eleven intervals of equal width specified in the preceding frequency distribution, from 50–59 to 150–159 μg/dl.
c. Describe the shape of the histogram.

7. The percentages of low birth weight infants in various countries around the world are contained in a data set called `unicef` [17]. The measurements themselves are saved under the variable name `lowbwt`.

a. Construct a box plot for these data.
b. Do the data appear to be skewed?
c. Do the data contain any outlying observations?

References

[1] Oken, M. M., Creech, R. H., Tormey, D. C., Horton, J., Davis, T. E., McFadden, E. T., and Carbone, P. P., "Toxicity and Response Criteria of the Eastern Cooperative Oncology Group," *American Journal of Clinical Oncology*, Volume 5, December 1982, 649–655.

[2] National Center for Health Statistics, "Advance Report of Final Mortality Statistics, 1988," *Monthly Vital Statistics Report*, Volume 39, November 28, 1990.

[3] Garfinkel, L., and Silverberg, E., "Lung Cancer and Smoking Trends in the United States over the Past 25 Years," *Ca–A Cancer Journal for Clinicians*, Volume 41, May/June 1991, 137–145.

[4] National Center for Health Statistics, Fulwood, R., Kalsbeek, W., Rifkind, B., Russell-Briefel, R., Muesing, R., LaRosa, J., and Lippel, K., "Total Serum Cholesterol Levels of Adults 20–74 Years of Age: United States, 1976–1980," *Vital and Health Statistics*, Series 11, Number 236, May 1986.

[5] Spear, M. E., *Charting Statistics*, New York: McGraw-Hill, 1952.

[6] Tukey, J. W., *Exploratory Data Analysis*, Reading, Mass.: Addison-Wesley, 1977.

[7] Bethel, R. A., Sheppard, D., Geffroy, B., Tam, E., Nadel, J. A., and Boushey, H. A., "Effect of 0.25 ppm Sulphur Dioxide on Airway Resistance in Freely Breathing, Heavily Exercising, Asthmatic Subjects," *American Review of Respiratory Disease*, Volume 31, April 1985, 659–661.

[8] Centers for Disease Control, "Summary of Notifiable Diseases, United States, 1989," *Morbidity and Mortality Weekly Report*, Volume 38, October 5, 1990.

[9] Schieber, G. J., and Poullier, J. P., "International Health Spending: Issues and Trends," *Health Affairs*, Volume 10, Spring 1991, 106–116.

[10] Tufte, E. R., *The Visual Display of Quantitative Information*, Cheshire, Conn.: Graphics Press, 1983.

[11] Waller, A. E., Baker, S. P., and Szocka, A., "Childhood Injury Deaths: National Analysis and Geographic Variations," *American Journal of Public Health*, Volume 79, March 1989, 310–315.

[12] National Center for Health Statistics, "Advance Reports, 1986," *Supplements to the Monthly Vital Statistics Report*, Series 24, March 1990.

[13] Rumeau-Rouquette, C., "The French Perinatal Program: 'Born in France'," *Child Health and Development*, Volume 3, *Prevention of Perinatal Mortality and Morbidity*, New York: Karger, 1984.

[14] National Center for Health Statistics, Nelson, C., "Office Visits to Cardiovascular Disease Specialists, 1985," *Vital and Health Statistics*, Advance Data Report Number 171, June 23, 1989.

[15] Yassi, A., Cheang, M., Tenenbein, M., Bawden, G., Spiegel, J., and Redekop, T., "An Analysis of Occupational Blood Lead Trends in Manitoba, 1979 Through 1987," *American Journal of Public Health*, Volume 81, June 1991, 736–740.

[16] National Center for Health Statistics, Fulwood, R., Johnson, C. L., Bryner, J. D., Gunter, E. W., and McGrath, C. R., "Hematological and Nutritional Biochemistry Reference Data for Persons 6 Months–74 Years of Age: United States, 1976–1980," *Vital and Health Statistics*, Series 11, Number 232, December 1982.

[17] United Nations Children's Fund, *The State of the World's Children 1991*, New York: Oxford University Press.

CHAPTER

3

Numerical Summary Measures

In the preceding chapter we studied tables and graphs as methods of organizing, visually summarizing, and displaying data. Although these techniques are extremely useful, they do not allow us to make concise, quantitative statements that characterize a distribution as a whole. In order to do this, we rely on *numerical summary measures*.

3.1 Measures of Central Tendency

The most commonly investigated characteristic of a set of data is its center, or the point about which the observations tend to cluster. Consider the data taken from a study that examines the response to ozone and sulphur dioxide among adolescents suffering from asthma [1]. Listed in Table 3.1 are the initial measurements of forced expiratory volume in 1 second for the 13 subjects involved in the study. Before investigating the effects of the pollutants, we might wish to determine the typical baseline FEV_1 for the individuals in this group.

3.1.1 Mean

The most frequently used measure of central tendency is the *arithmetic mean*, or *average*. The mean is calculated by summing all the observations in a set of data and dividing by the total number of measurements. In Table 3.1, for example, we have $n = 13$ observations. If x is used to represent FEV_1, then $x_1 = 2.30$ denotes the first in the series of observations; $x_2 = 2.15$, the second; and so on up through $x_{13} = 3.38$. In general, x_i refers to a single measurement where i can take on any

Table 3.1 Forced expiratory volumes in 1 second for 13 adolescents suffering from asthma

Subject	FEV_1 (liters)
1	2.30
2	2.15
3	3.50
4	2.60
5	2.75
6	2.82
7	4.05
8	2.25
9	2.68
10	3.00
11	4.02
12	2.85
13	3.38

value from 1 to n. The mean of the observations in the sample—represented by $\bar{x}$, or x-bar—is

$$\bar{x} = \frac{1}{n}\sum_{i=1}^{n} x_i.$$

For the FEV_1 data, therefore,

$$\begin{aligned}\bar{x} &= \left(\frac{1}{13}\right)(2.30 + 2.15 + 3.50 + 2.60 + 2.75 + 2.82 + 4.05 \\ &\quad + 2.25 + 2.68 + 3.00 + 4.02 + 2.85 + 3.38) \\ &= \frac{38.35}{13} \\ &= 2.95 \text{ liters.}\end{aligned}$$

Note that we have used some mathematical shorthand. The Greek letter sigma, $\sum$, is the symbol for summation. The expression $\sum_{i=1}^{n} x_i$ means that we are to add up the values of all of the observations in the sample, from x_1 to x_n. When $\sum$ appears in the text, the limits of summation are placed beside it; when it does not, the limits are above and below it. Nevertheless, both representations denote exactly the same thing.

The mean can be used as a summary measure for both discrete and continuous measurements. In general, however, it is not appropriate for either nominal or ordinal data. One exception to this rule applies when we have dichotomous data, and the two possible outcomes are represented by 0 and 1. In this situation, the mean of the observations is equal to the proportion of 1s in the data set. For example, suppose that we wish to know the proportion of the asthmatic adolescents in the previously mentioned study who are males. Listed in Table 3.2 are the

Table 3.2 Indicators of gender for 13 adolescents suffering from asthma

Subject	Gender
1	0
2	1
3	1
4	0
5	0
6	1
7	1
8	1
9	0
10	1
11	1
12	1
13	0

relevant dichotomous data; the value 1 represents a male, whereas 0 designates a female. If we compute the mean of these observations, we find that

$$\begin{aligned}\bar{x} &= \frac{1}{n}\sum_{i=1}^{n} x_i \\ &= \left(\frac{1}{13}\right)(0+1+1+0+0+1+1+1+0+1+1+1+0) \\ &= \frac{8}{13} \\ &= 0.615.\end{aligned}$$

Therefore, 61.5% of the study subjects are males.

The method by which the mean is calculated takes into consideration the magnitude of each and every observation in a set of data. What happens when one observation has a value that is very different from the others? Suppose, for instance, that we had recorded the data in Table 3.1 on disk and that this disk was accidentally Xrayed at the airport; as a result, the FEV_1 measurement of subject 11 is now recorded as 40.2 rather than 4.02. The mean FEV_1 of all 13 subjects would then be calculated as

$$\begin{aligned}\bar{x} &= \left(\frac{1}{13}\right)(2.30+2.15+3.50+2.60+2.75+2.82+4.05 \\ &\qquad +2.25+2.68+3.00+40.2+2.85+3.38) \\ &= \frac{74.53}{13} \\ &= 5.73 \text{ liters,}\end{aligned}$$

which is nearly twice as large as it was before. Clearly, the mean is extremely sensitive to unusual values. In this particular example, we would have rightfully questioned an FEV_1 measurement of 40.2 liters and would have either corrected the error or separated this observation from the others. In general, however, the error might not be as obvious, or the unusual observation might not be an error at all. Since it is our intent to characterize an entire group of individuals, we might prefer to use a summary measure that is not as sensitive to each and every observation.

3.1.2 Median

One measure of central tendency that is not as sensitive to the value of each measurement is the median. The median can be used as a summary measure for ordinal data as well as discrete and continuous data. The *median* is defined as the 50th percentile of a set of measurements; if a list of observations is ranked from the smallest to the largest, then half the values would be greater than or equal to the median, whereas the other half would be less than or equal to it. Therefore, if a set of data contains a total of n observations where n is odd, then the median is the middle value, or the $[(n+1)/2]^{\text{th}}$ largest measurement; if n is even, the median is usually taken to be the average of the two middlemost values, the $(n/2)^{\text{th}}$ and $[(n/2)+1]^{\text{th}}$ observations. If we were to rank the 13 FEV_1 measurements listed in Table 3.1, for example, the following sequence would result:

2.15, 2.25, 2.30, 2.60, 2.68, 2.75, 2.82, 2.85, 3.00, 3.38, 3.50, 4.02, 4.05.

Since there are an odd number of observations in the list, the median would be the $(13+1)/2 = 7^{\text{th}}$ observation, or 2.82. Six of the measurements are less than 2.82 liters, and six are greater.

The calculation of the median takes into consideration the ordering and the relative magnitude of the observations in a set of data. In the situation where the FEV_1 of subject 11 was recorded as 40.2 rather than 4.02, the ranking of the measurements would change only slightly:

2.15, 2.25, 2.30, 2.60, 2.68, 2.75, 2.82, 2.85, 3.00, 3.38, 3.50, 4.05, 40.2.

As a result, the median FEV_1 would remain 2.82 liters. The median is said to be *robust*; that is, it is much less sensitive to unusual data points than is the mean.

3.1.3 Mode

A third measure of central tendency is the mode; it can be used as a summary measure for all types of data. The *mode* of a set of values is the observation that occurs most frequently. The FEV_1 data in Table 3.1 do not have a unique mode since each of the values occurs only once. The mode for the dichotomous data in Table 3.2 is 1; this value appears eight times, whereas 0 appears only five times.

The best measure of central location for a given set of data often depends on the way in which the values are distributed. If they are symmetric and *unimodal*—meaning that there is only one peak, as in the idealized distribution pictured in Figure 3.1(a)—then the mean, the median, and the mode should all be roughly the same. If the distribution of values is symmetric but *bimodal*, as in Figure 3.1(b), then the mean and median should again be approximately the same. Note, however, that this common value could lie **between** the two peaks and hence be a measurement that is unlikely to occur. A bimodal distribution often indicates that the population from which the values are taken actually consists of two distinct subgroups that differ in the characteristic being measured; in this situation, it might be better to report two modes rather than the mean or the median. The data in Figure 3.1(c) are skewed to the right, and those in Figure 3.1(d) are skewed to the left. When the data are not symmetric, the median is often the best measure of central tendency. Because the mean is sensitive to extreme observations, it is pulled in the direction of the outlying data values and as a result might end up either excessively inflated or deflated. Note that when the data are skewed to the right, the mean lies to the right of the median; when they are skewed to the left, the mean lies to the left of the median.

Regardless of the measure of central tendency used in a particular situation, it can be misleading to assume that this value is representative of all observations. One example that illustrates this point was included in the November 17, 1991, episode of the popular news program "60 Minutes." The program contained a segment on diet and mortality that contrasted the French and American experiences. Although the French diet is extremely high in fat and cholesterol, France has a much lower rate of heart disease than the United States. This paradoxical difference was attributed to the French habit of drinking wine—red wine in particular—with meals. Studies have suggested that moderate alcohol consumption can lessen the risk of heart disease. While the per capita intake of wine in

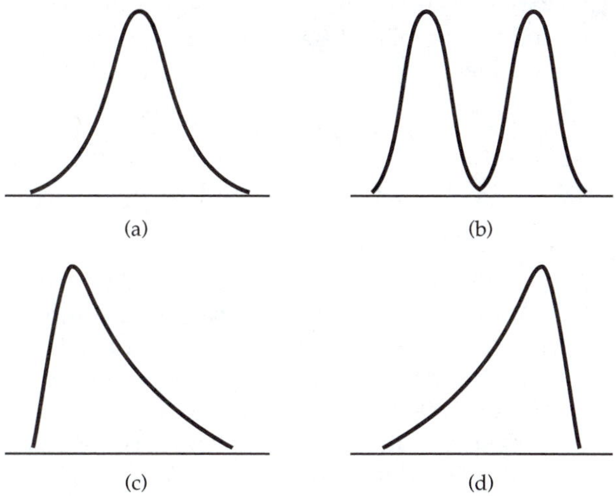

Figure 3.1. Possible distributions of data values

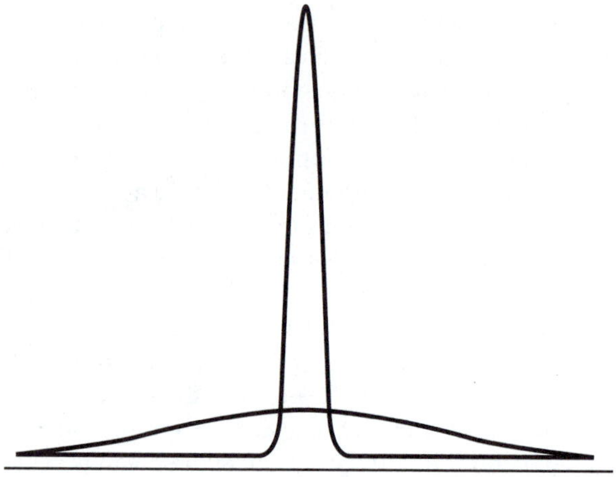

Figure 3.2. Two distributions with identical means, medians, and modes

France is one of the highest in the world, the per capita intake in the United States is one of the lowest; the program implied that the French drink a moderate amount of wine each day, perhaps two or three glasses. The reality may be quite different, however. According to a wine industry survey conducted in 1990, more than half of all French adults never drink wine at all [2]. Of those who do, only 28% of males and 11% of females drink it daily. Obviously the distribution is far more variable than the "typical value" suggests. Thus, although it is helpful to know where the center of a data set lies, this information is usually not sufficient to characterize a distribution.

As another example, the two very different distributions of data values pictured in Figure 3.2 have the same means, medians, and modes. To know how good our measure of central tendency actually is, we need to have some idea of the variation among the data values. Do all the observations tend to be quite similar and therefore lie close to the center, or are they spread out across a broad range of values?

3.2 Measures of Dispersion

3.2.1 Range

One number that can be used to describe the variability in a set of data values is known as the range. The *range* of a group of measurements is defined as the difference between the largest and the smallest observations. Although the range is easy to compute, its usefulness is limited; it considers only the extreme values of a data set rather than the majority of the observations. Therefore, like the mean, it is highly sensitive to exceptionally large or exceptionally small values. The range for the FEV_1 data in Table 3.1 is $4.05 - 2.15 = 1.90$ liters. The ranges of values for the average annual sulphur dioxide concentrations in the air of a number of cities around the world are displayed in Figure 3.3 [3].

Shown is the range of annual values at individual sites and the composite 5-year average for the city.

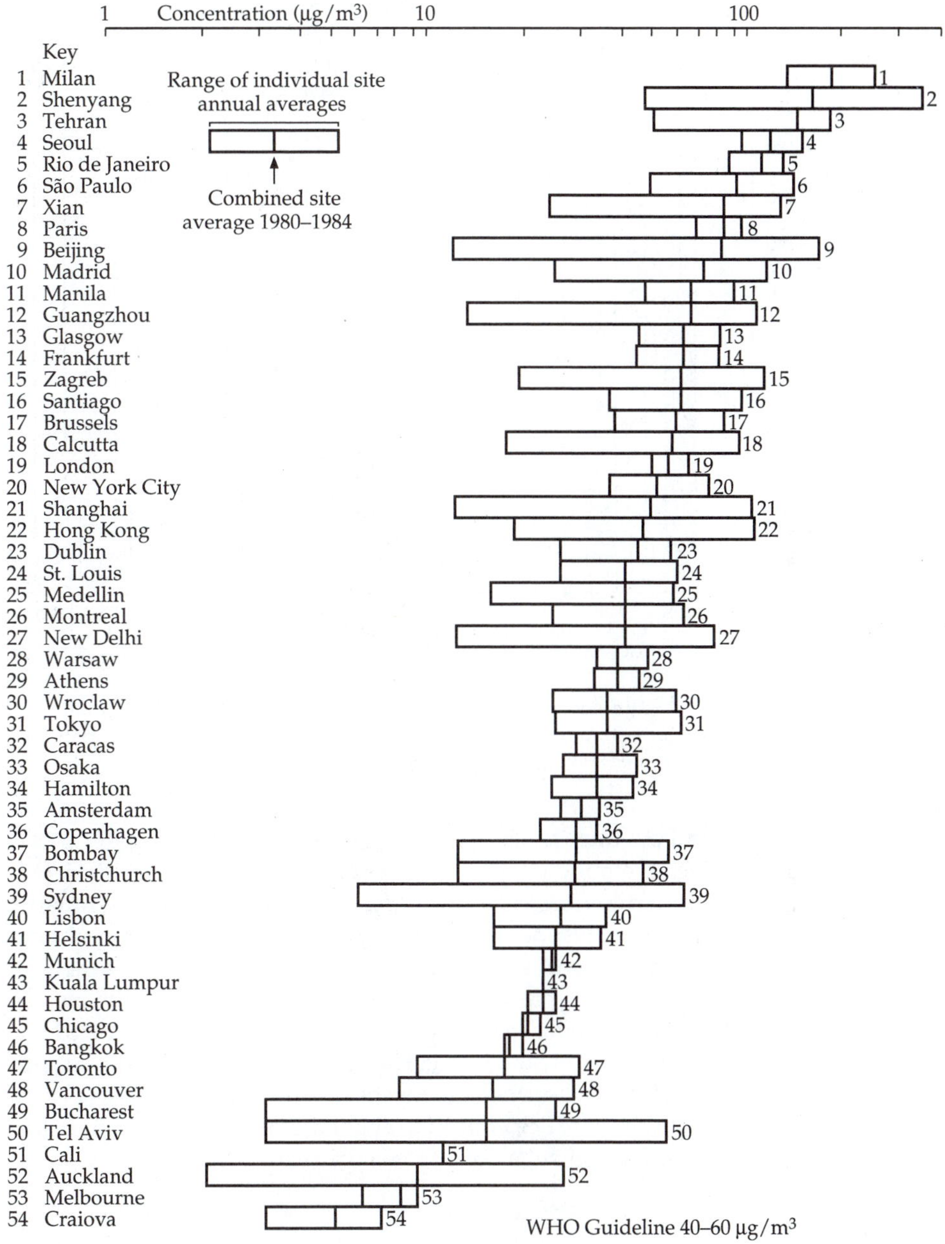

Figure 3.3. Summary of annual sulphur dioxide averages, 1980–1984

3.2.2 Interquartile Range

A second measure of variability—one that is not as easily influenced by extreme values—is called the *interquartile range*. The interquartile range is calculated

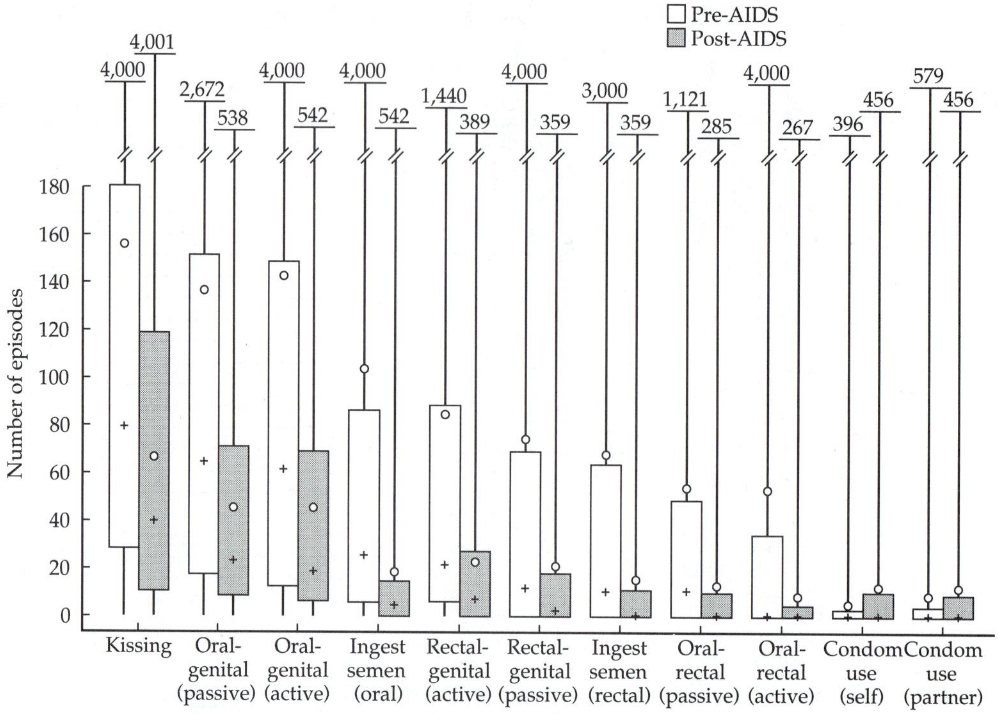

Figure 3.4. The median (+) and mean (○) annual frequency, interquartile range (boxes), and total range of engaging in specific acts during the year before hearing about AIDS and after hearing about AIDS

by subtracting the 25th percentile of the data from the 75th percentile; it encompasses the middle 50% of the observations. For the FEV_1 data in Table 3.1, the interquartile range is $3.38 - 2.60 = 0.78$ liters. The interquartile ranges of the numbers of episodes of various types of sexual behavior practiced by homosexual males before and after learning about AIDS—as well as their means, medians, and ranges—are presented in Figure 3.4 [4]. Note that the means are larger than the medians in all cases, indicating that the data are skewed and that there are a number of unusually large values that are causing the means to be inflated. The difference between the means and the medians is less evident after the men learned about AIDS; education about the virus appears to have had a constraining effect on sexual behavior, especially in extreme cases.

3.2.3 Variance and Standard Deviation

Another commonly used measure of dispersion for a set of data values is known as the variance. The *variance* quantifies the amount of variability or spread about the mean of a sample. To accomplish this, we might simply attempt to calculate the average distance of the individual observations from $\bar{x}$, or

$$\frac{1}{n}\sum_{i=1}^{n}(x_i - \bar{x}).$$

It can be shown mathematically, however, that $\sum_{i=1}^{n}(x_i - \bar{x})$ is always equal to zero. By definition, the sum of the deviations from the mean of all observations less than $\bar{x}$ is equal to the sum of the deviations of all observations greater than $\bar{x}$; consequently, these two sums cancel each other out. To eliminate this problem, we might try to square the deviations from the mean—recall that a squared quantity is always positive—and then find the average of these squared distances. This procedure gives us the variance of the observations.

More explicitly, the variance is calculated by subtracting the mean of a set of data values from each of the observations, squaring these deviations, adding them up, and dividing by one less than the number of observations in the data set. Representing the variance by s^2,

$$s^2 = \frac{1}{(n-1)}\sum_{i=1}^{n}(x_i - \bar{x})^2.$$

The reason for dividing by $n - 1$ rather than n when we calculate the variance will be discussed further in Chapter 9; we can, nonetheless, think of the variance as an average of squared deviations. For the FEV_1 measurements presented in Table 3.1, the variance is

$$\begin{aligned}
s^2 &= \frac{1}{(13-1)}\sum_{i=1}^{13}(x_i - 2.95)^2 \\
&= \left(\frac{1}{12}\right)[(-0.65)^2 + (-0.80)^2 + (0.55)^2 + (-0.35)^2 + (-0.20)^2 \\
&\quad + (-0.13)^2 + (1.10)^2 + (-0.70)^2 + (-0.27)^2 + (0.05)^2 \\
&\quad + (1.07)^2 + (-0.10)^2 + (0.43)^2] \\
&= \frac{4.66}{12} \\
&= 0.39 \text{ liters}^2.
\end{aligned}$$

The *standard deviation* of a set of data is the square root of the variance. Given the 13 FEV_1 measurements in Table 3.1, therefore, the standard deviation is equal to

$$\begin{aligned}
s &= \sqrt{s^2} \\
&= \sqrt{0.39 \text{ liters}^2} \\
&= 0.62 \text{ liters}.
\end{aligned}$$

In practice, the standard deviation is used more frequently than the variance. This is primarily because the standard deviation has the same units of measurement as the mean. In a comparison of two groups of data, the group with the smaller standard deviation has the more homogeneous observations; the group with the larger standard deviation exhibits a greater amount of variability. The actual magnitude of the standard deviation depends on the values in the data set; what is large for one data set may be small for another. In addition, because the standard deviation has units of measurement, it is meaningless to compare standard deviations for two unrelated quantities.

3.2.4 Coefficient of Variation

It is possible to make comparisons among data sets representing different quantities using a numerical summary measure known as the *coefficient of variation*. The coefficient of variation relates the standard deviation of a sample to its mean; it is the ratio of s to $\bar{x}$ multiplied by 100 and is, therefore, a measure of relative variability. Because the standard deviation and the mean share the same units of measurement, the units cancel out and leave the coefficient of variation a dimensionless number. Since it is independent of measurement units, it can be used to compare the relative variation between any two sets of values. For example, the coefficient of variation for the FEV_1 data in Table 3.1 is

$$\begin{aligned} CV &= \frac{s}{\bar{x}} \times 100\% \\ &= \frac{0.62}{2.95} \times 100\% \\ &= 21.0\%. \end{aligned}$$

Although the coefficient of variation is still used as a summary measure in some circles, its statistical properties are not very good. As a result, its use is diminishing and should be discouraged.

3.3 Grouped Data

If we wish to count the change that we have in our pocket, there are two ways that we can do it. The first is to consecutively sum the values of the coins as we retrieve them. The second is to begin by grouping the coins together by denomination, then multiplying the value of each coin by the number of coins of that denomination, and finally summing these values. For example, if we have 3 pennies, 4 nickels, 2 dimes, and 1 quarter, we have a total of

$$\begin{aligned} 3(1) + 4(5) + 2(10) + 1(25) &= 3 + 20 + 20 + 25 \\ &= 68 \text{ cents.} \end{aligned}$$

The same procedure can be used to sum any given set ofobservations. For

Table 3.3 Duration of transfusion therapy for ten patients with sickle cell disease

Subject	Duration (years)
1	12
2	11
3	12
4	6
5	11
6	11
7	8
8	5
9	5
10	5

instance, consider the data in Table 3.3 [5]. Among patients with sickle cell disease, transfusion therapy is often used to prevent recurrent strokes after an initial cerebrovascular event. Long-term transfusion therapy, however, has its own associated risks and is not always advisable. Table 3.3 lists the duration of therapy for ten patients enrolled in a study investigating the effects of stopping the transfusions. We might be interested in determining the mean of these values.

The standard technique for finding the mean of these observations is to simply add the values and divide by $n = 10$; in this case, we find that

$$\begin{aligned}\bar{x} &= \frac{1}{n}\sum_{i=1}^{n} x_i \\ &= \left(\frac{1}{10}\right)(12 + 11 + 12 + 6 + 11 + 11 + 8 + 5 + 5 + 5) \\ &= \frac{86}{10} \\ &= 8.6 \text{ years.}\end{aligned}$$

Alternatively, we could have found the sum of the measurements by first grouping the observations with equal values; note that there are three 5s, one 6, one 8, three 11s, and two 12s. In this case,

$$\begin{aligned}\sum_{i=1}^{10} x_i &= 3(5) + 1(6) + 1(8) + 3(11) + 2(12) \\ &= 15 + 6 + 8 + 33 + 24 \\ &= 86,\end{aligned}$$

and

$$\begin{aligned}\bar{x} &= \frac{86}{10} \\ &= 8.6 \text{ years.}\end{aligned}$$

We obtain the same mean regardless of the method we use.

The technique of grouping measurements that have equal values before calculating the mean has one distinct advantage over the standard method: this procedure can be applied to data that have been summarized in the form of a frequency distribution. Data that are organized in this way are often referred to as *grouped data.* Even if the original observations are no longer available, we might still be interested in calculating numerical summary measures for the data. An obstacle arises in that we do not know the values of the individual observations; however, we are able to determine the number of measurements that fall into each specified interval. This information can be used to calculate a *grouped mean.*

To compute the mean of a set of data arranged as a frequency distribution, we begin by assuming that all the values that fall into a particular interval are equal to the midpoint of that interval. Recall the serum cholesterol level data for the group of 25- to 34-year-old men that was presented in Table 2.5; these data are reproduced in Table 3.4. The first interval contains values that range from 80 to 119 mg/100 ml, with a midpoint of 99.5. We assume, therefore, that all 13 measurements within this interval take the value 99.5 mg/100 ml. Similarly, we assume that the 150 observations in the second interval all take the value 139.5 mg/100 ml. Because we are making these assumptions, our calculation is only approximate.

To find the mean of the grouped data, we first sum the measurements by multiplying the midpoint of each interval by the corresponding frequency and adding these products; we then divide by the total number of values. Therefore,

$$\bar{x} = \frac{\sum_{i=1}^{k} m_i f_i}{\sum_{i=1}^{k} f_i}$$

where k is the number of intervals in the table, m_i is the midpoint of the ith interval, and f_i is the frequency associated with the ith interval. Note that the sum of the

Table 3.4 Absolute frequencies of serum cholesterol levels for U.S. males, aged 25 to 34 years, 1976–1980

Cholesterol Level (mg/100 ml)	Number of Men
80–119	13
120–159	150
160–199	442
200–239	299
240–279	115
280–319	34
320–359	9
360–399	5
Total	1,067

frequencies $\sum_{i=1}^{k} f_i$ is equal to the total number of observations n. For the data in Table 3.4,

$$
\begin{aligned}
\bar{x} &= \frac{\sum_{i=1}^{8} m_i f_i}{\sum_{i=1}^{8} f_i} \\
&= \left(\frac{1}{1,067}\right)[99.5(13) + 139.5(150) + 179.5(442) + 219.5(299) \\
&\quad + 259.5(115) + 299.5(34) + 339.5(9) + 379.5(5)] \\
&= \frac{212,166.5}{1,067} \\
&= 198.8 \text{ mg/100 ml.}
\end{aligned}
$$

The grouped mean is actually a weighted average of the interval midpoints; each midpoint is weighted by the frequency of observations within the interval.

After we have calculated the mean of a set of grouped data, we might also wish to find its standard deviation. Again we assume that all the observations falling within a particular interval are equal to the midpoint of that interval. The variance of the data is

$$
s^2 = \frac{\sum_{i=1}^{k}(m_i - \bar{x})^2 f_i}{[\sum_{i=1}^{k} f_i] - 1}
$$

where all terms are defined as for the mean. The variance for the data in Table 3.4, therefore, is

$$
\begin{aligned}
s^2 &= \frac{\sum_{i=1}^{8}(m_i - 198.8)^2 f_i}{[\sum_{i=1}^{8} f_i] - 1} \\
&= \left(\frac{1}{1,067 - 1}\right)[(-99.3)^2(13) + (-59.3)^2(150) + (-19.3)^2(442) \\
&\quad + (20.7)^2(299) + (60.7)^2(115) + (100.7)^2(34) \\
&\quad + (140.7)^2(9) + (180.7)^2(5)] \\
&= \frac{2,058,342.8}{1,066} \\
&= 1,930.9 \text{ (mg/100 ml)}^2.
\end{aligned}
$$

Recall that the standard deviation is the square root of the variance; consequently, the standard deviation of the serum cholesterol level data is

$$
\begin{aligned}
s &= \sqrt{1,930.9 \text{ (mg/100 ml)}^2} \\
&= 43.9 \text{ mg/100 ml.}
\end{aligned}
$$

3.4 Chebychev's Inequality

Once the mean and the standard deviation of a given set of data have been calculated, these two numbers can be used to summarize the characteristics of the entire distribution. The mean tells us where the observations are centered; the standard deviation gives us an idea of the amount of dispersion about that center. This concept can be made more precise using *Chebychev's inequality*.

Chebychev's inequality allows us to say that for any number k that is greater than or equal to 1, at least $[1 - (1/k)^2]$ of the measurements in a set of data lie within k standard deviations of their mean [6]. Given that $k = 2$, for example, at least

$$\begin{aligned} 1 - \left(\frac{1}{2}\right)^2 &= 1 - \left(\frac{1}{4}\right) \\ &= \frac{3}{4} \end{aligned}$$

of the values lie within two standard deviations of the mean. Equivalently, we could say that $\bar{x} \pm 2s$ encompasses at least 75% of the observations in the group. This statement is true no matter what the values of $\bar{x}$ and s are. Similarly, if $k = 3$, at least

$$\begin{aligned} 1 - \left(\frac{1}{3}\right)^2 &= 1 - \left(\frac{1}{9}\right) \\ &= \frac{8}{9} \end{aligned}$$

of the observations lie within three standard deviations of the mean; therefore, $\bar{x} \pm 3s$ contains at least 88.9% of the measurements.

Chebychev's inequality is a conservative statement; it applies to the mean and standard deviation of **any** distribution, no matter what its shape. Returning to the FEV_1 data in Table 3.1, for instance, we can say that the interval

$$2.95 \pm (2 \times 0.62)$$

or

$$(1.71\,,\,4.19)$$

encompasses at least 75% of the observations, whereas the interval

$$2.95 \pm (3 \times 0.62)$$

or

$$(1.09\,,\,4.81)$$

contains at least 88.9%. In fact, both intervals contain all 13 of the measurements. Similarly, for the serum cholesterol level data in Table 3.4, we are able to state that the interval

$$198.8 \pm (2 \times 43.9)$$

or

$$(111.0\,,\ 286.6)$$

encloses at least 75% of the values, whereas the interval

$$198.8 \pm (3 \times 43.9)$$

or

$$(67.1\,,\ 330.5)$$

contains at least 88.9%. Thus, although conservative, Chebychev's inequality allows us to use the mean and the standard deviation of a set of data–just two numbers–to describe the entire group.

3.5 Further Applications

In a study investigating the causes of death among persons with severe asthma, data were recorded for ten patients who arrived at the hospital in a state of respiratory arrest; breathing had stopped, and the subjects were unconscious. Table 3.5 lists the heart rates of the ten patients upon admission [7]. How can we characterize this set of observations?

To begin, we might be interested in finding a typical heart rate for the ten individuals. The most commonly used measure of central tendency is the mean. To find the mean of these data, we simply sum all the observations and divide by $n = 10$. Therefore, for the data in Table 3.5,

$$\begin{aligned}\bar{x} &= \frac{1}{n}\sum_{i=1}^{n} x_i \\ &= \left(\frac{1}{10}\right)(167 + 150 + 125 + 120 + 150 + 150 + 40 \\ &\qquad + 136 + 120 + 150) \\ &= \frac{1,308}{10} \\ &= 130.8 \text{ beats per minute.}\end{aligned}$$

The mean heart rate upon admission to the hospital is 130.8 beats per minute.

In this data set, the heart rate of patient 7 is considerably lower than the heart rates of the other subjects. What would happen if this observation were removed from the group? In this case,

$$\begin{aligned}\bar{x} &= \left(\frac{1}{9}\right)(167 + 150 + 125 + 120 + 150 + 150 + 136 + 120 + 150) \\ &= \frac{1,268}{9} \\ &= 140.9 \text{ beats per minute.}\end{aligned}$$

The mean has increased by approximately ten beats per minute; this change demonstrates how much influence a single unusual observation can have on the mean.

A second measure of central tendency is the median or 50th percentile of the set of data. Ranking the measurements from the smallest to the largest, we have

40, 120, 120, 125, 136, 150, 150, 150, 150, 167.

Since there are an even number of observations, the median is taken to be the average of the two middlemost values. In this case, these values are the $10/2 = 5^{\text{th}}$ and the $(10/2) + 1 = 6^{\text{th}}$ observations. Consequently, the median of these data is

$(136 + 150)/2 = 143$ beats per minute, a number quite a bit larger than the mean. Five observations are smaller than the median, and five are larger.

The calculation of the median takes into account the ordering and the relative magnitude of the observations. If we were to again remove patient 7, the ranking of heart rates would be

120, 120, 125, 136, 150, 150, 150, 150, 167.

There are nine observations in the list; the median is the $[(9 + 1)/2] = 5^{\text{th}}$ largest measurement, or 150 beats per minute. Although the median has increased somewhat, it has not changed as much as the mean did when patient 7's observation was removed from the group.

The mode of a set of data is the observation that occurs most frequently. For the measurements in Table 3.5, the mode is 150 beats per minute; this is the only value that occurs four times.

Once we have found the center of a set of data, we often want to estimate the amount of variability among the observations as well; this allows us to quantify the degree to which the summary is representative of the entire group. One measure of dispersion that can be used is the range. The range of the data is the difference between the largest and smallest measurements. For the observations in Table 3.5, the range is $167 - 40 = 127$ beats per minute. Since the range considers only the two most extreme values, it is highly sensitive to outliers.

The interquartile range of a set of data is defined as the 75th percentile minus the 25th percentile. For the data in Table 3.5, the interquartile range is $150 - 120 = 30$ beats per minute; this is the range of the middle 50% of the observations.

The most commonly used measures of dispersion for a set of data values are the variance and the standard deviation. The variance quantifies the amount of variability around the mean of the data; it is calculated by subtracting the mean from each of the measurements, squaring these deviations, summing them, and

Table 3.5 Heart rates for ten patients in a state of respiratory arrest

Patient	Heart Rate (beats per minute)
1	167
2	150
3	125
4	120
5	150
6	150
7	40
8	136
9	120
10	150

dividing by the total number of observations minus 1. The variance of the heart rates in Table 3.5 is

$$\begin{aligned} s^2 &= \frac{1}{(10-1)} \sum_{i=1}^{10} (x_i - 130.8)^2 \\ &= \left(\frac{1}{9}\right) [(36.2)^2 + (19.2)^2 + (-5.8)^2 + (-10.8)^2 + (19.2)^2 \\ &= \quad + (19.2)^2 + (-90.8)^2 + (5.2)^2 + (-10.8)^2 + (19.2)^2] \\ &= \frac{11,323.6}{9} \\ &= 1,258.2 \text{ (beats per minute)}^2. \end{aligned}$$

The standard deviation is the square root of the variance; it is used more frequently in practice because it has the same units of measurement as the mean. For the ten measures of heart rate, the standard deviation is

$$\begin{aligned} s &= \sqrt{1,258.2 \text{ (beats per minute)}^2} \\ &= 35.5 \text{ beats per minute.} \end{aligned}$$

Now consider the frequency distribution of birth weights in Table 3.6; these data were first presented in Table 2.9. Are we able to further summarize these data and make a concise statement about their distribution? Although we do not know the actual values of the 3,751,275 measurements of birth weight, we do know the numbers of observations that fall into each interval. We can, therefore, apply

Table 3.6 Absolute frequencies of birth weights for infants born in the United States, 1986

Birth Weight (grams)	Number of Infants
0–499	4,843
500–999	17,487
1,000–1,499	23,139
1,500–1,999	49,112
2,000–2,499	160,919
2,500–2,999	597,738
3,000–3,499	1,376,008
3,500–3,999	1,106,634
4,000–4,499	344,390
4,500–4,999	62,769
5,000–5,499	8,236
Total	3,751,275

the techniques for grouped data to obtain numerical summary measures for these observations.

To find the mean, we begin by assuming that all the observations within a particular interval are equal to the midpoint of that interval. For example, we assume that the 4,843 measurements in the first interval all take the value 249.5 grams and that the 17,487 measurements in the second interval take the value 749.5 grams. We then multiply each midpoint by the corresponding frequency in the interval, sum these products, and divide by the total number of observations. For the data in Table 3.6,

$$\begin{aligned}
\bar{x} &= \frac{\sum_{i=1}^{11} m_i f_i}{\sum_{i=1}^{11} f_i} \\
&= \left(\frac{1}{3751275}\right)[249.5(4843) + 749.5(17,487) + 1249.5(23,139) \\
&\quad + 1749.5(49,112) + 2249.5(160,919) + 2749.5(597,738) \\
&\quad + 3249.5(1,376,008) + 3749.5(1,106,634) + 4249.5(344,390) \\
&\quad + 4749.5(62,769) + 5249.5(8236)] \\
&= \frac{12,560,121,114.5}{3751275} \\
&= 3348.2 \text{ grams.}
\end{aligned}$$

The grouped mean is a weighted average of the interval midpoints.

In addition to calculating a measure of central tendency, we are also able to calculate a measure of dispersion for the frequency distribution. The variance of the data in Table 3.6 is

$$\begin{aligned}
s^2 &= \frac{\sum_{i=1}^{11}(m_i - 3348.2)^2 f_i}{[\sum_{i=1}^{11} f_i] - 1} \\
&= \frac{1}{(3751275 - 1)}[(-3098.7)^2(4843) + (-2598.7)^2(17487) \\
&\quad + (-2098.7)^2(23139) + (-1598.7)^2(49112) + (-1098.7)^2(160919) \\
&\quad + (-598.7)^2(597738) + (-98.7)^2(1376008) + (401.3)^2(1106634) \\
&\quad + (901.3)^2(344390) + (1401.3)^2(62769) + (1901.3)^2(8236)] \\
&= \frac{1,423,951,273,348.3}{3751274} \\
&= 379,591.4 \text{ grams}^2.
\end{aligned}$$

Table 3.7 Stata output displaying numerical summary measures

heartrate				
	Percentiles	Smallest		
1%	40	40		
5%	40	120		
10%	80	120	Obs	10
25%	120	125	Sum of Wgt.	10
50%	143		Mean	130.8
		Largest	Std. Dev.	35.4708
75%	150	150		
90%	158.5	150	Variance	1258.178
95%	167	150	Skewness	−1.772591
99%	167	167	Kurtosis	5.479789

The standard deviation, which is the square root of the variance, is

$$\begin{aligned} s &= \sqrt{379,591.4 \text{ grams}^2} \\ &= 616.1 \text{ grams.} \end{aligned}$$

Instead of working out all these numerical summary measures by hand, we could have used a computer to do the calculations for us. Table 3.7 shows the relevant output from Stata for the heart rate data in Table 3.5. Selected percentiles of the data are displayed on the left-hand side of the table. Using these values, we can determine the median and the interquartile range. The middle column contains the four smallest and the four largest measurements; these allow us to calculate the range. The information on the right-hand side of the table includes the number of observations, the mean of the data, the standard deviation, and the variance.

3.6 Review Exercises

1. A study was conducted investigating the long-term prognosis of children who have suffered an acute episode of bacterial meningitis. Listed below are the times to the onset of seizure for 13 children who took part in the study [8]. In months, the measurements are:

0.10 0.25 0.50 4 12 12 24 24 31 36 42 55 96

a. Find the following numerical summary measures of the data.

i. mean
ii. median
iii. mode

iv. range
v. standard deviation

b. Show that $\sum_{i=1}^{13}(x_i - \bar{x})$ is equal to 0.

2. Recall the serum zinc level data for 462 males between the ages of 15 and 17 that were examined in Chapter 2. The observations are stored on your disk in a data set called `serzinc`; the serum zinc measurements are saved under the variable name `zinc`.

a. Find the mean, median, standard deviation, range, and interquartile range of the data.
b. Use Chebychev's inequality to describe the distribution of values.
c. Do you think that these summary measures have done an adequate job of describing the data?

3. The percentages of low birth weight infants—defined as those weighing less than 2,500 grams—for a number of nations around the world are saved under the variable name `lowbwt` in the data set `unicef` [9].

a. Compute the mean and the median of these observations.
b. For this data set, which of these numbers would you prefer as a measure of central tendency? Why?

4. Displayed below is a frequency distribution containing a summary of the resting systolic blood pressures for a sample of 35 patients with ischemic heart disease [10].

Blood Pressure (mm Hg)	Number of Patients
115–124	4
125–134	5
135–144	5
145–154	7
155–164	5
165–174	4
175–184	5
Total	35

a. Compute the grouped mean and the grouped standard deviation of the data.
b. The 35 measurements of systolic blood pressure are stored on your disk in a file called `ischemic`; the values are saved under the variable name `sbp`. Compute the ungrouped mean and the standard deviation of these data.

c. Are the grouped and the ungrouped numerical summary measures the same? Why or why not?

5. Figures 3.5 and 3.6 display the infant mortality rates for 98 nations on three continents: Africa, Asia, and Europe [9]. Figure 3.5 provides histograms that illustrate the distribution of infant mortality rates for each of the continents. Figure 3.6 displays the same data using one-way scatter plots and box plots.

a. Without doing any calculations, which continent would you expect to have the smallest mean? the largest median? the smallest standard deviation? Explain your reasoning.
b. For Africa, would you expect the mean and the median infant mortality rates to be equal? Would you expect the mean and the median to be equal for Asia? Why or why not?

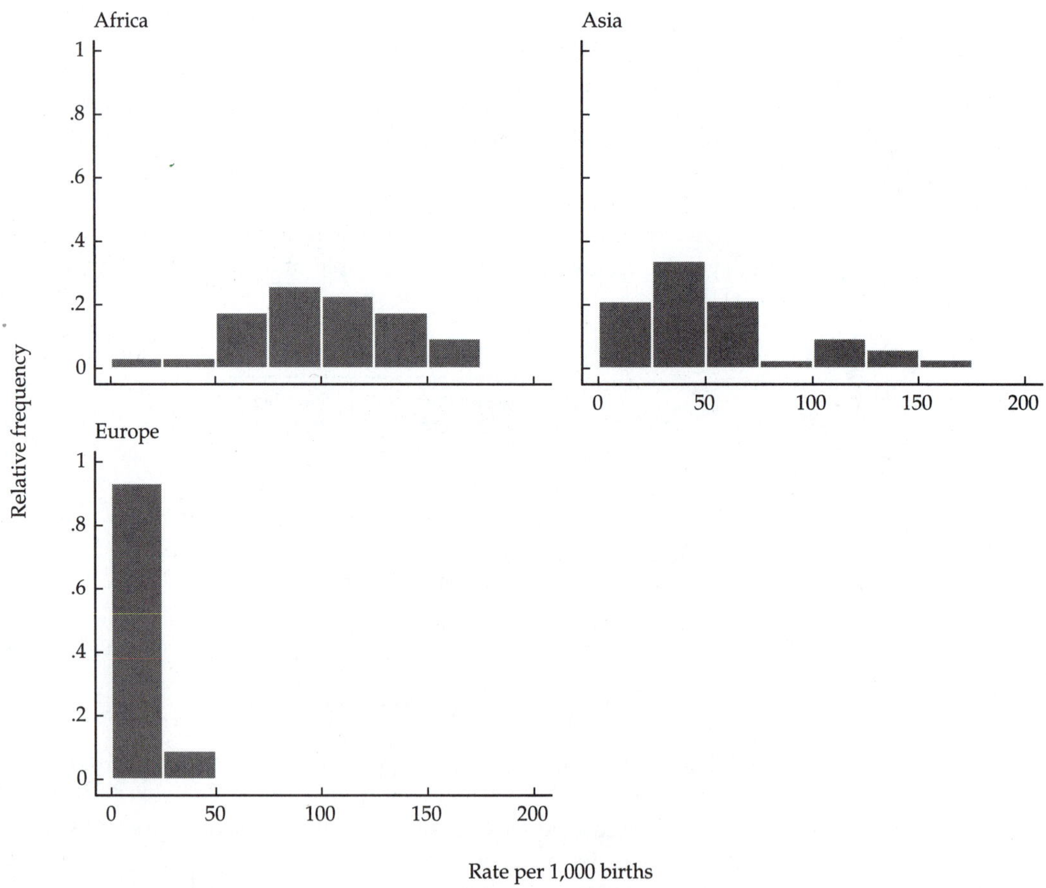

Figure 3.5. Histograms of infant mortality rates for Africa, Asia, and Europe, 1989

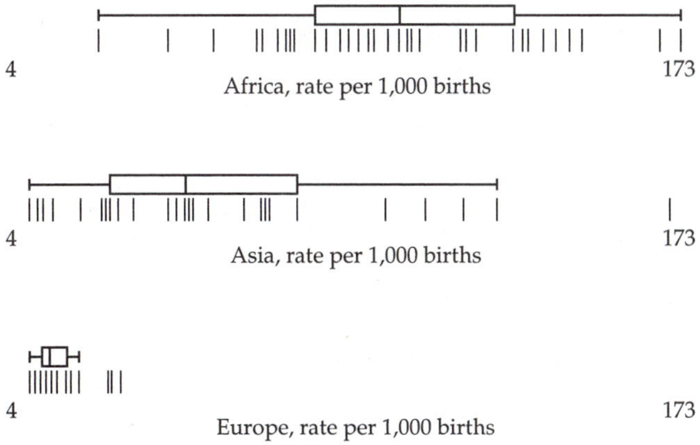

Figure 3.6. One-way scatter plots and box plots of infant mortality rates for Africa, Asia, and Europe, 1989

References

[1] Koenig, J. Q., Covert, D. S., Hanley, Q. S., Van Belle, G., and Pierson, W. E., "Prior Exposure to Ozone Potentiates Subsequent Response to Sulfur Dioxide in Adolescent Asthmatic Subjects," *American Review of Respiratory Disease*, Volume 141, February 1990, 377–380.

[2] Prial, F. J., "Wine Talk," *The New York Times*, December 25, 1991, 29.

[3] United Nations Environment Programme, World Health Organization, "Urban Air Quality on Average," *Global Pollution and Health: Results of Health-Related Environmental Monitoring*, 1987.

[4] Martin, J. L., "The Impact of AIDS on Gay Male Sexual Behavior Patterns in New York City," *American Journal of Public Health*, Volume 77, May 1987, 578–581.

[5] Wang, W. C., Kovnar, E. H., Tonkin, I. L., Mulhern, R. K., Langston, J. W., Day, S. W., Schell, M. J., and Wilimas, J. A., "High Risk of Recurrent Stroke after Discontinuance of Five to Twelve Years of Transfusion Therapy in Patients with Sickle Cell Disease," *Journal of Pediatrics*, Volume 118, March 1991, 377–382.

[6] Parzen, E., *Modern Probability Theory and Its Applications*, New York: Wiley, 1960.

[7] Molfino, N. A., Nannini, L. J., Martelli, A. N., and Slutsky, A. S., "Respiratory Arrest in Near-Fatal Asthma," *The New England Journal of Medicine*, Volume 324, January 31, 1991, 285–288.

[8] Pomeroy, S. L., Holmes, S. J., Dodge, P. R., and Feigin, R. D., "Seizures and Other Neurologic Sequelae of Bacterial Meningitis in Children," *The New England Journal of Medicine*, Volume 323, December 13, 1990, 1651–1656.

[9] United Nations Children's Fund, *The State of the World's Children 1991*, New York: Oxford University Press.

[10] Miller, P. F., Sheps, D. S., Bragdon, E. E., Herbst, M. C., Dalton, J. L., Hinderliter, A. L., Koch, G. G., Maixner, W., and Ekelund, L. G., "Aging and Pain Perception in Ischemic Heart Disease," *American Heart Journal*, Volume 120, July 1990, 22–30.

CHAPTER 4

Rates and Standardization

Demographic data and *vital statistics* are numbers that are used to characterize or describe a population. Demographic data include such information as the size of the population and its composition by sex, race, and age. Vital statistics describe the life of the population; they deal with births, deaths, marriages, divorces, and occurrences of disease. Researchers and public health professionals use both types of data to describe the health status of a population, to spot trends and make projections, and to plan for necessary services such as housing and medical care.

Vital statistics are also used to make comparisons among groups. For example, we might want to compare the number of deaths in the United States in 1986 to the number of deaths in 1987. If we consider only the raw numbers of deaths in each year—2,105,361 in 1986 and 2,123,323 in 1987—it would be difficult to interpret the observed increase [1]. It is possible that there were fewer deaths in 1986 simply because the population base was smaller in that year.

4.1 Rates

To make comparisons more meaningful, rates may be used instead of raw numbers. A *rate* is defined as the number of cases of a particular outcome of interest that occur over a given time period divided by a population that is specified during that time period. For example, we might be interested in the number of ear infections per 1,000 elementary school students during a two-month period. Although they are often used interchangeably, the terms "rate" and "proportion" are not synonymous. A *proportion* is a ratio in which all individuals included in the numerator must also be included in the denominator, such as the fraction of women over the age of 60 who have had a mastectomy. It has no units of measurement. A rate is intrinsically dependent on a measure of time and does incorporate units of measurement.

Since the number of occurrences of a specified outcome depends upon the size of the population being considered, dividing by their population sizes makes any two groups more comparable. For example, instead of comparing the total numbers of deaths in 1986 and 1987, we could compare the death rates for those years. A death rate, or *mortality rate*, is the number of deaths that occur during some time period, such as a calendar year, divided by the total population at risk during that time period. This rate is often expressed in terms of deaths per 1,000 population or deaths per 100,000 population; the multiplier for a rate–whether 1,000, 10,000, or 100,000–is usually chosen to minimize the number of decimal places in a display. If we were to compute the mortality rates for the United States, we would find the death rate in 1986 to be 873.2 per 100,000 population and the death rate in 1987 to be 872.4 per 100,000 population [1]. Although a greater number of people died in 1987, the death rate actually decreased slightly.

One example of a commonly used mortality rate is the *infant mortality rate*. This quantity is defined as the number of deaths during a calendar year among infants under one year of age divided by the total number of live births during that year. The infant mortality rate is one of the most important representative

Table 4.1 Infant mortality rates for selected countries, 1989

Nation	Mortality Rate per 1,000 Live Births
Argentina	31
Australia	8
Brazil	61
Canada	7
China	31
Egypt	67
Ethiopia	133
Finland	6
France	8
Greece	11
India	96
Israel	10
Italy	10
Japan	4
Mexico	41
Poland	16
Saudi Arabia	67
Spain	9
Switzerland	7
USSR	25
United Kingdom	8
United States	10
Venezuela	35
Yugoslavia	24

measures of the health status of a nation; even though the number of births varies considerably from country to country, it is informative to compare infant mortality rates. The rates for selected nations appear in Table 4.1 [2].

The rates that we have considered up to this point have been crude rates. A *crude rate* is a single number computed as a summary measure for an entire population; it disregards differences caused by age, sex, race, and other characteristics. Figure 4.1 (on p. 64) displays trends in the crude marriage, divorce, death, and infant mortality rates in the United States for the years 1984 through 1988 [3].

Factors such as age, sex, and race may have a significant effect on the rates that describe vital statistics. Consider the mortality rates in Table 4.2 [1]. In the United States, females tend to have lower death rates than males, white individuals have lower death rates than members of other races, and, after age 5, death rate increases with age. Rates that are computed within relatively small, well-defined subgroups are called *specific rates*. Mortality rates calculated for individual age groups, for instance, are known as *age-specific death rates*.

4.2 Standardization of Rates

The National Health Interview Survey is an ongoing household survey of the civilian, noninstitutionalized population of the United States. It provides the following data enumerating hearing impairments due to injury reported by individuals 17 years of age and older in 1980–1981 [4].

Employment Status	Population	Impairments	Rate per 1,000
Currently employed	98,917	552	5.58
Currently unemployed	7,462	27	3.62
Not in the labor force	56,778	368	6.48
Total	163,157	947	5.80

The employment status categories are defined by the National Center for Health Statistics as follows: "Currently employed" individuals are persons who were working at any time during the two-week interview period; the "currently unemployed" comprise individuals looking for work and those who have been laid off; persons classified as "not in the labor force" include homemakers, volunteer workers, and persons who have retired.

Judging from the crude data, the rate of hearing impairment due to injury seems to vary with employment status. The rate among individuals who are currently employed is 5.58 per 1,000 population, whereas the rate among those not in the labor force is 6.48 per 1,000 population. Of these two groups, it appears that individuals who are not in the labor force are at greater risk of hearing impairment due to injury than those who are currently employed. Is this a valid conclusion?

Table 4.2 Total deaths and death rates by age, race, and sex, United States, 1988

(Rates per 100,000 population in specified group)

Age	All Races			White			All Other					
							Total			Black		
	Both Sexes	Male	Female	Both Sexes	Male	Female	Both Sexes	Male	Female	Both Sexes	Male	Female
						Number						
All ages	2,167,999	1,125,540	1,042,459	1,876,906	965,419	911,487	291,093	160,121	130,972	264,019	144,228	119,791
Under 1 year	38,910	22,007	16,903	25,925	14,878	11,047	12,985	7,129	5,856	11,840	6,503	5,337
1–4 years	7,429	4,221	3,208	5,380	3,073	2,307	2,049	1,148	901	1,784	1,016	768
5–9 years	4,357	2,574	1,783	3,193	1,889	1,304	1,164	685	479	997	592	405
10–14 years	4,568	2,907	1,661	3,460	2,248	1,212	1,108	659	449	947	561	386
15–19 years	16,031	11,687	4,344	12,538	9,033	3,505	3,493	2,654	839	2,998	2,312	686
20–24 years	22,136	16,844	5,292	16,502	12,589	3,913	5,634	4,255	1,379	4,977	3,744	1,233
25–29 years	26,809	19,653	7,156	19,510	14,485	5,025	7,299	5,168	2,131	6,606	4,677	1,929
30–34 years	32,328	23,332	8,996	23,031	16,869	6,162	9,297	6,463	2,834	8,556	5,967	2,589
35–39 years	36,616	25,433	11,183	26,043	18,285	7,758	10,573	7,148	3,425	9,763	6,632	3,131
40–44 years	40,838	27,001	13,837	30,717	20,454	10,263	10,121	6,547	3,574	9,262	6,011	3,251
45–49 years	49,684	31,597	18,087	38,143	24,324	13,819	11,541	7,273	4,268	10,481	6,641	3,840
50–54 years	67,788	42,318	25,470	53,296	33,489	19,807	14,492	8,829	5,663	13,151	8,040	5,111
55–59 years	105,363	65,194	40,169	86,289	53,910	32,379	19,074	11,284	7,790	17,384	10,280	7,104
60–64 years	164,386	100,174	64,212	138,357	85,141	53,216	26,029	15,033	10,996	23,769	13,727	10,042
65–69 years	221,504	131,931	89,573	190,770	114,474	76,296	30,734	17,457	13,277	28,084	15,850	12,234
70–74 years	267,041	151,970	115,071	235,866	135,187	100,679	31,175	16,783	14,392	28,377	15,139	13,238
75–79 years	302,580	158,662	143,918	269,922	142,069	127,853	32,658	16,593	16,065	29,622	14,811	14,811
80–84 years	299,334	136,059	163,275	272,232	123,668	148,564	27,102	12,391	14,711	24,399	10,811	13,588
85 years and over	459,710	151,559	308,151	425,306	139,042	286,264	34,404	12,517	21,887	30,875	10,818	20,057
Not stated	587	417	170	426	312	114	161	105	56	147	96	51

(continued)

Table 4.2 (continued)

Age	All Races			White			All Other: Total			All Other: Black		
	Both Sexes	Male	Female	Both Sexes	Male	Female	Both Sexes	Male	Female	Both Sexes	Male	Female
	Rate											
All ages	882.0	940.0	826.9	905.1	952.2	860.0	757.5	872.7	652.2	874.2	1,006.8	754.5
Under 1 year	1,008.3	1,113.7	897.7	832.0	930.5	728.2	1,745.3	1,891.0	1,600.0	1,996.6	2,167.7	1,821.5
1–4 years........	50.9	56.5	45.0	45.7	51.0	40.2	72.3	79.8	64.6	80.8	90.5	70.7
5–9 years........	24.2	27.9	20.3	22.0	25.4	18.5	33.0	38.4	27.5	35.8	41.9	29.5
10–14 years	27.5	34.1	20.5	25.9	32.8	18.7	33.8	39.5	27.8	36.2	42.3	30.0
15–19 years	88.0	125.8	48.7	85.1	120.0	48.6	100.5	150.4	49.1	107.5	164.3	49.6
20–24 years	115.4	175.3	55.3	104.4	158.3	49.8	166.7	257.3	79.9	184.1	286.9	88.1
25–29 years	122.5	179.5	65.5	106.7	156.8	55.5	203.7	301.3	114.0	233.6	348.5	129.8
30–34 years	148.3	214.0	82.6	125.7	182.2	67.9	267.7	392.6	155.1	319.6	475.1	182.3
35–39 years	191.3	268.3	115.8	160.2	224.5	95.7	366.5	535.4	221.0	446.6	662.5	264.2
40–44 years	253.3	341.1	168.6	220.3	295.7	146.1	463.8	656.7	301.6	570.3	817.8	365.7
45–49 years	381.4	496.9	271.3	339.4	438.9	242.6	645.8	891.3	439.5	769.5	1,085.1	511.3
50–54 years	608.7	784.7	443.4	555.1	712.8	404.0	944.1	1,270.4	674.2	1,101.4	1,508.4	773.2
55–59 years	966.9	1,254.9	704.6	908.8	1,181.7	656.5	1,360.5	1,782.6	1,013.0	1,559.1	2,027.6	1,168.4
60–64 years	1,503.4	1,965.7	1,100.1	1,432.7	1,881.6	1,036.9	2,039.9	2,632.7	1,559.7	2,321.2	2,971.2	1,786.8
65–69 years	2,216.6	2,903.4	1,643.8	2,146.1	2,820.3	1,580.0	2,783.9	3,599.4	2,144.9	3,137.9	4,053.7	2,432.2
70–74 years	3,378.6	4,469.7	2,554.9	3,320.2	4,410.7	2,493.3	3,892.0	5,009.9	3,088.4	4,352.3	5,627.9	3,456.4
75–79 years	5,125.9	6,833.0	4,018.9	5,075.6	6,794.3	3,962.0	5,582.6	7,152.2	4,551.0	6,223.1	8,093.4	5,054.9
80–84 years	8,271.2	10,781.2	6,927.2	8,222.0	10,763.1	6,871.6	8,799.4	10,965.5	7,583.0	9,759.6	12,147.2	8,439.8
85 years and over	15,594.0	18,370.8	14,508.1	15,875.6	18,814.9	14,755.9	12,789.6	14,554.7	11,960.1	13,482.5	15,454.3	12,694.3

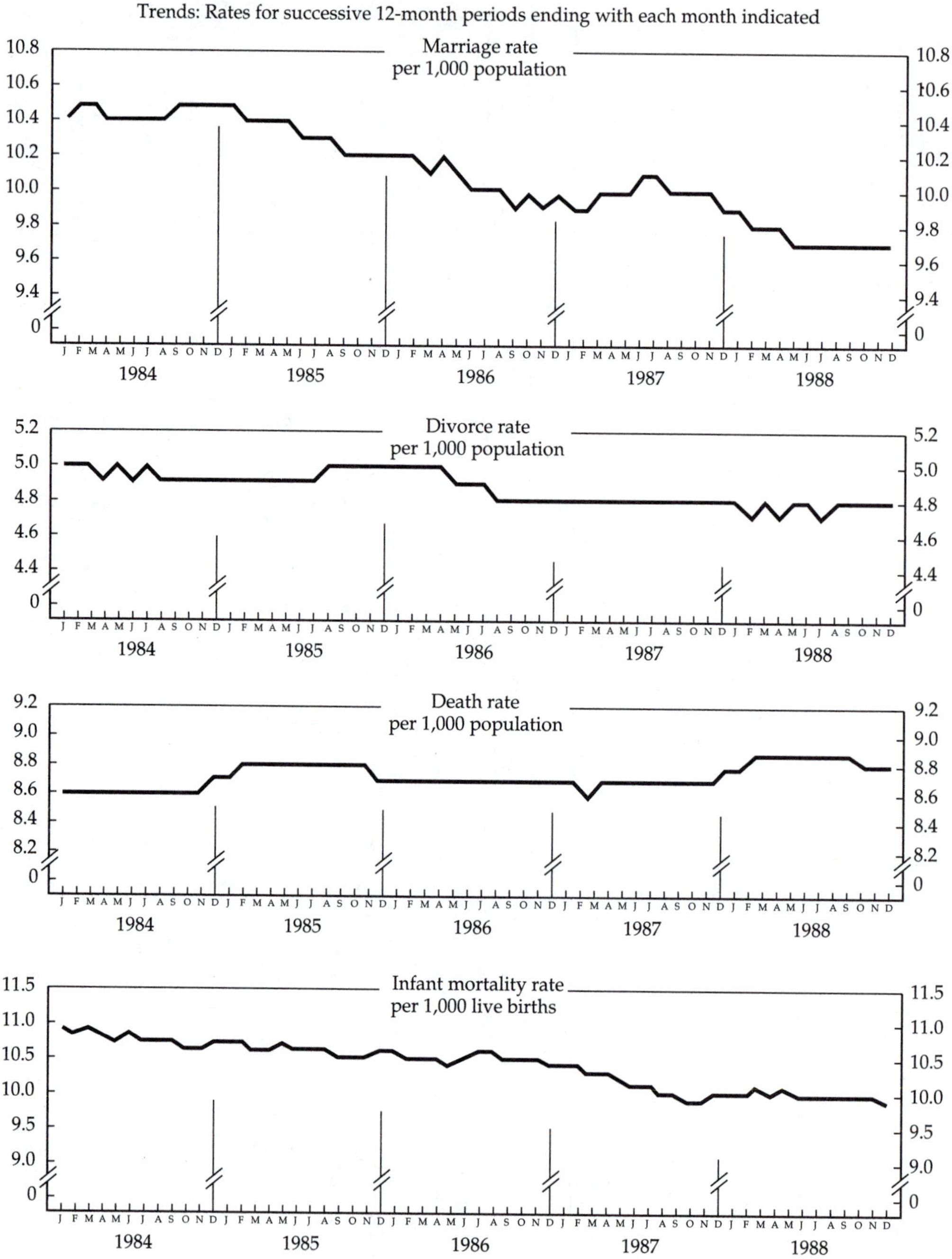

Figure 4.1. Trends in the crude marriage, divorce, death, and infant mortality rates, United States, 1984–1988

One problem that often arises in the comparison of crude rates is that the populations may differ substantially with respect to important characteristics such as age and sex. For example, if we have two populations from different geographical areas–one composed entirely of males and the other entirely of females–we could never be sure whether a difference in their mortality rates is due to location or to some effect of gender. In this situation, gender is referred to as a *confounder*. Because it is associated with both geographical area and death rate, it obscures the true relationship between these factors.

To answer the question about the hearing impairment data, it would be wise to check whether the two subpopulations of interest have similar underlying structures. We therefore break down each subpopulation by age group.

	Currently Employed		Not in the Labor Force	
Age	**Population**	**Percent**	**Population**	**Percent**
17–44	67,987	68.7	20,760	36.6
45–64	27,592	27.9	15,108	26.6
65+	3,338	3.4	20,910	36.8
Total	98,917	100.0	56,778	100.0

As we can see, the two groups differ in age composition; the individuals not included in the labor force are considerably older than those who are currently employed.

Now consider the age-specific impairment rates in the surveyed population as a whole.

Age	Population	Impairments	Rate per 1,000
17–44	94,930	441	4.65
45–64	43,857	308	7.02
65+	24,370	198	8.12
Total	163,157	947	5.80

Note that the rate of hearing impairment due to injury in the entire surveyed population is actually a weighted average of the age-specific rates; in particular,

$$5.80 = \frac{(94,930)(4.65) + (43,857)(7.02) + (24,370)(8.12)}{163,157}.$$

Also observe that the rates increase with age. Age is a confounder in the relationship between hearing impairment and employment status; it is independently associated with both of these quantities. As a result, we cannot be sure whether the higher rate of impairment among individuals not in the labor force is the result of some inherent characteristic of the individuals in this group or whether it is simply the effect of age.

To make a more accurate comparison between the two subpopulations, we should consider their age-specific impairment rates rather than their overall crude rates.

	Currently Employed			Not in the Labor Force		
Age	Population	Impair-ments	Rate per 1,000	Population	Impair-ments	Rate per 1,000
17–44	67,987	346	5.09	20,760	80	3.85
45–64	27,592	179	6.49	15,108	117	7.74
65+	3,338	27	8.09	20,910	171	8.18
Total	98,917	552	5.58	56,778	368	6.48

However, it is difficult to draw an overall conclusion based on these two sets of rates. Among individuals over the age of 45, the rate of hearing impairment due to injury is higher for persons not in the labor force than it is for the currently employed; among those between the ages of 17 and 44, however, the rate is much lower for persons not in the labor force. Comparing the crude impairment rates of the two subpopulations provided an incomplete picture of the true situation.

Although the calculation of subgroup-specific rates gives us a more accurate comparison among populations than the calculation of crude rates does, we could, if we have many subgroups, end up with an overwhelming number of rates to compare. It would be convenient to be able to summarize an entire situation with a single number calculated for each subpopulation, a number that adjusts for differences in composition. In practice, there are two ways to do this. The first is known as the *direct method of standardization*; the second is called the *indirect method of standardization*. Both strategies focus on the two components that factor into the calculation of a crude rate–the composition of the population and its subgroup-specific rates–and attempt to overcome the problem of confounding by holding one of these components constant across populations. Indices such as the consumer price index have a similar objective.

4.2.1 Direct Method of Standardization

The *direct method* of adjusting for differences among populations involves computing the overall rates that would result if, instead of having different distributions, all populations were to have the same standard composition. The first step in applying this technique is to select the standard distribution. For the hearing impairment example, we use the total population questioned in the National Health Interview Survey. We then calculate the numbers of impairments that would have occurred in the two employment status subgroups, assuming that each had this standard population distribution while retaining its own individual age-specific impairment rates.

	Total	Currently Employed		Not in the Labor Force	
Age	(1) Population	(2) Rate per 1,000	(3) Expected Impairments	(4) Rate per 1,000	(5) Expected Impairments
17–44	94,930	5.09	483.2	3.85	365.5
45–64	43,857	6.49	284.6	7.74	339.5
65+	24,370	8.09	197.2	8.18	199.3
Total	163,157		965.0		904.3

The expected number of impairments for the group of currently employed individuals is calculated by multiplying column (1) by column (2) and dividing by 1,000; the expected number of impairments for those not in the labor force is obtained by multiplying column (1) by column (4) and dividing by 1,000.

The age-adjusted impairment rate for each group is calculated by dividing its expected number of hearing impairments due to injury by the total standard population.

$$\text{Currently employed:} \quad \frac{965.0}{163,157} = 5.91 \text{ per } 1,000$$

$$\text{Not in the labor force:} \quad \frac{904.3}{163,157} = 5.54 \text{ per } 1,000$$

These age-adjusted rates are the impairment rates that would apply if both the currently employed and those not in the labor force had the same age distribution as the total surveyed population. After we control for the effect of age, the adjusted impairment rate for individuals who are currently employed is higher than the adjusted rate for individuals not in the labor force.

Note that the choice of a different standard age distribution would lead to different adjusted impairment rates. This is not important, however, since an adjusted rate has no meaning by itself. It is merely a construct that is calculated based on a hypothetical standard distribution; it does not reflect the true impairment rate of any population. Adjusted rates, therefore, have meaning only when we are comparing two or more groups. Since they do depend on the standard distribution selected, this population must be chosen carefully. If we were to choose a different but reasonable standard age distribution—for example, the U.S. population in 1988—the magnitude of the **difference** between the adjusted impairment rates of those currently employed and those not in the labor force should not change drastically; the currently employed should still have a slightly higher adjusted rate of hearing impairment.

4.2.2 Indirect Method of Standardization

The *indirect method* of adjusting for differences in composition involves the use of a set of standard age-specific impairment rates along with the actual age composition of each population. We again use the total surveyed population as the standard. This time, however, we calculate the number of impairments that would have occurred in the two population subgroups if each took on the age-specific impairment rates of the survey population while retaining its own age distribution.

	Total	Currently Employed		Not in the Labor Force	
Age	(1) Rate per 1,000	(2) Population	(3) Expected Impairments	(4) Population	(5) Expected Impairments
17–44	4.65	67,987	316.1	20,760	96.5
45–64	7.02	27,592	193.7	15,108	106.1
65+	8.12	3,338	27.1	20,910	169.8
Total	5.80	98,917	536.9	56,778	372.4

The expected number of impairments among those currently employed is calculated by multiplying column (1) by column (2) and dividing by 1,000; the expected impairments for those not in the labor force are obtained by multiplying column (1) by column (4) and dividing by 1,000.

We next divide the observed number of hearing impairments in each group by the expected number of impairments. The quantity that results is known as the *standardized morbidity ratio*. If the data pertained to deaths rather than impairments, dividing the observed number of deaths by the expected number would give us the *standardized mortality ratio*.

$$\text{Currently employed:} \quad \frac{552}{536.9} = 1.03$$
$$= 103\%$$

$$\text{Not in the labor force:} \quad \frac{368}{372.4} = 0.99$$
$$= 99\%$$

These standardized morbidity ratios indicate that the group of currently employed individuals has a 3% higher impairment rate than the surveyed population as a whole, whereas the group not included in the labor force has an impairment rate that is 1% lower than the total population. Recall that the total surveyed population also includes the group of individuals who are not currently employed.

The application of the indirect method often concludes with a comparison of the standardized ratios. We could, however, continue on and compute the actual age-adjusted impairment rates for each group. These rates are derived by multiplying the crude impairment rate for the total surveyed population by the appropriate standardized ratios.

$$\text{Currently employed:} \quad \frac{5.80}{1,000} \times 1.03 = 5.97 \text{ per } 1,000$$

$$\text{Not in the labor force:} \quad \frac{5.80}{1,000} \times 0.99 = 5.74 \text{ per } 1,000$$

With the effect of age removed, the group of currently employed individuals is once again seen to have a slightly higher adjusted impairment rate than the individuals not included in the labor force. Although the actual numbers are different, this is the same conclusion that was reached when the direct method of standardization was applied.

4.2.3 Use of Standardized Rates

Standardized rates, age-adjusted rates in particular, are encountered frequently in the study of vital statistics. One interesting example involves data taken from three different studies that investigate the relationship between death rate and smoking status [5]. Each study compares three groups of men: nonsmokers, cigarette smokers, and smokers of cigars and pipes. The crude death rates per 1,000 person-years of experience are displayed below. These rates were calculated by dividing the total number of deaths in each group by the corresponding person-years of exposure and multiplying by 1,000.

	Death Rate per 1,000 Person-Years		
Smoking Group	**Canada**	**Great Britain**	**United States**
Nonsmokers	20.2	11.3	13.5
Cigarettes	20.5	14.1	13.5
Cigars/pipes	35.5	20.7	17.4

The studies conducted in Canada, Great Britain, and the United States all seem to carry the same message: individuals who smoke either cigars or pipes should give up smoking or, at the very least, switch to cigarettes.

If we consider the age compositions of the various smoking groups, we find that they differ considerably. To illustrate this, the mean ages for the three groups in each of the studies are displayed below.

	Mean Age (years)		
Smoking Group	**Canada**	**Great Britain**	**United States**
Nonsmokers	54.9	49.1	57.0
Cigarettes	50.5	49.8	53.2
Cigars/pipes	65.9	55.7	59.7

In general, smokers of cigars and pipes tend to be older than nonsmokers and cigarette smokers.

Suppose that we now standardize for the differences in age composition by separating the men into three different subclasses; the categories of age are chosen so that roughly the same number of men fall into each class. If the group of nonsmokers is chosen as the standard age distribution, the corresponding adjusted death rates are as shown below. (Note that these rates cannot be calculated from the information provided.)

	Death Rate per 1,000 Person-Years		
Smoking Group	Canada	Great Britain	United States
Nonsmokers	20.2	11.3	13.5
Cigarettes	28.3	12.8	17.7
Cigars/pipes	21.2	12.0	14.2

The adjusted rates for the nonsmokers are identical to the crude rates; this is to be expected since this group was used as the standard distribution. Note that cigarette smoking appears much more dangerous than it did before. Also, the adjusted death rates for cigar and pipe smokers are considerably lower than the crude rates. Thus, after adjusting for differences in age, we are led to a very different interpretation of the data.

As another example of standardized rates that are encountered in the study of vital statistics, Figure 4.2 depicts the trends in age-adjusted mortality rates for 14 of the 15 leading causes of death in the United States [1]. The vertical bars represent changes in disease classification. The standardized rates were calculated by means of the direct method; they use the U.S. population in 1940 as the standard distribution. Although the adjusted rates have been decreasing for the most part, there are cases in which they are on the rise. Figures 4.3(a) and (b) illustrate the dramatic increase in the death rate due to AIDS relative to other causes among men and women 25 to 44 years of age [6]. Table 4.3 (on pp. 73–75) displays the overall age-adjusted death rates for the United States by race and sex [1]. Again, these adjusted rates have been decreasing over time; note, however, that in all years females have lower death rates than males, and white individuals have lower death rates than persons in other racial groups.

In practice we must know when to use an adjusted rate rather than a crude rate or subgroup-specific rates. In general, if there are no confounding factors such as age or sex then the crude rates are adequate. When one or more confounders are present, subgroup-specific rates are always sufficient for comparisons. Adjusted rates—single numbers that summarize the situation in each population being considered—should be used only if they are meaningful. If the distribution of the standard population is radically different from the distributions of the populations being compared, for instance, then standardization is inappropriate. Also, when direct standardization is applied, the subgroup-specific rates should have the same general trends in all the groups being compared as well as in the standard

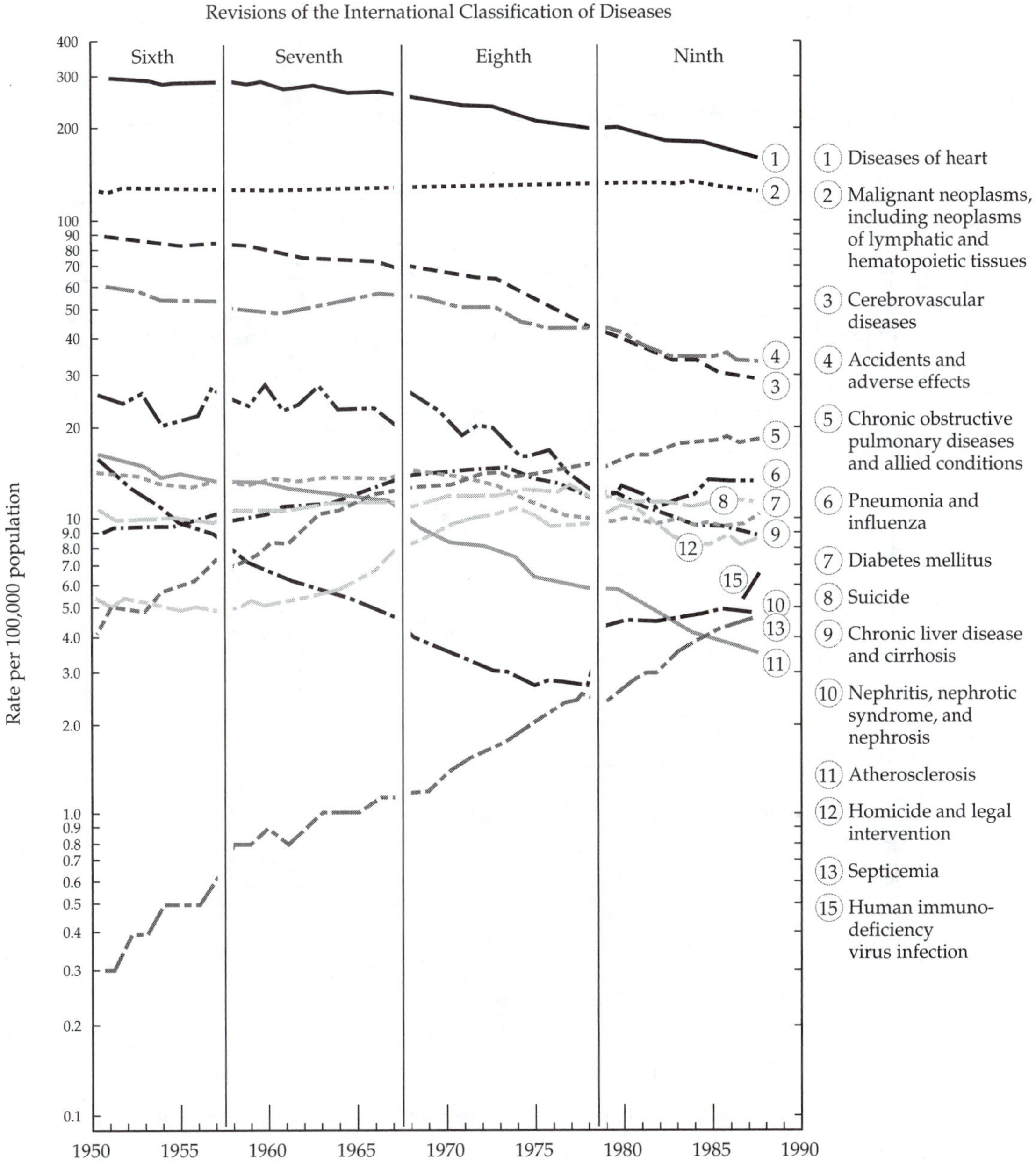

Figure 4.2. Age-adjusted death rates for 14 of the 15 leading causes of death, United States, 1950–1988

population. More explicitly, if we were to graph the subgroup-specific rates for two different groups, the trends would ideally be parallel, as in Figure 4.4(a) (see p. 76). A somewhat less ideal situation occurs when the trends are not parallel

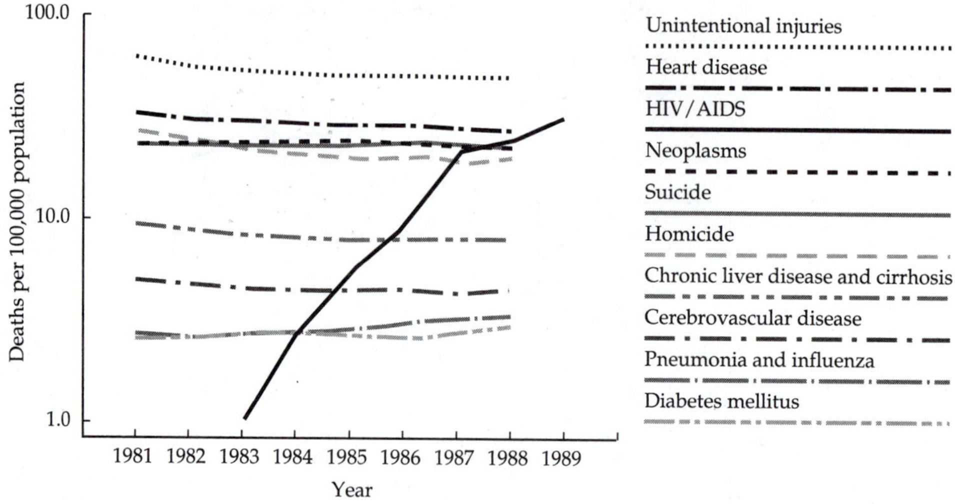

(a) Men

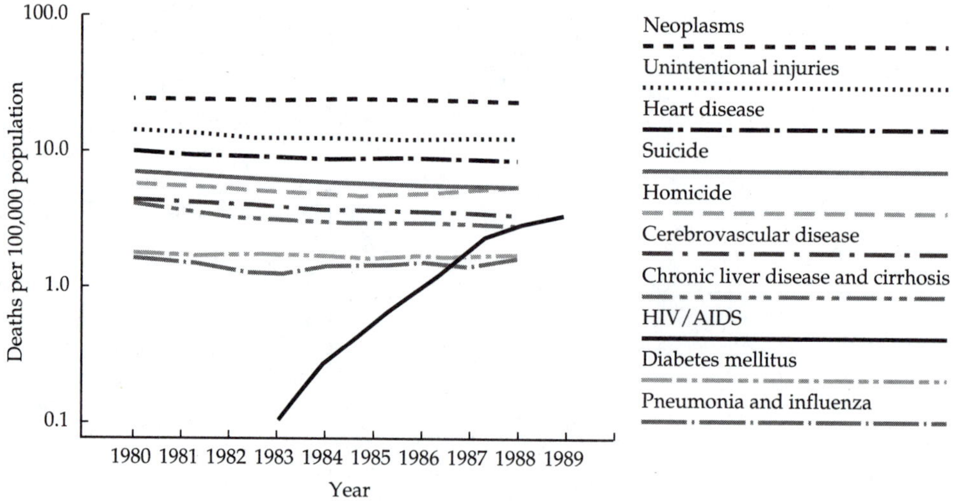

(b) Women

Figure 4.3. Leading causes of death for men and women 25 to 44 years of age, United States, 1981–1988

but are comparable, as in Figure 4.4(b). When the subgroup-specific rates follow very different patterns in the two populations, as in Figure 4.4(c), direct standardization should not be attempted. In a situation such as this, a single number is not able to capture the complex behavior of the rates; by manipulating the standard, they could be made to show just about anything that we choose.

Table 4.3 Total deaths, death rates, and age-adjusted death rates by race and sex, United States, 1940, 1950, 1960, 1970, and 1975–1988

(Rates per 100,000 population in specified group. Rates are based on populations enumerated as of April 1 for census years and estimated as of July 1 for all other years. Beginning 1970, excludes deaths of nonresidents of the United States.)

Year	All Races			White			All Other					
							Total			Black		
	Both Sexes	Male	Female	Both Sexes	Male	Female	Both Sexes	Male	Female	Both Sexes	Male	Female
					Number							
1988	2,167,999	1,125,540	1,042,459	1,876,906	965,419	911,487	291,093	160,121	130,972	264,019	144,228	119,791
1987	2,123,323	1,107,958	1,015,365	1,843,067	953,382	889,685	280,256	154,576	125,680	254,814	139,551	115,263
1986	2,105,361	1,104,005	1,001,356	1,831,083	952,554	878,529	274,278	151,451	122,827	250,326	137,214	113,112
1985	2,086,440	1,097,758	988,682	1,819,054	950,455	868,599	267,386	147,303	120,083	244,207	133,610	110,597
1984	2,039,369	1,076,514	962,855	1,781,897	934,529	847,368	257,472	141,985	115,487	235,884	129,147	106,737
1983	2,019,201	1,071,923	947,278	1,765,582	931,779	833,803	253,619	140,144	113,475	233,124	127,911	105,213
1982	1,974,797	1,056,440	918,357	1,729,085	919,239	809,846	245,712	137,201	108,511	226,513	125,610	100,903
1981	1,977,981	1,063,772	914,209	1,731,233	925,490	805,743	246,748	138,282	108,466	228,560	127,296	101,264
1980	1,989,841	1,075,078	914,763	1,738,607	933,878	804,729	251,234	141,200	110,034	233,135	130,138	102,997
1979	1,913,841	1,044,959	868,882	1,676,145	910,137	766,008	237,696	135,822	102,874	220,818	124,433	96,385
1978	1,927,788	1,055,290	872,498	1,689,722	920,123	769,599	238,066	135,167	102,899	221,340	124,663	96,677
1977	1,899,597	1,046,243	853,354	1,664,100	912,670	751,430	235,497	133,573	101,924	220,076	123,894	96,182
1976	1,909,440	1,051,983	857,457	1,674,989	918,589	756,400	234,451	133,394	101,057	219,442	123,977	95,465
1975	1,892,879	1,050,819	842,060	1,660,366	917,804	742,562	232,513	133,015	99,498	217,932	123,770	94,162
1970	1,921,031	1,078,478	842,553	1,682,096	942,437	739,659	238,935	136,041	102,894	225,647	127,540	98,107
1960	1,711,982	975,648	736,334	1,505,335	860,857	644,478	206,647	114,791	91,856	196,010	107,701	88,309
1950	1,452,454	827,749	624,705	1,276,085	731,366	544,719	176,369	96,383	79,986	169,606	92,004	77,602
1940	1,417,269	791,003	626,266	1,231,223	690,901	540,322	186,046	100,102	85,944	178,743	95,517	83,226

(continued)

Table 4.3 (continued)

Year	All Races			White			All Other					
							Total			Black		
	Both Sexes	Male	Female	Both Sexes	Male	Female	Both Sexes	Male	Female	Both Sexes	Male	Female
					Death Rate							
1988	882.0	940.0	826.9	905.1	952.2	860.0	757.5	872.7	652.2	874.2	1,006.8	754.5
1987	872.4	934.7	813.1	895.5	947.8	845.5	745.8	861.5	640.0	856.9	989.5	737.3
1986	873.2	940.7	809.3	896.2	954.4	840.7	745.6	862.7	638.8	854.2	987.7	733.9
1985	873.9	945.0	806.6	897.1	960.0	837.1	743.3	858.6	638.1	845.7	976.8	727.7
1984	862.3	935.9	792.7	885.2	951.1	822.3	731.5	846.4	627.0	828.5	958.1	712.0
1983	862.8	942.6	787.4	884.6	957.4	815.3	736.6	854.9	629.2	830.5	963.3	711.2
1982	852.0	938.0	770.7	872.9	951.8	797.9	728.9	854.8	614.5	819.2	960.4	692.4
1981	862.4	954.5	775.4	880.3	965.1	799.6	755.0	889.4	633.0	841.7	991.6	707.3
1980	878.3	976.9	785.3	892.5	983.3	806.1	791.7	936.5	660.6	875.4	1,034.1	733.3
1979	852.2	957.5	752.7	865.2	963.3	771.8	771.0	920.3	635.7	839.3	999.6	695.3
1978	868.0	977.5	764.5	880.2	982.7	782.7	790.0	943.2	651.0	855.1	1,016.8	709.5
1977	864.4	978.9	756.0	874.6	983.0	771.3	798.6	951.7	659.6	864.0	1,026.0	718.0
1976	877.6	993.8	767.6	887.7	997.3	783.1	812.1	969.8	668.6	875.0	1,041.6	724.5
1975	878.5	1,002.0	761.4	886.9	1,004.1	775.1	823.1	987.6	673.1	882.5	1,055.4	726.1
1970	945.3	1,090.3	807.8	946.3	1,086.7	812.6	938.4	1,115.9	775.3	999.3	1,186.6	829.2
1960	954.7	1,104.5	809.2	947.8	1,098.5	800.9	1,008.5	1,152.0	872.6	1,038.6	1,181.7	905.0
1950	963.8	1,106.1	823.5	945.7	1,089.5	803.3	1,119.4	1,251.1	993.5	...	...	...
1940	1,076.4	1,197.4	954.6	1,041.5	1,162.2	919.4	1,382.8	1,513.7	1,256.2	...	...	...

(continued)

Table 4.3 (continued)

							All Other					
	All Races			White			Total			Black		
Year	Both Sexes	Male	Female	Both Sexes	Male	Female	Both Sexes	Male	Female	Both Sexes	Male	Female
					Age-Adjusted Death Rate							
1988.........	535.5	696.7	404.4	509.8	664.3	384.4	692.5	904.3	523.5	788.8	1,037.8	593.1
1987.........	535.5	698.6	403.3	511.1	668.2	384.1	688.0	898.1	520.1	778.6	1,023.2	586.2
1986.........	541.7	709.1	406.6	518.0	679.8	387.7	693.1	905.7	524.0	781.0	1,026.9	588.2
1985.........	546.1	716.8	409.4	523.1	688.7	390.6	697.8	910.4	530.0	779.9	1,024.0	589.1
1984.........	545.9	716.7	409.6	523.6	689.9	391.3	694.5	904.4	529.0	771.9	1,011.7	585.3
1983.........	550.5	725.3	411.5	528.0	698.4	392.7	703.1	916.3	535.2	777.9	1,019.6	590.4
1982.........	553.8	733.1	411.2	531.8	706.0	393.3	704.6	928.4	529.3	778.7	1,035.0	581.4
1981.........	568.2	753.3	420.4	544.6	724.4	401.4	732.6	964.3	549.3	803.9	1,067.7	599.1
1980.........	585.8	777.2	432.6	559.4	745.3	411.1	774.2	1,015.1	582.6	842.5	1,112.8	631.1
1979.........	577.0	768.6	423.1	551.9	738.4	402.5	757.1	994.8	566.9	812.1	1,073.3	605.0
1978.........	595.0	791.4	437.4	569.5	761.1	416.4	780.0	1,021.0	585.8	831.8	1,093.9	622.7
1977.........	602.1	801.3	441.8	575.7	770.6	419.6	796.7	1,036.1	601.8	849.3	1,112.1	639.6
1976.........	618.5	820.9	455.0	591.3	789.3	432.5	818.5	1,063.4	618.3	870.5	1,138.3	654.5
1975.........	630.4	837.2	462.5	602.2	804.3	439.0	840.6	1,090.1	634.5	890.8	1,163.0	670.6
1970.........	714.3	931.6	532.5	679.6	893.4	501.7	983.4	1,231.4	770.8	1,044.0	1,318.6	814.4
1960.........	760.9	949.3	590.6	727.0	917.7	555.0	1,046.1	1,211.0	893.3	1,073.3	1,246.1	916.9
1950.........	841.5	1,001.6	688.4	800.4	963.1	645.0	1,225.7	1,358.5	1,095.7	...	...	...
1940.........	1,076.1	1,213.0	938.9	1,017.2	1,155.1	879.0	1,634.7	1,764.4	1,504.7	...	...	...

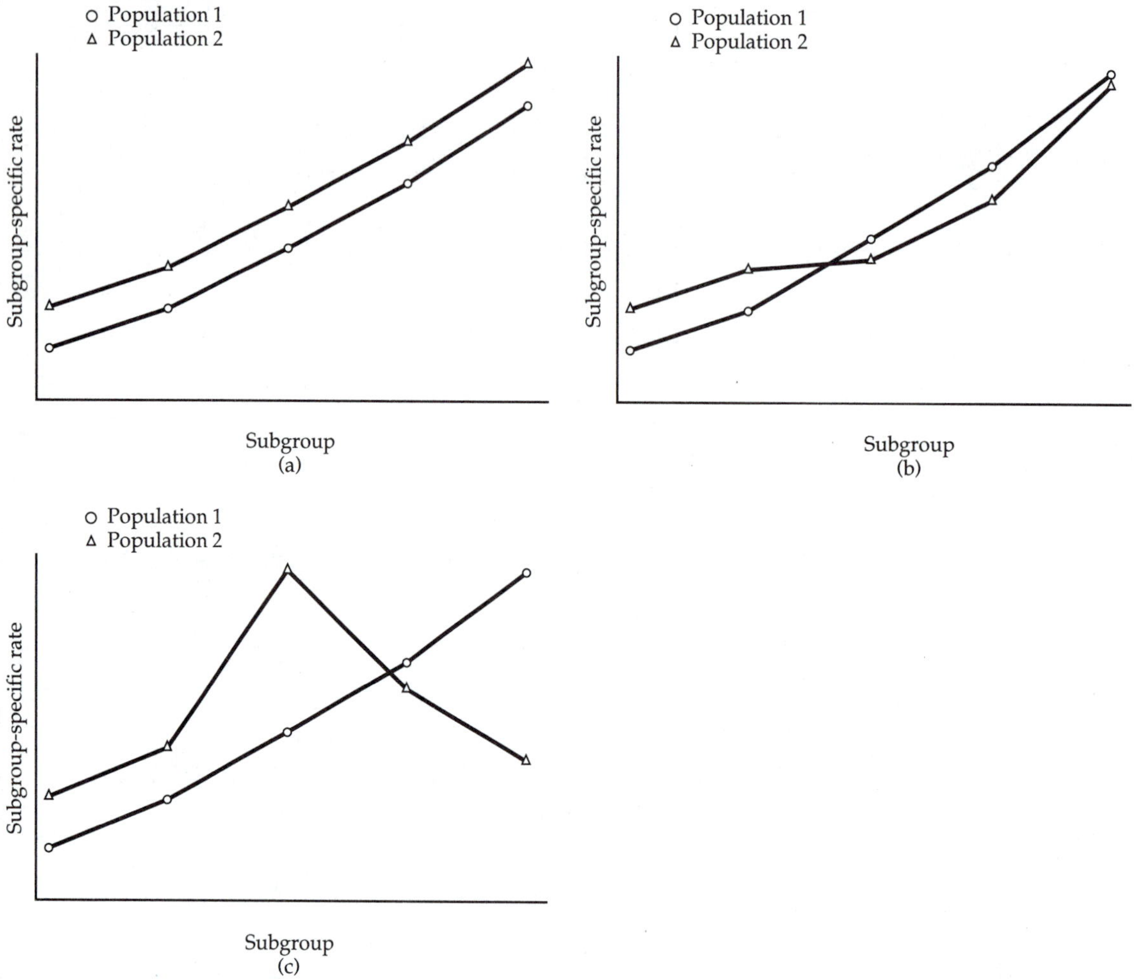

Figure 4.4. Trends in subgroup-specific rates for two populations

The age-specific hearing impairment rates for the subpopulations of currently employed individuals and those not in the labor force are displayed in Figure 4.5. The graph most closely resembles Figure 4.4(b); the trends are similar although not parallel. We conclude that it is not inappropriate to use the direct method to standardize for age.

In practice, the direct method of standardization is used more frequently than the indirect method. However, the direct method requires that subgroup-specific rates be available for all populations being compared; in many situations this is not feasible. If the subgroup-specific rates are available but have been calculated based on very small numbers—and are therefore subject to a great deal of fluctuation—then the indirect method is preferred. In any event, the application

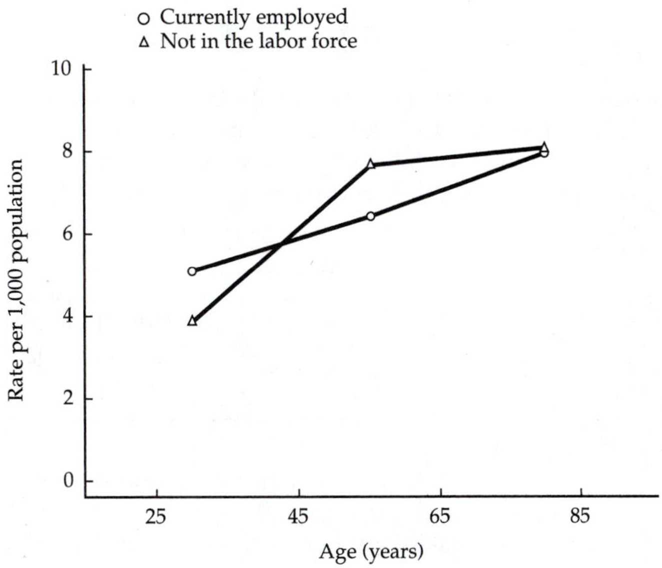

Figure 4.5. Age-specific hearing impairment rates for the currently employed and those not in the labor force, 1980–1981

of either method should lead to the same conclusions. If the conclusions differ, then the situation needs to be investigated further.

4.3 Further Applications

Suppose that we are interested in comparing the infant mortality rates in two states, Colorado and Louisiana. How may this be accomplished? To begin, consider the following data, which specify the numbers of live births and deaths of infants less than 1 year of age that occurred in each state in 1987 [7, 8].

State	Live Birth	Infant Deaths	Rate per 1,000
Colorado	53,808	527	9.8
Louisiana	73,967	872	11.8

The crude data show that the infant mortality rate in Colorado is 9.8 per 1,000 live births, whereas the rate in Louisiana is 11.8 per 1,000 live births. As a result, it appears that infants born in Louisiana are more likely to die before they are 1 year old than those born in Colorado.

We might suspect that race is a confounder in the relationship between state and infant mortality rate; hence we examine the underlying distributions of race in the two populations.

	Colorado		Louisiana	
Race	Live Births	Percent	Live Births	Percent
Black	3,166	5.9	29,670	40.1
White	48,805	90.7	42,749	57.8
Other	1,837	3.4	1,548	2.1
Total	53,808	100.0	73,967	100.0

The states do differ in racial composition. In Colorado, most of the live births are white children; in Louisiana, a much larger percentage of the infants are black. We focus on these two subgroups.

First we examine the race-specific infant mortality rates across the entire United States in 1987.

Race	Live Births	Infant Deaths	Rate per 1,000
Black	641,567	11,461	17.9
White	2,992,488	25,810	8.6
Other	175,339	1,137	6.5
Total	3,809,394	38,408	10.1

The crude infant mortality rate for the United States is a weighted average of the race-specific rates;

$$10.1 = \frac{641,567(17.9) + 2,992,488(8.6) + 175,339(6.5)}{3,809,394}.$$

Note that the infant mortality rate is considerably higher among black children than it is among white. Since race is associated with both state and infant mortality rate, it is a confounder in the relationship between these two quantities. Perhaps the higher crude infant mortality rate in Louisiana is a consequence of the greater proportion of black children who are born there.

A more accurate comparison between states can be made by looking at the race-specific infant mortality rates rather than the crude rates.

	Colorado			Louisiana		
Race	**Live Births**	**Infant Deaths**	**Rate per 1,000**	**Live Births**	**Infant Deaths**	**Rate per 1,000**
Black	3,166	52	16.4	29,670	525	17.7
White	48,805	469	9.6	42,749	344	8.0
Other	1,837	6	3.3	1,548	3	1.9
Total	53,808	527	9.8	73,967	872	11.8

To draw any type of conclusion based on these two sets of rates is difficult. It is true that the infant mortality rate among black children is higher in Louisiana than it is in Colorado; among white children, however, the infant mortality rate is higher in Colorado.

Direct Method of Standardization

To apply the direct method of standardization, we first select a standard race distribution; in this example, we use all the live births in the United States in 1987. We then calculate the number of infant deaths that would have occurred in the two states, assuming that each possessed this standard race composition while retaining its own individual race-specific mortality rates.

	U.S.	Colorado		Louisiana	
Race	**Live Births**	**Rate per 1,000**	**Expected Deaths**	**Rate per 1,000**	**Expected Deaths**
Black	641,567	16.4	10,521.7	17.7	11,355.7
White	2,992,488	9.6	28,727.9	8.0	23,939.9
Other	175,339	3.3	578.6	1.9	333.1
Total	3,809,394		39,828.2		35,628.7

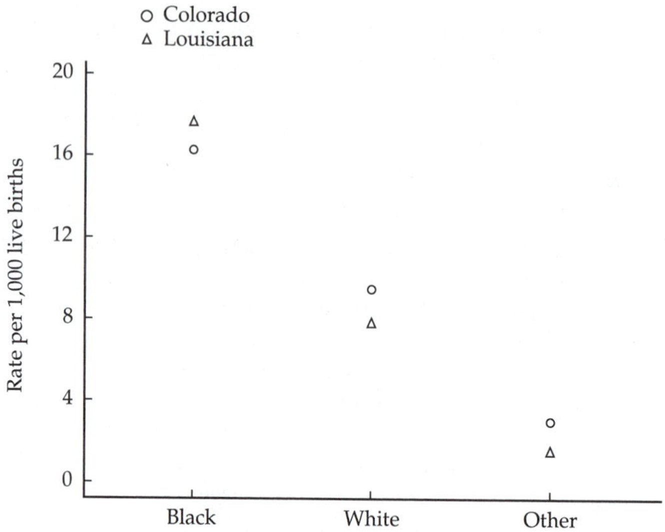

Figure 4.6. Race-specific infant mortality rates for Colorado and Louisiana, 1987

The race-adjusted infant mortality rate for each state is calculated by dividing its expected number of infant deaths by the total number of live births in the standard population:

$$\text{Colorado}: \quad \frac{39,828.2}{3,809,394} = 10.5 \text{ per } 1,000$$

$$\text{Louisiana}: \quad \frac{35,628.7}{3,809,394} = 9.4 \text{ per } 1,000$$

These race-adjusted rates are the infant mortality rates that would apply if the live births in both Colorado and Louisiana had the same distribution of race as the United States as a whole in 1987. Although the crude infant mortality rate for Louisiana is higher than the crude rate for Colorado, after we control for the effect of race, the adjusted infant mortality rate in Colorado is higher.

In this example, was it appropriate to calculate the race-adjusted infant mortality rates for these two populations? To attempt to answer this question, consider the plot of the race-specific mortality rates for Colorado and Louisiana presented in Figure 4.6. Since race is not a continuous measurement, we cannot connect the points with straight lines. Observe, however, that the two sets of rates follow the same trend; for both states, black children have the highest infant mortality rate, followed by white children, and then by children from other groups. Therefore, we conclude that it is appropriate to use the direct method to standardize for race.

Indirect Method of Standardization

To apply the indirect method of standardization, we select a standard set of race-specific infant mortality rates and apply them to the actual numbers of live births in each state. As the standard, we again choose the live births in the United States in 1987. We then calculate the numbers of infant deaths that would have occurred in each state if they were to take on the race-specific infant mortality rates of the United States as a whole while retaining their own distributions of race.

	U.S.	Colorado		Louisiana	
Race	Rate per 1,000	Live Births	Expected Deaths	Live Births	Expected Deaths
Black	17.9	3,166	56.7	29,670	531.1
White	8.6	48,805	419.7	42,749	367.6
Other	6.5	1,837	11.9	1,548	10.1
Total	10.1	53,808	488.3	73,967	908.8

The standardized mortality ratio for each state is obtained by dividing the observed number of infant deaths by the expected number of deaths:

$$\begin{aligned}\text{Colorado}: \quad \frac{527}{488.3} &= 1.08\\ &= 108\%\end{aligned}$$

$$\begin{aligned}\text{Louisiana}: \quad \frac{872}{908.8} &= 0.96\\ &= 96\%\end{aligned}$$

These standardized mortality ratios indicate that Colorado has an 8% higher infant mortality rate than the United States as a whole, whereas Louisiana has a 4% lower infant mortality rate. We now calculate the race-adjusted infant mortality rate for each state by multiplying its standardized mortality ratio by the crude infant mortality rate in the standard population, 10.1 per 1,000 live births:

$$\text{Colorado}: \quad \frac{10.1}{1,000} \times 1.08 = 10.9 \text{ per } 1,000$$

$$\text{Louisiana}: \quad \frac{10.1}{1,000} \times 0.96 = 9.7 \text{ per } 1,000$$

After we control for the effect of race, Colorado has a higher adjusted infant mortality rate than Louisiana. This is the same conclusion that was reached when the direct method of standardization was applied.

4.4 Review Exercises

1. The total numbers of deaths in the United States in various years are presented below [1].

Year	Number of Deaths
1980	1,989,841
1970	1,921,031
1960	1,711,982
1950	1,452,454
1940	1,417,269

The statement is made that since the number of deaths has been increasing over the years, the population as a whole must be growing less healthy. Do you agree with this statement? Why or why not?

2. The following data were reported for the state of Massachusetts in 1986 [9].

	Number
Population	5,829,504
Live births	82,163
Deaths	
Total	55,860
Under 1 year	695

Compute the following rates:

a. crude birth rate
b. crude death rate
c. infant mortality rate

3. Between the years 1984 and 1987, the crude death rate for females in the United States increased steadily. At the same time, the age-adjusted death rate decreased [1]. The relevant data are shown below.

Year	Crude Death Rate per 100,000 Population	Age-Adjusted Death Rate per 100,000 Population
1987	813.1	403.3
1986	809.3	406.6
1985	806.6	409.4
1984	792.7	409.6

Explain how the age-adjusted rate could decrease while the crude rate is increasing.

4. Consider the following data pertaining to cancer mortality in the United States in 1940 and 1986 [10, 11].

	1940		1986	
Age	**Population (thousands)**	**Deaths**	**Population (thousands)**	**Deaths**
0–4	10,541	494	18,152	666
5–14	22,431	667	33,860	1,165
15–24	23,922	1,287	39,021	2,115
25–34	21,339	3,696	42,779	5,604
35–44	18,333	11,198	33,070	14,991
45–54	15,512	26,180	22,815	37,800
55–64	10,572	39,071	22,232	98,805
65–74	6,377	44,328	17,332	146,803
75+	2,643	31,279	11,836	161,381
Total	131,670	158,200	241,097	469,330

a. Compute the crude cancer mortality rates for 1940 and 1986, and compare these rates.
b. For both 1940 and 1986, calculate the proportion of the total population in each age group. Describe how the two populations differ with respect to age composition.
c. Compute the age-specific cancer death rates for each population. Is there a relationship between age and death rate?
d. Does it appear to be necessary to control for the effect of age when comparing cancer death rates in the two populations? Why or why not?
e. Using the U.S. population in 1940 as the standard distribution, apply the direct method of standardization. What are the age-adjusted cancer mortality rates for 1940 and 1986?
f. How does the age-adjusted death rate differ from the crude death rate in each of these populations? Explain these results.
g. Using suitably scaled axes, plot the age-specific cancer mortality rates versus age for both 1940 and 1986. Comment on the extent to which it was appropriate to standardize for age using the direct method.
h. Using the age-specific cancer mortality rates for 1940 as the standard, apply the indirect method to compute the standard mortality ratios for 1940 and 1986.
i. How does the 1986 population compare with the 1940 population in terms of cancer mortality rate?
j. Calculate the age-adjusted cancer mortality rates for 1940 and 1986. Do the results obtained using the indirect method correspond to those obtained when the direct method was applied?

References

[1] National Center for Health Statistics, "Advance Report of Final Mortality Statistics, 1988," *Monthly Vital Statistics Report*, Volume 39, November 28, 1990.

[2] United Nations Children's Fund, *The State of the World's Children 1991*, New York: Oxford University Press.

[3] National Center for Health Statistics, "Births, Marriages, Divorces, and Deaths for 1988," *Monthly Vital Statistics Report*, Volume 37, April 1989.

[4] National Center for Health Statistics, Collins, J. G., "Types of Injuries and Impairments Due to Injuries, United States," *Vital and Health Statistics*, Series 10, Number 159, November 1986.

[5] Cochran, W. G., "The Effectiveness of Adjustment by Subclassification in Removing Bias in Observational Studies," *Biometrics*, Volume 24, June 1968, 295–313.

[6] Centers for Disease Control, "Mortality Attributable to HIV Infection/AIDS—United States, 1981–1990," *Morbidity and Mortality Weekly Report*, Volume 40, Number 3, January 25, 1991.

[7] National Center for Health Statistics, *Vital Statistics of the United States, 1987*, Volume I—Natality, 1989.

[8] National Center for Health Statistics, *Vital Statistics of the United States, 1987*, Volume II—Mortality, Part A, 1990.

[9] Massachusetts Department of Public Health, *1986 Annual Report, Vital Statistics of Massachusetts*, Public Document 1, January 1989.

[10] National Center for Health Statistics, *Vital Statistics of the United States, 1986*, Volume II—Mortality, Part B, 1988.

[11] United States Department of Commerce, *Vital Statistics of the United States, 1940*, Part II—Natality and Mortality Data Tabulated by Place of Residence, 1943.

CHAPTER

5

Life Tables

It is almost always impossible to predict how long a particular individual will live, let alone predict the span of life for each person in a population of millions. In spite of this, policy planners must be able to assess and describe the health of a nation. For many years, life tables have been used as a means of summarizing the health status of a group of individuals. Like the techniques that we have studied in previous chapters, they are descriptive in nature. *Life tables* identify the death rates experienced by a particular population over a given period of time. They have many practical applications; life tables are used to analyze the mortality of a population, to make international comparisons, to compute insurance premiums and annuities, and to predict survival. They have even been used, rather inanely, in a debate regarding the existence of a biological limit to human life [1].

By following a fictional cohort of individuals–usually a group of 100,000–from birth until the last individual in the cohort has died, a life table describes the mortality experience over a specified period of time. For example, Table 5.1 shows the 1988 life table for the United States [2]. It assumes that we observe 100,000 persons from birth until death and that, as they pass through each year of life, the individuals die at the rates experienced by the U.S. population in 1988. The number 100,000 is chosen to simplify calculations; it is arbitrary, and we would end up with the same results no matter what number we chose.

5.1 Computation of the Life Table

5.1.1 Column 1

Before describing some of its applications, we begin by examining the way in which the life table is computed. The first column of the life table is the age interval. The age interval represents the period of life between age x and age $x + n$, where n is the length of the interval. Therefore, 0–1 represents the one-year period

Table 5.1 Abridged life table for the total population, United States, 1988

Age Interval	Proportion Dying	Of 100,000 Born Alive		Stationary Population		Average Remaining Lifetime
Period of Life Between Two Exact Ages Stated in Years (1)	Proportion of Persons Alive at Beginning of Age Interval Dying During Interval (2)	Number Living at Beginning of Age Interval (3)	Number Dying During Age Interval (4)	In the Age Interval (5)	In This and All Subsequent Age Intervals (6)	Average Number of Years of Life Remaining at Beginning of Age Interval (7)
x to $x+n$	${}_nq_x$	l_x	${}_nd_x$	${}_nL_x$	T_x	$\mathring{e}_x$
0–1	0.0100	100,000	999	99,147	7,494,642	74.9
1–5	0.0020	99,001	198	395,540	7,395,495	74.7
5–10	0.0012	98,803	120	493,688	6,999,955	70.8
10–15	0.0014	98,683	134	493,155	6,506,267	65.9
15–20	0.0044	98,549	431	491,767	6,013,112	61.0
20–25	0.0058	98,118	565	489,206	5,521,345	56.3
25–30	0.0061	97,553	596	486,274	5,032,139	51.6
30–35	0.0074	96,957	717	483,035	4,545,865	46.9
35–40	0.0096	96,240	924	479,021	4,062,830	42.2
40–45	0.0126	95,316	1,204	473,785	3,583,809	37.6
45–50	0.0189	94,112	1,777	466,443	3,110,024	33.0
50–55	0.0300	92,335	2,766	455,194	2,643,581	28.6
55–60	0.0473	89,569	4,238	437,859	2,188,387	24.4
60–65	0.0728	85,331	6,208	411,976	1,750,528	20.5
65–70	0.1055	79,123	8,344	375,656	1,338,552	16.9
70–75	0.1568	70,779	11,096	327,120	962,896	13.6
75–80	0.2288	59,683	13,654	265,113	635,776	10.7
80–85	0.3445	46,029	15,858	190,715	370,663	8.1
85 and over	1.0000	30,171	30,171	179,948	179,948	6.0

of life from birth until an individual's first birthday. Interval 1–5 represents the time from the first birthday until the fifth birthday, a period of four years. All other age intervals span five years, except for the last. This is an open-ended interval representing the entire period of life beyond the 85th birthday.

For the sake of convenience, abridged life tables, such as Table 5.1, are used most often in practice. A complete life table would have an entry for each individual year; as an example, a complete life table calculated for the U.S. population in 1979–1981 is included at the end of this chapter [3]. Note from Table 5.8 that, within each of the five-year intervals between ages 5 and 85, mortality is relatively constant. This is not true for the very young; mortality in the first year of life is quite different from that in years 2 through 4. At the other end of the scale, the combination of everyone over 85 years of age into a single group is influenced in part by tradition. In the past, life expectancy was considerably less than 85 years, and only a small proportion of individuals lived beyond that age.

5.1.2 Column 2

The second column of the life table, represented by ${}_nq_x$, lists the proportion of individuals alive at the beginning of the interval x to $x + n$ who die some time during the interval. This quantity is also called the *hazard function*; it can be calculated from the age-specific death rates for the United States population in 1988, such as those provided in Table 4.2. For example,

$$\begin{aligned}
{}_1q_0 &= \text{proportion of individuals alive at birth who die before} \\
&\qquad \text{their first birthday} \\
&= \frac{1,008.3}{100,000} \qquad \text{(from Table 4.2)} \\
&= 0.010083 \\
&\approx 0.0100.
\end{aligned}$$

The numbers included in the two tables do not match exactly because Table 4.2 is abridged and lists only the estimated age-specific death rate for a given interval, whereas Table 5.1 was calculated using more precise data. Note that

$$\begin{aligned}
{}_\infty q_{85} &= \text{proportion of individuals alive on their } 85^{\text{th}} \text{ birthday who} \\
&\qquad \text{die during the time period after their } 85^{\text{th}} \text{ birthday} \\
&= 1.0000,
\end{aligned}$$

since death is inevitable. The other proportions in column 2 are somewhat more difficult to calculate.

If we were to plot the actual age-specific death rates—otherwise known as the hazard function—from birth until age 90 for the U.S. population in 1979–1981, the result would be Figure 5.1. A plot of the age-specific death rates in 1988 would be similar. From the graph, a single rate may be chosen to summarize each age

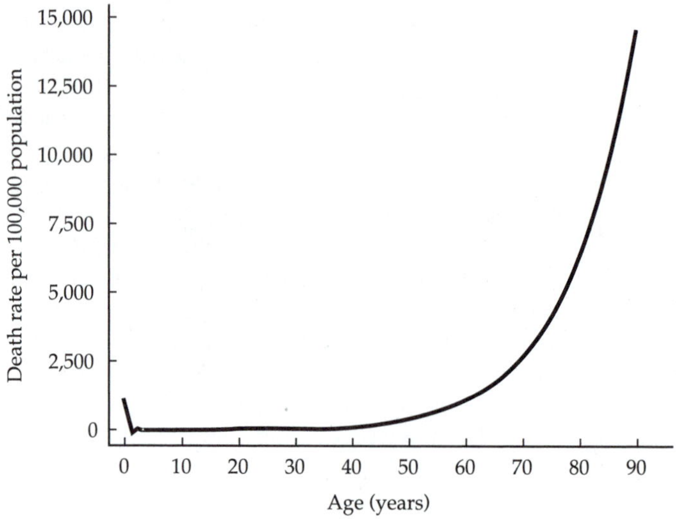

Figure 5.1. Age-specific death rates from birth until age 90 for the U.S. population, 1979–1981

interval in column 1. For an abridged table, this value is used to approximate the proportion of individuals dying within the interval. For example, consider the age group 1–5. From Table 4.2, the approximate age-specific death rate for this interval is 50.9 per 100,000 population, or 0.000509. Since this is the estimated mortality rate for **each** year within the four-year period, the total proportion of individuals dying between their first and fifth birthdays is roughly

$$\begin{aligned} {}_4q_1 &= 4 \times 0.000509 \\ &= 0.002036 \\ &\approx 0.0020. \end{aligned}$$

Similarly,

$$\begin{aligned} {}_5q_5 &= 5 \times 0.000242 \\ &= 0.001210 \\ &\approx 0.0012, \end{aligned}$$

$$\begin{aligned} {}_5q_{10} &= 5 \times 0.000275 \\ &= 0.001375 \\ &\approx 0.0014, \end{aligned}$$

and so on. The estimates calculated using data from the abridged table are accurate if the age-specific death rates are relatively constant throughout an interval. In theory, we should actually compound the mortality each year rather than simply multiply by the number of years; in practice, however, the approximation is usually adequate.

5.1.3 Columns 3 and 4

The third column of the life table, l_x, displays the number of individuals out of the original 100,000 who are still alive on their x^{th} birthday. Note that column headings with a single subscript refer to conditions at the beginning of the interval; those with two subscripts specify conditions within the interval. Hence the fourth column, ${}_n d_x$, lists the number out of the l_x alive at the beginning of the interval x to $x + n$ who die some time during the interval. The computations of these two columns are interdependent. For example, l_0, the number of individuals born into the cohort, is equal to 100,000. The number dying before their first birthday is the total number born alive multiplied by the proportion of individuals who die during the 0–1 age interval, or

$$\begin{aligned} {}_1 d_0 &= 100,000 \times {}_1 q_0 \\ &= 100,000 \times 0.0100 \\ &= 1,000 \\ &\approx 999. \end{aligned}$$

Therefore, the number of individuals out of the original 100,000 who live to see their first birthday is

$$\begin{aligned} l_1 &= 100,000 - 999 \\ &= 99,001. \end{aligned}$$

Similarly,

$$\begin{aligned} {}_4 d_1 &= 99,001 \times {}_4 q_1 \\ &= 99,001 \times 0.0020 \\ &= 198, \end{aligned}$$

$$\begin{aligned} l_5 &= 99,001 - 198 \\ &= 98,803, \end{aligned}$$

$$\begin{aligned} {}_5 d_5 &= 98,803 \times {}_5 q_5 \\ &= 98,803 \times 0.0012 \\ &= 118.6 \\ &\approx 120, \end{aligned}$$

and

$$\begin{aligned} l_{10} &= 98,803 - 120 \\ &= 98,683. \end{aligned}$$

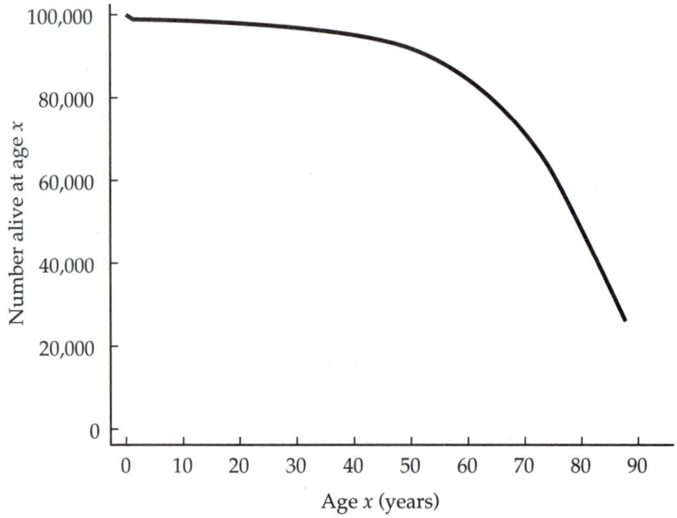

Figure 5.2. Number of individuals alive at age x from birth until age 85 for the U.S. population, 1988

In general, the number of individuals alive at the beginning of a particular interval is equal to the number who were alive at the beginning of the previous interval minus the number who died during that interval, or

$$l_{x+n} = l_x - {}_nd_x.$$

If we plot l_x against age x, as in Figure 5.2, we observe the manner in which the number of survivors decreases over time. The number of individuals dying during an age interval is derived by multiplying the number alive at the beginning of the interval by the proportion of persons dying during the interval, or

$$_nd_x = l_x \times {}_nq_x.$$

The results of our calculations do not match the entries in Table 5.1 exactly because we are using data from an abridged table; Table 5.1 was computed using more precise values.

5.1.4 Column 5

Column 5 of the life table, represented by $_nL_x$, is known as the *stationary population* within an interval. Demographers find this concept useful; it may be interpreted in the following way. Suppose that a cohort of 100,000 individuals is born each year. Furthermore, assume that in each cohort the proportion of individuals dying within the age interval x to $x+n$ is given by $_nq_x$ in column 2. As a result, the 1988 age-specific death rates apply to every cohort. If no migration occurs and a sufficient number of years are allowed to pass, we would reach a stationary population; the number of persons alive in any given age group would never change. As soon as one individual leaves an interval—either by dying or by growing older and entering the subsequent interval—his or her place would be taken by someone from the previous age group. Consequently, a census taken at any arbitrary point in time would enumerate the same population with the same distribution among age groups.

The stationary population ${}_nL_x$ has a second interpretation as well. It may be thought of as the total time in years lived during the age interval x to $x + n$ by the l_x individuals alive at the beginning of the interval. Consider, for example, the age interval 1–5. There are $l_1 = 99,001$ individuals alive at the beginning of this interval. Of these 99,001, $l_5 = 98,803$ are still alive on their fifth birthday. Since each of these 98,803 individuals survives throughout the entire four-year period, they contribute $98,803 \times 4 = 395,212$ *person-years* to the total time lived during the interval. There are also ${}_4d_1 = 198$ persons who die at some point during the interval. Each of these individuals contributes some amount of person-time before he or she dies, anywhere from zero to four years. If we sum the person-years contributed by each of the individuals who die during the interval—information that is not provided in the abridged life table—and add this to the 395,212 person-years lived by the individuals who survive the entire four years, we have a total of ${}_4L_1 = 395,540$ years lived during the age interval 1–5.

Note that the procedure used to derive this total is identical to the technique for summing a set of observations that was presented in Chapter 3. We begin by grouping individuals according to the number of years that they have contributed to the interval. For each group, we then multiply the number of individuals by the associated number of years. Finally, we sum these quantities to obtain the total time lived during the interval. Once again, all calculations are based on the assumption that the 1988 age-specific death rates hold throughout the entire lifetime of the cohort.

5.1.5 Column 6

The sixth column of the life table, T_x, displays the total stationary population in the age interval x to $x + n$ and all subsequent intervals. In other words, it is the total number of person-years lived beyond their x^{th} birthday by the l_x individuals alive on that birthday. It is obtained by summing column 5 from the bottom up; for example,

$$\begin{aligned} T_{65} &= {}_5L_{65} + {}_5L_{70} + {}_5L_{75} + {}_5L_{80} + {}_\infty L_{85} \\ &= 375,656 + 327,120 + 265,113 + 190,715 + 179,948 \\ &= 1,338,552. \end{aligned}$$

5.1.6 Column 7

Finally, column 7 of the life table, $\overset{\circ}{e}_x$, is the average number of years of life remaining for an individual who is alive at age x. It is calculated by dividing the total number of person-years lived beyond the xth birthday, T_x, by the number of individuals who survive to age x or beyond; dividing by l_x eliminates the dependency on the size of the original cohort. As an example,

$$\begin{aligned}\overset{\circ}{e}_0 &= \text{average number of years of life remaining at birth}\\ &= \frac{T_0}{l_0}\\ &= \frac{7,494,642}{100,000}\\ &= 74.9\end{aligned}$$

and

$$\begin{aligned}\overset{\circ}{e}_1 &= \text{average number of years of life remaining after the first birthday}\\ &= \frac{T_1}{l_1}\\ &= \frac{7,395,495}{99,001}\\ &= 74.7.\end{aligned}$$

If $\overset{\circ}{e}_1$ is larger than $\overset{\circ}{e}_0$—which was the case in the United States as recently as 1976—this reflects a high infant mortality rate. It implies that if an infant is strong enough to survive the first year of life, then the average remaining lifetime on his or her first birthday is greater than it was at birth.

5.2 Applications

Life tables have a number of practical applications. For example, they enable us to predict the chance that an individual will live to a particular age x. Suppose we wish to know the probability that a person will survive from birth until age 65 or, equivalently, the proportion of persons who do so. Of the 100,000 individuals who were born into the cohort, 79,123 are still alive on their 65th birthday. Therefore, the proportion of persons surviving until age 65 is

$$\begin{aligned}\frac{l_{65}}{l_0} &= \frac{79,123}{100,000}\\ &= 0.79123\end{aligned}$$

or about 79.1%. The probability that a 50-year-old will reach his or her 65th birthday is

$$\frac{l_{65}}{l_{50}} = \frac{79,123}{92,335}$$
$$= 0.85691$$

or approximately 85.7%. This increase in probability–from 79.1% to 85.7%–is important to an individual calculating insurance rates. The concept of probability will be discussed further in Chapter 6; for now, we consider "probability" to be synonymous with "proportion."

Average remaining lifetime, $\mathring{e}_0$ in particular, is often used to describe the health status of a population [4]. The average life expectancies at birth in the United States are broken down by race and sex in Table 5.2 [2]. Because of its application as a population summary measure, average remaining lifetime is often used to make international comparisons. The average life expectancies at birth for selected nations are provided in Table 5.3 [5]. In all the countries included,

Table 5.2 Life expectancy at birth by race and sex, United States, 1940, 1950, 1960, 1970–1988

							All Other					
	All Races			White			Total			Black		
Age	Both Sexes	Male	Female	Both Sexes	Male	Female	Both Sexes	Male	Female	Both Sexes	Male	Female
						Years						
1988	74.9	71.5	78.3	75.6	72.3	78.9	71.2	67.1	75.1	69.2	64.9	73.4
1987	75.0	71.5	78.4	75.6	72.2	78.9	71.3	67.3	75.2	69.4	65.2	73.6
1986	74.8	71.3	78.3	75.4	72.0	78.8	71.2	67.2	75.1	69.4	65.2	73.5
1985	74.7	71.2	78.2	75.3	71.9	78.7	71.2	67.2	75.0	69.5	64.3	73.5
1984	74.7	71.2	78.2	75.3	71.8	78.7	71.3	67.4	75.0	69.7	65.6	73.7
1983	74.6	71.0	78.1	75.2	71.7	78.7	71.1	67.2	74.9	69.6	65.4	73.6
1982	74.5	70.9	78.1	75.1	71.5	78.7	71.0	66.8	75.0	69.4	65.1	73.7
1981	74.2	70.4	77.8	74.8	71.1	78.4	70.3	66.1	74.4	68.9	64.5	73.2
1980	73.7	70.0	77.4	74.4	70.7	78.1	69.5	65.3	73.6	68.1	63.8	72.5
1979	73.9	70.0	77.8	74.6	70.8	78.4	69.8	65.4	74.1	68.5	64.0	72.9
1978	73.5	69.6	77.3	74.1	70.4	78.0	69.3	65.0	73.5	68.1	63.7	72.4
1977	73.3	69.5	77.2	74.0	70.2	77.9	68.9	64.7	73.2	67.7	63.4	72.0
1976	72.9	69.1	76.8	73.6	69.9	77.5	68.4	64.2	72.7	67.2	62.9	71.6
1975	72.6	68.8	76.6	73.4	69.5	77.3	68.0	63.7	72.4	66.8	62.4	71.3
1974	72.0	68.2	75.9	72.8	69.0	76.7	67.1	62.9	71.3	66.0	61.7	70.3
1973	71.4	67.6	75.3	72.2	68.5	76.1	66.1	62.0	70.3	65.0	60.9	69.3
1972	71.2	67.4	75.1	72.0	68.3	75.9	65.7	61.5	70.1	64.7	60.4	69.1
1971	71.1	67.4	75.0	72.0	68.3	75.8	65.6	61.6	69.8	64.6	60.5	68.9
1970	70.8	67.1	74.7	71.7	68.0	75.6	65.3	61.3	69.4	64.1	60.0	68.3
1960	69.7	66.6	73.1	70.6	67.4	74.1	63.6	61.1	66.3	...	...	...
1950	68.2	65.6	71.1	69.1	66.5	72.2	60.8	59.1	62.9	...	...	...
1940	62.9	60.8	65.2	64.2	62.1	66.6	53.1	51.5	54.9	...	...	...

females have a longer life expectancy than males; there are only a few countries in the world for which the opposite is true.

Average remaining lifetime is also used as a basis for the computation of life insurance premiums. Recall, however, that the formula $\overset{\circ}{e}_x = T_x/l_x$ is derived based on the stationary population of columns 5 and 6. All calculations depend on the assumption that the age-specific death rates for 1988 remain constant throughout the entire lifetime of a cohort born in that year. In reality, the cohort members would be subject to the ${}_1q_0$ for 1988 during their first year of life, the ${}_1q_1$ for 1989 during their second year of life, the ${}_1q_2$ for 1990 during their third year of life, and so on. Since the age-specific death rates for future years are unknown, in the life table they are estimated by the 1988 death rates. As illustrated in Figure 5.3, however, the death rates for all age groups have been decreasing in recent years [2]. If this trend continues, the average remaining lifetime calculated in column 7 is actually an underestimate of the true remaining lifetime. Consequently, if individuals live longer than expected and continue to pay premiums throughout their lifetimes, insurance companies which use $\overset{\circ}{e}_x$ to predict survival will end up increasing their profits. In contrast, government and

Table 5.3 Average life expectancies at birth for selected countries, 1989

Nation	$\overset{\circ}{e}_0$	**Females as a Percentage of Males**
Argentina	71	108.8
Australia	76	109.6
Brazil	65	107.9
Canada	77	109.5
China	70	105.9
Egypt	60	103.4
Ethiopia	45	106.8
Finland	75	111.3
France	76	111.1
Greece	76	105.4
India	59	101.7
Israel	76	105.4
Italy	76	108.2
Japan	79	107.9
Mexico	69	110.6
Poland	72	111.8
Saudi Arabia	64	106.5
Spain	77	108.1
Switzerland	77	109.5
USSR	71	113.6
United Kingdom	76	106.8
United States	76	109.7
Venezuela	70	109.0
Yugoslavia	72	108.7

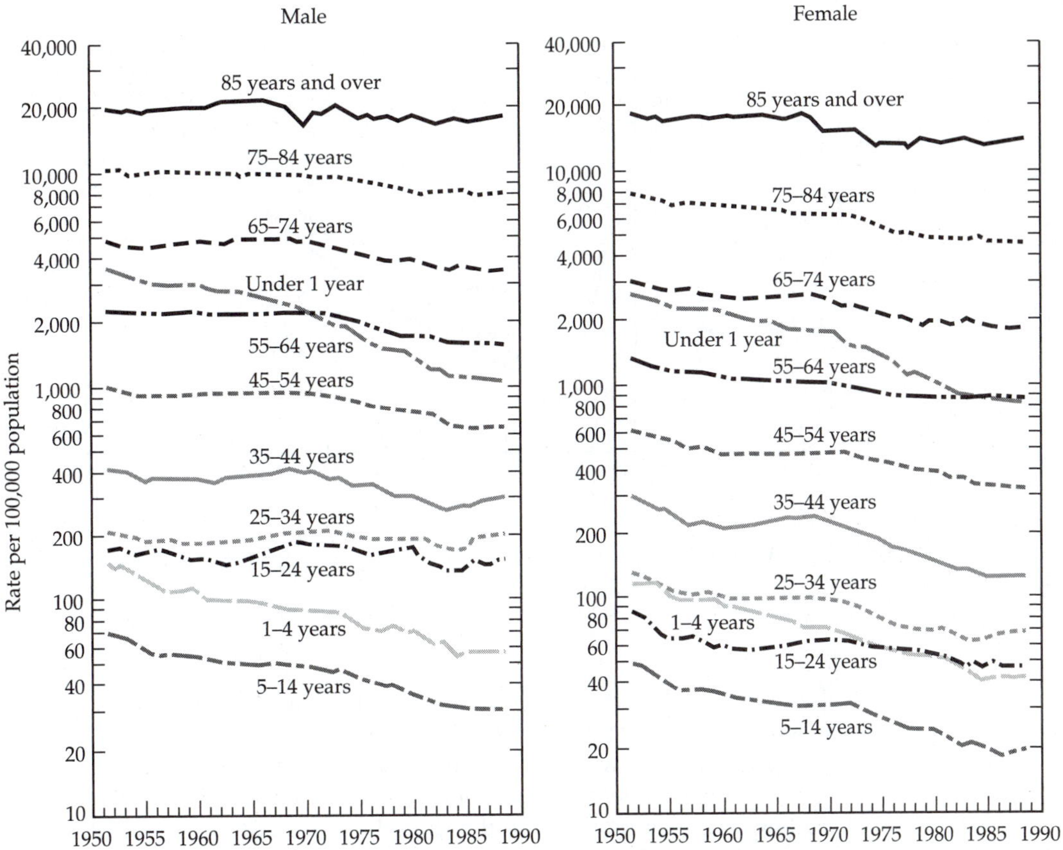

Figure 5.3. Death rates by age and sex, United States, 1950–1988

public health organizations that use $\mathring{e}_x$ to plan for future services could run into serious problems.

5.3 Years of Potential Life Lost

We may wish to use the methodology of life tables to quantify premature mortality. The improvement in survival during the last century in England and Wales is illustrated in Figure 5.4 [6]. Displayed are the age- and sex-specific mortality rates in both 1851 and 1951. The experience in the United States has been similar; in particular, the most significant reductions in mortality have been made in the younger age groups. The older groups do not show much improvement. This is largely a reflection of the advances brought about by the eradication of a large number of infectious, or "childhood," diseases, as well as great improvements in nutrition, housing, and sanitation. In other words, the reductions in mortality are

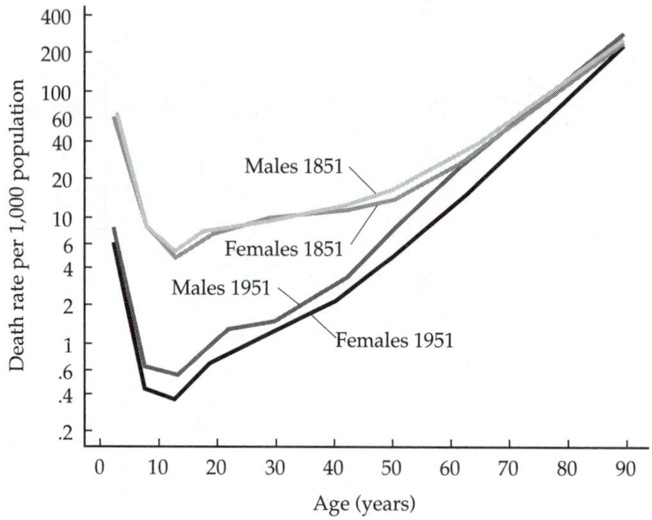

Figure 5.4. Age- and sex-specific death rates per 1,000 population, England and Wales, 1851 and 1951

the result of advances in the prevention of disease. Note that the difference in survival between genders is a relatively recent phenomenon.

Health policy workers must be able to determine the relative importance of various causes of death. Table 5.4 displays the mean ages at death for the seven leading causes of mortality in the United States in the 1930s and 1940s [7]. We can see that, for the causes that tend to affect younger persons, the average age at death improves more over the 15-year period than it does for causes that primarily affect the elderly. This difference is not incorporated in the calculation of overall mortality rates for those years, however, since each death counts the same as any other. The majority of deaths occur among older persons; therefore, mortality data are dominated by diseases of the elderly.

Incidentally, it should be noted that studies of mortality usually rely on information provided by death certificates. Such data must be treated with caution. Death certificates are legal, not medical, documents. Furthermore, procedures for filling out death certificates vary with time and across different cultures and are often prone to error when the deaths of older people are involved.

When used for a relatively stable group over a short period of time, the average age at death can be quite informative. Like other crude summary measures, however, it may be affected by a number of factors. As a general descriptive measure of survival, it was discredited long ago in favor of the life table method. A related concept, the *years of potential life lost*, or *YPLL*, was introduced in the early part of this century [8]. Rather than taking into account the years of life lived by an individual, the YPLL considers the amount of life that is lost as the result of a premature death. This has the effect of placing more importance on life that is lost owing to diseases of the young.

Table 5.4 Mean age at death in years for the seven leading causes of mortality, United States, 1930–1945

Cause of Death	1930	1935	1940	1945
Cerebrovascular*	67.5	67.7	68.5	69.1
Diseases of the heart	64.5	65.9	67.1	67.6
Nephritis	64.0	65.5	66.7	66.8
Cancer	61.4	62.1	62.5	63.0
Pneumonia/influenza	40.1	43.7	47.2	46.2
Accidents	41.3	43.8	45.5	45.8
Tuberculosis	37.4	39.9	41.7	43.7
All others	35.1	38.5	42.0	43.5
All causes combined	48.8	52.2	55.8	57.4

*Intracranial lesions of vascular origin

There are a number of different ways of defining YPLL. The one that is used most commonly, and that is reported by the Centers for Disease Control, excludes from consideration the deaths of all individuals over the age of 65 [9]. Other persons are allotted 65 years at birth. The years of potential life lost by an individual is the number of years fewer than 65 that the person actually lives. If a 50-year-old man dies of heart disease, for instance, then he has lost 15 years of potential life before the age of 65. The total YPLL for a population is obtained by summing the contributions from each individual.

Years of potential life lost may be contrasted with the more common crude mortality rate that we discussed in the previous chapter. Recall that the crude mortality rate assigns equal weight to each death and averages over all individuals. YPLL, in contrast, places more value on the deaths of younger persons; in fact, deaths occurring beyond a certain age are excluded from consideration.

A second method of calculating YPLL is based on the claim that the potential life lost by an individual is equal to his or her average remaining lifetime at the age at which death occurs [9]. According to the U.S. life table for 1988, a 50-year-old man is expected to have an average of 28.5 years of life remaining. If this individual were to die of heart disease, he would be losing 28.5 years of potential life rather than 15. Again, this method counts the quantity of life lost, but it does not arbitrarily assign each person a maximum of 65 years. As a result, it continues to value individuals who live beyond this age.

To compare each of the three methods described—the crude mortality rate, the YPLL based on average remaining lifetime, and the YPLL prior to age 65—the ten leading causes of death in the United States in 1984 are ranked in Table 5.5 [9]. According to the crude death rates, heart disease is the leading cause of mortality in the United States; however, if we concentrate only on deaths occurring before the age of 65, unintentional injury appears to be the leading cause. Note that, in general, the ranking generated by the YPLL based on life expectancy lies somewhere between the rankings produced by the other two methods.

As previously mentioned, the Centers for Disease Control choose to report the years of potential life lost based on the allotment of 65 years, thus discounting

Table 5.5 Rankings of the ten leading causes of death by method of calculation, United States, 1984

Rank	Crude Mortaility Rate	YPLL (Life Expectancy)	YPLL (Age 65)
1	Heart disease	Heart disease	Unintentional injuries
2	Malignant neoplasms	Malignant neoplasms	Malignant neoplasms
3	Cerebrovascular disease	Unintentional injuries	Heart disease
4	Unintentional injuries	Suicide/homicide	Suicide/homicide
5	COPD*	Cerebrovascular disease	Congenital anomalies
6	Pneumonia/influenza	COPD*	Prematurity
7	Suicide/homicide	Congenital anomalies	SIDS†
8	Diabetes	Pneumonia/influenza	Cerebrovascular disease
9	Cirrhosis	Cirrhosis	Cirrhosis
10	Congenital anomalies	Diabetes	Pneumonia/influenza

*Chronic obstructive pulmonary diseases and allied conditions
† Sudden Infant Death Syndrome

lives lost beyond that age. Table 5.6 displays the years of potential life lost for persons who died in the United States in 1986 and 1987 [10]. The total YPLL dropped from 5,016 years per 100,000 U.S. residents in 1986 to 4,949 years per 100,000 residents in 1987. Overall, the years of life lost to most major causes of death declined. There were small increases in the number of years lost to malignant neoplasms and cerebrovascular disease. The number of years of potential life lost as a result of HIV infection, however, increased an amazing 45%.

With a little algebra we are able to show that, for a particular cause of mortality, the mean age at death and the years of potential life lost before age 65 are closely related. Suppose that in a given population there are N_{65} individuals who die before age 65 from the specified cause. Denote by m_{65} the average age at death of these persons. In this case,

$$\text{YPLL} = N_{65}\,(65 - m_{65}).$$

Note that YPLL increases if m_{65} decreases, meaning that individuals dying from the particular cause are doing so at an earlier age or, if N_{65} increases, implying that more persons are dying as a result of the cause. If we divide YPLL by N_{65} to obtain the average number of years lost by each individual, the resulting quantity is simply

$$\frac{\text{YPLL}}{N_{65}} = 65 - m_{65},$$

or 65 minus the mean age at death.

When we compare years of potential life lost across different groups or populations, the same problems exist as when we compare either crude rates or the mean ages at death; in particular, both the sizes of the populations and their age structures may vary. To compensate for differing population sizes, we may calculate a YPLL rate per 1,000 persons. In addition, a technique similar to the direct

Table 5.6 Years of potential life lost before age 65 by cause of death, United States, 1986 and 1987

Cause of Death	YPLL 1986	YPLL 1987
All causes	12,093,486	12,045,778
Unintentional injuries	2,358,426	2,295,710
Malignant neoplasms	1,832,210	1,837,742
Heart disease	1,557,041	1,494,227
Suicide/homicide	1,360,508	1,289,223
Congenital anomalies	661,117	642,551
Prematurity	428,796	422,813
HIV infection	246,823	357,536
SIDS	340,431	286,733
Cerebrovascular disease	246,131	246,479
Chronic liver disease	231,558	228,145
Pneumonia/influenza	175,386	166,775
Pulmonary disease	128,590	123,260
Diabetes mellitus	121,117	119,155

method of standardization has been proposed in which a standard population is used to adjust for differing age compositions [11]. Although these approaches are sensible, they do not seem to have found favor and are not frequently applied.

5.4 Further Applications

Table 5.7 (on p. 102) is an abridged life table for the U.S. population in 1985 [12]. Suppose that we wish to fill in the missing entries.

Recall that column 3 of the life table, l_x, lists the number of individuals out of the original cohort of 100,000 who are still alive on their x^{th} birthday. Column 4, ${}_nd_x$, displays the number out of the l_x alive at the beginning of the age interval x to $x+n$ who died some time during the interval. Thus l_{70}—the number of individuals out of the original 100,000 who are alive at age 70—is equal to the number of persons who were alive at age 65 minus the number who died between their 65th and 70th birthdays; in particular,

$$\begin{aligned} l_{70} &= l_{65} - {}_5d_{65} \\ &= 78,678 - 8,588 \\ &= 70,090. \end{aligned}$$

The number of persons dying during the age interval 70–75, ${}_5d_{70}$, is equal to the number of individuals alive on their 70th birthday multiplied by the proportion of persons dying during the interval; therefore,

$$\begin{aligned} {}_5d_{70} &= l_{70} \times {}_5q_{70} \\ &= 70,090 \times 0.1625 \\ &= 11,390. \end{aligned}$$

Similarly, we can fill in the remainder of columns 3 and 4 by computing

$$\begin{aligned} l_{75} &= l_{70} - {}_5d_{70} \\ &= 70,090 - 11,390 \\ &= 58,700, \end{aligned}$$

$$\begin{aligned} {}_5d_{75} &= l_{75} \times {}_5q_{75} \\ &= 58,700 \times 0.2349 \\ &= 13,789, \end{aligned}$$

$$\begin{aligned} l_{80} &= l_{75} - {}_5d_{75} \\ &= 58,700 - 13,789 \\ &= 44,911, \end{aligned}$$

$$\begin{aligned} {}_5d_{80} &= l_{80} \times {}_5q_{80} \\ &= 44,911 \times 0.3480 \\ &= 15,629, \end{aligned}$$

$$\begin{aligned} l_{85} &= l_{80} - {}_5d_{80} \\ &= 44,911 - 15,629 \\ &= 29,282, \end{aligned}$$

and

$$\begin{aligned} {}_\infty d_{85} &= l_{85} \times {}_\infty q_{85} \\ &= 29,282 \times 1.0000 \\ &= 29,282. \end{aligned}$$

Column 6 of the life table, T_x, specifies the total number of person-years lived beyond the xth birthday by all members of the cohort alive on that birthday. It is calculated by summing column 5—the total time in years lived during the age interval x to $x + n$—from the bottom up. Therefore, we have

$$\begin{aligned} T_{85} &= {}_\infty L_{85} \\ &= 175,908, \end{aligned}$$

$$\begin{aligned} T_{80} &= {}_5L_{80} + {}_\infty L_{85} \\ &= {}_5L_{80} + T_{85} \\ &= 185,697 + 175,908 \\ &= 361,605, \end{aligned}$$

$$\begin{aligned} T_{75} &= {}_5L_{75} + T_{80} \\ &= 259,866 + 361,605 \\ &= 621,471, \end{aligned}$$

and

$$\begin{aligned} T_{70} &= {}_5L_{70} + T_{75} \\ &= 322,963 + 621,471 \\ &= 944,434. \end{aligned}$$

Column 7 of the life table, $\overset{\circ}{e}_x$, shows the average number of years of life remaining to an individual who is alive at age x. It is calculated as the total number

Table 5.7 Abridged life table for the total population, United States, 1985

Age Interval	Proportion Dying	Of 100,000 Born Alive		Stationary Population		Average Remaining Lifetime
Period of Life Between Two Exact Ages Stated in Years (1)	Proportion of Persons Alive at Beginning of Age Interval Dying During Interval (2)	Number Living at Beginning of Age Interval (3)	Number Dying During Age Interval (4)	In the Age Interval (5)	In This and All Subsequent Age Intervals (6)	Average Number of Years of Life Remaining at Beginning of Age Interval (7)
x to $x+n$	${}_nq_x$	l_x	${}_nd_x$	${}_nL_x$	T_x	$\mathring{e}_x$
0–1........................	0.0107	100,000	1,069	99,079	7,472,607	74.7
1–5........................	0.0020	98,931	200	395,255	7,373,528	74.5
5–10	0.0012	98,731	123	493,320	6,978,273	70.7
10–15	0.0014	98,608	135	492,778	6,484,953	65.8
15–20	0.0040	98,473	397	491,465	5,992,175	60.9
20–25	0.0054	98,076	534	489,072	5,500,710	56.1
25–30	0.0057	97,542	553	486,326	5,011,638	51.4
30–35	0.0067	96,989	648	483,363	4,525,312	46.7
35–40	0.0086	96,341	831	479,745	4,041,949	42.0
40–45	0.0126	95,510	1,199	474,766	3,562,204	37.3
45–50	0.0197	94,311	1,854	467,259	3,087,438	32.7
50–55	0.0316	92,457	2,924	455,434	2,620,179	28.3
55–60	0.0497	89,533	4,447	437,186	2,164,745	24.2
60–65	0.0753	85,086	6,408	410,278	1,727,559	20.3
65–70	0.1092	78,678	8,588	372,847	1,317,281	16.7
70–75	0.1625			322,963		
75–80	0.2349			259,866		
80–85	0.3480			185,697		
85 and over................	1.0000			175,908		

of person-years lived beyond the xth birthday divided by the number of persons still alive on that birthday; thus,

$$\overset{\circ}{e}_{70} = \frac{T_{70}}{l_{70}} = \frac{944,434}{70,090} = 13.5,$$

$$\overset{\circ}{e}_{75} = \frac{T_{75}}{l_{75}} = \frac{621,471}{58,700} = 10.6,$$

$$\overset{\circ}{e}_{80} = \frac{T_{80}}{l_{80}} = \frac{361,605}{44,911} = 8.1,$$

and

$$\overset{\circ}{e}_{85} = \frac{T_{85}}{l_{85}} = \frac{175,908}{29,282} = 6.0.$$

Note that the average remaining lifetime at birth, $\overset{\circ}{e}_0$, is not the same as the average age at death. Consider the following data comparing England with France and Sweden [13].

Country		Mean Duration of Life	Mean Age at Death	One Death in
England	(1841)	41 years	29 years	46 living
France	(1817–1831)	40 years	34 years	42 living
Sweden	(1801–1805)	39 years	31 years	41 living

The table was explained in the following way:

> The average age of the persons who died, or the "mean age at death," was 34 years in France, 31 years in Sweden, and 29 years in England; yet we know that the "expectation of life" is greater in England than in Sweden or France. A Society that granted life annuities to children in England would have to make 40 annual payments on an average, and only 38 in Sweden.

The mean duration of life is calculated using the life table. The mean age at death, however, is simply the average age of those persons dying in the specified time period; its value depends on the age composition of the population involved.

Suppose that we now wish to know the probability that a newborn infant will live to see his or her 75th birthday. Of the $l_0 = 100,000$ persons born into the cohort, $l_{75} = 58,700$ are still alive at age 75. Therefore, the probability of living from birth until age 75 is

$$\begin{aligned} \frac{l_{75}}{l_0} &= \frac{58,700}{100,000} \\ &= 0.587 \\ &= 58.7\%. \end{aligned}$$

If the cohort member has already reached 30 years of age, the probability of living until age 75 becomes

$$\begin{aligned} \frac{l_{75}}{l_{30}} &= \frac{58,700}{96,989} \\ &= 0.605 \\ &= 60.5\%. \end{aligned}$$

If he or she is 70 years old, then the probability of surviving another five years is

$$\begin{aligned} \frac{l_{75}}{l_{70}} &= \frac{58,700}{70,090} \\ &= 0.837 \\ &= 83.7\%. \end{aligned}$$

5.5 Review Exercises

1. A study exploring the lives of Canadian males in various states of health presents an interesting application of the life table [14]. When constructing their table, the authors divided the person-years lived during a given age interval, ${}_nL_x$, into person-years lived inside institutions and person-years lived outside

institutions. The life table can now be used to determine the average remaining years of life outside institutions.

(1) Age	(2) Number Alive	(3) Person-Years of Life Lived	(4) Rate of Institutionalization	(5) Person-Years in Institutions
0	100,000	1,475,419	0.0020387	3,008
15	98,004	972,823	0.0029173	2,838
25	96,408	1,896,306	0.0027395	5,195
45	92,509	1,699,064	0.0059150	10,050
65	72,274	1,037,206	0.0526520	54,611

(6) Person-Years Outside Institutions	(7) Total Person-Years Outside Institutions from Age x on	(8) Average Remaining Years Outside Institutions
1,472,411		
969,985		
1,891,111		
1,689,014		
982,595		

Columns 2 and 3 list the l_x and ${}_nL_x$ values from the 1978 life table for Canadian males. The rates of institutionalization displayed in column 4 were obtained from the Canadian Health Survey.

a. Explain how columns 5 and 6 are derived.
b. Complete columns 7 and 8 of the table. Note that column 8 represents the average remaining years of life outside institutions for **all** men involved in the study.

2. In young adult life, males have a higher rate of accidental and violent death than do females.

a. Explain the effect that this would have on a male life table relative to the corresponding female life table with respect to the age-specific death rates for the young adult years.
b. How might this affect the average life expectancies for males relative to females in these age groups?

3. Listed below are the average remaining lifetimes at various ages for males from two countries—Malawi, which has one of the lowest average life expectancies at birth, and Iceland, which has one of the highest [15].

	Average Remaining Lifetime (years)	
Age	Malawi	Iceland
0	40.9	73.7
1	47.9	73.3
5	60.8	69.5
15	55.0	59.8
25	47.3	50.6
35	40.3	41.2
45	34.0	31.9
55	27.7	23.4

a. In Malawi, the average life expectancy at age 1 is seven years greater than it is at birth. What phenomenon is causing this?
b. On the basis of comparing the average remaining lifetimes at birth, you might assume that the age-specific death rates in Malawi are higher than those in Iceland at all ages. This, however, is not the case. Explain how you know this by comparing the average life expectancies at subsequent ages.

4. Listed below are selected values of l_x taken from life tables for white females in the U.S. population in 1909–1911 and 1969–1971.

	Number of Survivors out of 100,000 Live Births	
Age	1909–1911	1969–1971
0	100,000	100,000
15	83,093	97,902
45	69,341	94,649
75	26,569	63,290

a. Compute the probabilities of surviving from birth to age 15, from age 15 to 45, and from age 45 to 75 in each cohort.
b. If the probability of surviving from age x to age $x+n$ is p_1 in 1909–1911 and p_2 in 1969–1971, then the relative percentage of improvement in survival is $(p_2 - p_1)/p_1$. Which age group has the greatest relative percentage of improvement in survival over the 60-year period?

5. Refer to Table 5.1, the 1988 life table for the United States.

a. What is the chance of surviving from birth until age 80?
b. If an individual reaches his or her 50th birthday, what is the probability of that person surviving to age 80?
c. What is the probability that a 25-year-old will survive ten years? A 45-year-old? A 65-year-old?

6. The number of deaths caused by cancer in 1984 and the corresponding years of potential life lost are displayed below for both men and women [16]. In this example, the years of potential life lost for a specified individual are defined as the remaining life expectancy at the age at which death occurs.

	Number of Deaths	YPLL
Men	242,763	3,284,558
Women	210,687	3,596,723

Although a greater number of men died as a result of cancer, women lost more years of potential life. Explain how this could occur.

7. To compute the U.S. life table for 1940, we assume that the 1940 age-specific death rates remain constant throughout the entire lifetime of a cohort born in that year. These death rates are listed in column 2 of the table below. In theory, however, any set of age-specific mortality rates could be used. The true mortality rates experienced by the cohort born in 1940 through age 49 are provided in column 3 of the table. For ages 50 and above, the 1989 rates are used. Note the differences in the two columns.

Age Interval x **to** $x + n$	**Proportion Dying** ${}_nq_x$	**Proportion Dying** ${}_nq_x$
0–1	0.0549	0.0549
1–5	0.0115	0.0101
5–10	0.0055	0.0038
10–15	0.0050	0.0028
15–20	0.0085	0.0056
20–25	0.0119	0.0053
25–30	0.0139	0.0077
30–35	0.0169	0.0077
35–40	0.0218	0.0123
40–45	0.0301	0.0106
45–50	0.0427	0.0247
50–55	0.0624	0.0300
55–60	0.0896	0.0473
60–65	0.1270	0.0728
65–70	0.1812	0.1055
70–75	0.2704	0.1568
75–80	0.3946	0.2288
80–85	0.5941	0.3445
85+	1.0000	1.0000

a. Without doing any calculations, predict which set of mortality rates will yield the longer life expectancy at birth.

(Exercise 7 continues on p. 112.)

Table 5.8 Complete life table for the total population, United States, 1979–1981

Age Interval	Proportion Dying	Of 100,000 Born Alive		Stationary Population		Average Remaining Lifetime
Period of Life Between Two Exact Ages (1)	Proportion of Persons Alive at Beginning of Age Interval Dying During Interval (2)	Number Living at Beginning of Age Interval (3)	Number Dying During Age Interval (4)	In the Age Interval (5)	In This and All Subsequent Age Intervals (6)	Average Number of Years of Life Remaining at Beginning of Age Interval (7)
x to $x + n$	$_nq_x$	l_x	$_nd_x$	$_nL_x$	T_x	$\mathring{e}_x$
Days						
0–1	.00463	100,000	463	273	7,387,758	73.88
1–7	.00246	99,537	245	1,635	7,387,485	74.22
7–28	.00139	99,292	138	5,708	7,385,850	74.38
28–365	.00418	99,154	414	91,357	7,380,142	74.43
Years						
0–1	.01260	100,000	1,260	98,973	7,387,758	73.88
1–2	.00093	98,740	92	98,694	7,288,785	73.82
2–3	.00065	98,648	64	98,617	7,190,091	72.89
3–4	.00050	98,584	49	98,560	7,091,474	71.93
4–5	.00040	98,535	40	98,515	6,992,914	70.97
5–6	.00037	98,495	36	98,477	6,894,399	70.00
6–7	.00033	98,459	33	98,442	6,795,922	69.02
7–8	.00030	98,426	30	98,412	6,697,480	68.05
8–9	.00027	98,396	26	98,383	6,599,068	67.07
9–10	.00023	98,370	23	98,358	6,500,685	66.08
10–11	.00020	98,347	19	98,338	6,402,327	65.10
11–12	.00019	98,328	19	98,319	6,303,989	64.11
12–13	.00025	98,309	24	98,297	6,205,670	63.12
13–14	.00037	98,285	37	98,266	6,107,373	62.14
14–15	.00053	98,248	52	98,222	6,009,107	61.16

(continued)

Table 5.8 (continued)

15–16	.00069	98,196	67	98,163	5,910,885	60.19
16–17	.00083	98,129	82	98,087	5,812,722	59.24
17–18	.00095	98,047	94	98,000	5,714,635	58.28
18–19	.00105	97,953	102	97,902	5,616,635	57.34
19–20	.00112	97,851	110	97,796	5,518,733	56.40
20–21	.00120	97,741	118	97,682	5,420,937	55.46
21–22	.00127	97,623	124	97,561	5,323,255	54.53
22–23	.00132	97,499	129	97,435	5,225,694	53.60
23–24	.00134	97,370	130	97,306	5,128,259	52.67
24–25	.00133	97,240	130	97,175	5,030,953	51.74
25–26	.00132	97,110	128	97,046	4,933,778	50.81
26–27	.00131	96,982	126	96,919	4,836,732	49.87
27–28	.00130	96,856	126	96,793	4,739,813	48.94
28–29	.00130	96,730	126	96,667	4,643,020	48.00
29–30	.00131	96,604	127	96,541	4,546,353	47.06
30–31	.00133	96,477	127	96,414	4,449,812	46.12
31–32	.00134	96,350	130	96,284	4,353,398	45.18
32–33	.00137	96,220	132	96,155	4,257,114	44.24
33–34	.00142	96,088	137	96,019	4,160,959	43.30
34–35	.00150	95,951	143	95,880	4,064,940	42.36
35–36	.00159	95,808	153	95,731	3,969,060	41.43
36–37	.00170	95,655	163	95,574	3,873,329	40.49
37–38	.00183	95,492	175	95,404	3,777,755	39.56
38–39	.00197	95,317	188	95,224	3,682,351	38.63
39–40	.00213	95,129	203	95,027	3,587,127	37.71
40–41	.00232	94,926	220	94,817	3,492,100	36.79
41–42	.00254	94,706	241	94,585	3,397,283	35.87
42–43	.00279	94,465	264	94,334	3,302,698	34.96
43–44	.00306	94,201	288	94,057	3,208,364	34.06
44–45	.00335	93,913	314	93,756	3,114,307	33.16
45–46	.00366	93,599	343	93,427	3,020,551	32.27
46–47	.00401	93,256	374	93,069	2,927,124	31.39
47–48	.00442	92,882	410	92,677	2,834,055	30.51
48–49	.00488	92,472	451	92,246	2,741,378	29.65
49–50	.00538	92,021	495	91,773	2,649,132	26.79

(continued)

Table 5.8 (continued)

Age Interval	Proportion Dying	Of 100,000 Born Alive		Stationary Population		Average Remaining Lifetime
Period of Life Between Two Exact Ages (1)	Proportion of Persons Alive at Beginning of Age Interval Dying During Interval (2)	Number Living at Beginning of Age Interval (3)	Number Dying During Age Interval (4)	In the Age Interval (5)	In This and All Subsequent Age Intervals (6)	Average Number of Years of Life Remaining at Beginning of Age Interval (7)
x to $x + n$	$_nq_x$	l_x	$_nd_x$	$_nL_x$	T_x	$\mathring{e}_x$
50–51	.00589	91,526	540	91,256	2,557,359	27.94
51–52	.00642	90,986	584	90,695	2,466,103	27.10
52–53	.00699	90,402	631	90,086	2,375,408	26.28
53–54	.00761	89,771	684	89,430	2,285,322	25.46
54–55	.00830	89,087	739	88,717	2,195,892	24.65
55–56	.00902	88,348	797	87,950	2,107,175	23.85
56–57	.00978	87,551	856	87,122	2,019,225	23.06
57–58	.01059	86,695	919	86,236	1,932,103	22.29
58–59	.01151	85,776	987	85,283	1,845,867	21.52
59–60	.01254	84,789	1,063	84,258	1,760,584	20.76
60–61	.01368	83,726	1,145	83,153	1,676,326	20.02
61–62	.01493	82,581	1,233	81,965	1,593,173	19.29
62–63	.01628	81,348	1,324	80,686	1,511,208	18.58
63–64	.01767	80,024	1,415	79,316	1,430,522	17.88
64–65	.01911	78,609	1,502	77,859	1,351,206	17.19
65–66	.02059	77,107	1,587	76,314	1,273,347	16.51
66–67	.02216	75,520	1,674	74,683	1,197,033	15.85
67–68	.02389	73,846	1,764	72,964	1,122,350	15.20
68–69	.02585	72,082	1,864	71,150	1,049,386	14.56
69–70	.02806	70,218	1,970	69,233	978,236	13.93
70–71	.03052	68,248	2,083	67,206	909,003	13.32
71–72	.03315	66,165	2,193	65,069	841,797	12.72
72–73	.03593	63,972	2,299	62,823	776,728	12.14
73–74	.03882	61,673	2,394	60,476	713,905	11.58
74–75	.04184	59,279	2,480	58,039	653,429	11.02

(continued)

Table 5.8 (continued)

75–76	.04507	56,799	2,560	55,520	595,390	10.48
76–77	.04867	54,239	2,640	52,919	539,870	9.95
77–78	.05274	51,599	2,721	50,238	486,951	9.44
78–79	.05742	48,878	2,807	47,475	436,713	8.93
79–80	.06277	46,071	2,891	44,626	389,238	8.45
80–81	.06882	43,180	2,972	41,694	344,612	7.98
81–82	.07552	40,208	3,036	38,689	302,918	7.53
82–83	.08278	37,172	3,077	35,634	264,229	7.11
83–84	.09041	34,095	3,083	32,553	228,595	6.70
84–85	.09842	31,012	3,052	29,486	196,042	6.32
85–86	.10725	27,960	2,999	26,461	166,556	5.96
86–87	.11712	24,961	2,923	23,500	140,095	5.61
87–88	.12717	22,038	2,803	20,636	116,595	5.29
88–89	.13708	19,235	2,637	17,917	95,959	4.99
89–90	.14728	16,598	2,444	15,376	78,042	4.70
90–91	.15868	14,154	2,246	13,031	62,666	4.43
91–92	.17169	11,908	2,045	10,886	49,635	4.17
92–93	.18570	9,863	1,831	8,948	38,749	3.93
93–94	.20023	8,032	1,608	7,228	29,801	3.71
94–95	.21495	6,424	1,381	5,733	22,573	3.51
95–96	.22976	5,043	1,159	4,463	16,840	3.34
96–97	.24338	3,884	945	3,412	12,377	3.19
97–98	.25637	2,939	754	2,562	8,965	3.05
98–99	.26868	2,185	587	1,892	6,403	2.93
99–100	.28030	1,598	448	1,374	4,511	2.82
100–101	.29120	1,150	335	983	3,137	2.73
101–102	.30139	815	245	692	2,154	2.64
102–103	.31089	570	177	481	1,462	2.57
103–104	.31970	393	126	330	981	2.50
104–105	.32786	267	88	223	651	2.44
105–106	.33539	179	60	150	428	2.38
106–107	.34233	119	41	99	278	2.33
107–108	.34870	78	27	64	179	2.29
108–109	.35453	51	18	42	115	2.24
109–110	.35988	33	12	27	73	2.20

b. Complete the life tables corresponding to each set of rates. Assume that all deaths within an interval occur at its midpoint, except for the last interval, where all deaths occur at age 91. Thus show that, even with the simplifications and approximations mentioned, the discrepancy between the two life expectancies at birth is more than five years.
c. What might be the ramifications of this difference in life expectancies?

References

[1] Fries, J. F., "Aging, Natural Death, and the Compression of Morbidity," *The New England Journal of Medicine*, Volume 303, July 17, 1980, 130–135.
[2] National Center for Health Statistics, "Advance Report of Final Mortality Statistics, 1988," *Monthly Vital Statistics Report*, Volume 39, November 28, 1990.
[3] National Center for Health Statistics, *United States Decennial Life Tables for 1979–1980*, Volume I, Number 1, August 1985.
[4] Vandenbroucke, J. P., "Survival and Expectation of Life from the 1400s to the Present: A Study of the Knighthood Order of the Golden Fleece," *American Journal of Epidemiology*, Volume 122, December 1985, 1007–1015.
[5] United Nations Children's Fund, *The State of the World's Children 1991*, New York: Oxford University Press.
[6] Taylor, I., and Knowelden, J., *Principles of Epidemiology*, London: Churchill, 1957.
[7] Dickinson, F. G., and Welker, E. L., *What Is the Leading Cause of Death? Two New Measures*, Bulletin 64, Chicago: American Medical Association, 1948.
[8] Dempsey, M., "Decline in Tuberculosis: The Death Rate Fails to Tell the Entire Story," *American Review of Tuberculosis*, Volume 86, August 1947, 157–164.
[9] Centers for Disease Control, "Premature Mortality in the United States: Public Health Issues in the Use of Years of Potential Life Lost," *Morbidity and Mortality Weekly Report*, Volume 35, Number 2S, December 19, 1986.
[10] Centers for Disease Control, "YPLL Before Age 65: United States, 1987," *Morbidity and Mortality Weekly Report*, Volume 38, Number 2, January 20, 1989.
[11] Haenszel, W., "A Standardized Rate for Mortality Defined in Units of Lost Years of Life," *American Journal of Public Health*, Volume 40, January 1950, 17–26.
[12] National Center for Health Statistics, "Advance Report of Final Mortality Statistics, 1985," *Monthly Vital Statistics Report*, Volume 36, September 1987.
[13] Farr, W., *Vital Statistics: A Memorial Volume of Selections from the Reports and Writings of William Farr*, London: The Sanitary Institute of Great Britain, 1883.
[14] Wilkins, R., and Adams, O. B., "Health Expectancy in Canada, Late 1970s: Demographic, Regional, and Social Dimensions," *American Journal of Public Health*, Volume 73, September 1983, 1073–1080.

[15] United Nations, "Expectation of Life at Specified Ages for Each Sex," *Demographic Yearbook 1982*, New York, 1984.

[16] Horm, J. W., and Sondik, E. J., "Person-Years of Life Lost Due to Cancer in the United States, 1970 and 1984," *American Journal of Public Health*, Volume 79, November 1989, 1490–1493.

CHAPTER

6

Probability

In the previous chapters, we studied the ways in which descriptive statistics may be used to organize and summarize data. However, in addition to describing a set of observations, we might also be interested in investigating how the information contained in the sample can be used to infer the characteristics of the population from which it was drawn. Before we can do this, we must first lay the groundwork. The foundation for statistical inference is the theory of probability.

6.1 Operations on Events

In Chapter 5, we used the term *probability* as a synonym for "proportion." Before we can give a more precise definition, we must first explain the concept of an event. An *event* is the basic element to which probability can be applied; it is the result of an observation or experiment, or the description of some potential outcome. For example, we might consider the event that a 30-year-old woman lives to see her 70th birthday or the event that the same woman is diagnosed with cervical cancer before she reaches the age of 40. An event either occurs, or it does not occur. In the study of probability, events are represented by uppercase letters such as A, B, and C.

A number of operations may be performed on events. The *intersection* of two events A and B, denoted $A \cap B$, is defined as the event "both A and B." For example, let A represent the event that a 30-year-old woman lives to see her 70th birthday and B the event that her 30-year-old husband is still alive at age 70. The intersection of A and B would be the event that both the 30-year-old woman and her husband are alive at age 70.

The *union* of A and B, denoted $A \cup B$, is the event "either A or B, or both." In the example mentioned above, the union of A and B would be the event that either the 30-year-old woman or her 30-year-old husband lives to age 70, or that they both live to be 70 years of age.

The *complement* of an event A, denoted A^c or $\overline{A}$, is the event "not A." Consequently, A^c is the event that the 30-year-old woman does not reach the age of 70. These three operations—the intersection, the union, and the complement—can be used to describe even the most complicated situations in terms of simple events.

We are now ready to discuss the concept of probability. As a mathematical system, probability theory is well-defined. Since we wish to apply this theory, however, we need a practical working definition. Many definitions of probability have been proposed; the one presented here is called the *frequentist definition*. It states:

> If an experiment is repeated n times under essentially identical conditions, and if the event A occurs m times, then as n grows large the ratio m/n approaches a fixed limit that is the probability of A;
>
> $$P(A) = \frac{m}{n}.$$

In other words, the probability of an event A is its relative frequency of occurrence in a large number of trials repeated under virtually identical conditions. The practical nature of this definition causes it to be somewhat vague, although it does seem to work well.

The numerical value of a probability lies between 0 and 1. If a particular event happens with certainty, then it has probability $n/n = 1$. For example,

$$\begin{aligned} P(A \cup A^c) &= P(A \text{ or } A^c) \\ &= P(\text{a 30-year-old woman lives to age 70 or} \\ &\qquad \text{she does not live to age 70)} \\ &= 1, \end{aligned}$$

since it is certain that the woman will either live or die. Note that it is impossible for A and A^c to occur simultaneously. If an event can never happen, it has probability $0/n = 0$; hence

$$\begin{aligned} P(A \cap A^c) &= P(A \text{ and } A^c) \\ &= P(\text{a 30-year-old woman lives to age 70 and} \\ &\qquad \text{does not live to age 70)} \\ &= 0. \end{aligned}$$

An event that can never occur is called the *null event* and is represented by the symbol ϕ. Therefore, $A \cap A^c = \phi$.

Using the frequentist definition of the probability of an event A, we can calculate the probability of the complementary event A^c in a straightforward manner. If an experiment is repeated n times under essentially identical conditions and the event A occurs m times, then the event A^c occurs $n - m$ times. Therefore, for large n,

$$\begin{aligned} P(A^c) &= \frac{n-m}{n} \\ &= 1 - \frac{m}{n} \\ &= 1 - P(A). \end{aligned}$$

Two events A and B that cannot occur simultaneously are said to be *mutually exclusive*. If A is the event that a woman is pregnant and B the event that she is not, for example, then the events A and B are mutually exclusive. By definition, $A \cap B = \phi$ and $P(A \cap B) = 0$.

When two events are mutually exclusive, the *additive rule of probability* states that the probability that either of the two events will occur is equal to the sum of the probabilities of the individual events; more explicitly,

$$P(A \cup B) = P(A) + P(B).$$

This rule can be extended to the case of three or more mutually exclusive events. If $A_1, A_2, \ldots,$ and A_n are n events such that $A_1 \cap A_2 = \phi$, $A_1 \cap A_3 = \phi$, $A_2 \cap A_3 = \phi$, and so on for all possible pairs, then

$$P(A_1 \cup A_2 \cup \cdots \cup A_n) = P(A_1) + P(A_2) + \cdots + P(A_n).$$

If the events A and B are not mutually exclusive, then this form of the additive rule no longer applies. Since the two events are able to occur simultaneously, there is some area in which they will overlap. If we were to simply sum the probabilities of the individual events, then this area of overlap would be counted twice. When two events are not mutually exclusive, therefore, the probability that either of the events will occur is equal to the sum of the individual probabilities minus the probability of their intersection;

$$P(A \cup B) = P(A) + P(B) - P(A \cap B).$$

6.2 Conditional Probability

We are often interested in determining the probability that an event B will occur given that another event A has already taken place. Does the prior occurrence of A cause the probability of B to change? For instance, instead of finding the probability that a person will live to the age of 65, we might wish to know the probability that the individual will survive for the next five years given that he or she has already reached the age of 60. If this is the case, then we are dealing with a *conditional probability*. The notation $P(B \mid A)$ is used to represent the probability of the event B given that event A has already occurred.

The *multiplicative rule of probability* states that the probability that two events A and B will both occur is equal to the probability of A multiplied by the probability of B given that A has already occurred. In notation,

$$P(A \cap B) = P(A)\, P(B \mid A).$$

Since it is arbitrary which event we call A and which we call B, we may also write

$$P(A \cap B) = P(B)\, P(A \mid B).$$

Dividing both sides of the first equation by $P(A)$, we find the formula for a conditional probability to be

$$P(B \mid A) = \frac{P(A \cap B)}{P(A)},$$

given that $P(A) \neq 0$. Similarly, we have

$$P(A \mid B) = \frac{P(A \cap B)}{P(B)},$$

given that $P(B) \neq 0$.

If A is the event that an individual is 60 years of age, and B is the event that he or she lives to reach 65, then $A \cap B$ is the event that the person is alive at age 60 and also at 65. Therefore, it is simply the event that the individual survives to see his or her 65th birthday. According to Table 5.1, the 1988 life table for the U. S. population,

$$\begin{aligned} P(A) &= \text{P(an individual lives to be 60 years of age)} \\ &= \frac{85,331}{100,000} \\ &= 0.85331, \end{aligned}$$

and

$$\begin{aligned} P(A \cap B) &= \text{P(an individual lives to be 65 years of age)} \\ &= \frac{79,123}{100,000} \\ &= 0.79123. \end{aligned}$$

Therefore,

$$\begin{aligned} P(B \mid A) &= \text{P(an individual lives to be 65 years of age} \mid \\ &\qquad \text{he or she is alive at 60)} \\ &= \frac{P(A \cap B)}{P(A)} \\ &= \frac{0.79123}{0.85331} \\ &= 0.9272. \end{aligned}$$

If a person lives to be 60, then his or her chance of surviving to age 65 is greater than it was at birth.

When we are concerned with two events such that the outcome of one event has no effect on the occurrence or nonoccurrence of the other, then the events are said to be *independent*. If A and B are independent events, then

$$P(A \mid B) = P(A),$$

and

$$P(B \mid A) = P(B).$$

In this special case, the multiplicative rule of probability may be written

$$P(A \cap B) = P(A)\,P(B).$$

The terms *independent* and *mutually exclusive* do not mean the same thing. If A and B are independent and event A occurs, the outcome of B is not affected. Event B might occur or it might not occur, and $P(B \mid A) = P(B)$. If A and B are mutually exclusive, however, and event A occurs, then event B cannot occur. By definition, $P(B \mid A) = 0$.

6.3 Bayes' Theorem

Chapter 4 included a presentation of data collected in the National Health Interview Survey of 1980–1981. The data pertain to hearing impairments due to injury reported by individuals 17 years of age and older. The 163,157 persons included in the study were subdivided into three mutually exclusive categories: the currently employed, the currently unemployed, and those not in the labor force.

Employment Status	Population	Impairments
Currently employed	98,917	552
Currently unemployed	7,462	27
Not in the labor force	56,778	368
Total	163,157	947

Let E_1 be the event that an individual included in the survey is currently employed, E_2 the event that he or she is currently unemployed, and E_3 the event that the individual is not in the labor force. If we assume that these numbers are large enough to satisfy the frequentist definition of probability, then, from the data provided, we find that

$$\begin{aligned} P(E_1) &= \frac{98,917}{163,157} \\ &= 0.6063, \end{aligned}$$

$$\begin{aligned} P(E_2) &= \frac{7,462}{163,157} \\ &= 0.0457, \end{aligned}$$

and

$$\begin{aligned} P(E_3) &= \frac{56,778}{163,157} \\ &= 0.3480. \end{aligned}$$

If S is the event that an individual in the study is either currently employed or currently unemployed or not in the labor force, then

$$S = E_1 \cup E_2 \cup E_3.$$

Since the three categories are mutually exclusive, the additive rule of probability may be applied;

$$\begin{aligned} P(S) &= P(E_1 \cup E_2 \cup E_3) \\ &= P(E_1) + P(E_2) + P(E_3) \\ &= 0.6063 + 0.0457 + 0.3480 \\ &= 1.0000. \end{aligned}$$

When the probabilities of mutually exclusive events sum to 1, the events are said to be *exhaustive*; there are no other possible outcomes. Therefore, every person included in the survey must fall into one of the three groups.

Now, let H be the event that an individual has a hearing impairment due to injury. Overall,

$$\begin{aligned} P(H) &= \frac{947}{163,157} \\ &= 0.0058. \end{aligned}$$

Looking at each employment status subgroup separately,

$$\begin{aligned} P(H \mid E_1) &= \text{P(an individual has a hearing impairment} \mid \\ &\qquad \text{he or she is currently employed)} \\ &= \frac{552}{98,917} \\ &= 0.0056, \end{aligned}$$

$$\begin{aligned} P(H \mid E_2) &= \text{P(an individual has a hearing impairment} \mid \\ &\qquad \text{he or she is currently unemployed)} \\ &= \frac{27}{7,462} \\ &= 0.0036, \end{aligned}$$

and

$$\begin{aligned} P(H \mid E_3) &= \text{P(an individual has a hearing impairment} \mid \\ &\qquad \text{he or she is not in the labor force)} \\ &= \frac{368}{56,778} \\ &= 0.0065. \end{aligned}$$

Note that H, the event that an individual has a hearing impairment due to injury, may be expressed as the union of three mutually exclusive events: $E_1 \cap H$, the event that an individual is currently employed and has a hearing impairment; $E_2 \cap H$, the event that he or she is currently unemployed and has a hearing impairment; and $E_3 \cap H$, the event that the individual is not in the labor force and has a hearing impairment. Thus,

$$H = (E_1 \cap H) \cup (E_2 \cap H) \cup (E_3 \cap H).$$

Everyone with a hearing impairment can be placed into one and only one of these three categories. Since the categories are mutually exclusive, we may apply the additive rule; therefore,

$$\begin{aligned} P(H) &= P[(E_1 \cap H) \cup (E_2 \cap H) \cup (E_3 \cap H)] \\ &= P(E_1 \cap H) + P(E_2 \cap H) + P(E_3 \cap H). \end{aligned}$$

Applying the multiplicative rule to each term separately,

$$\begin{aligned} P(H) &= P(E_1 \cap H) + P(E_2 \cap H) + P(E_3 \cap H) \\ &= P(E_1)\, P(H \mid E_1) + P(E_2)\, P(H \mid E_2) + P(E_3)\, P(H \mid E_3) \\ &= (0.6063)(0.0056) + (0.0457)(0.0036) + (0.3480)(0.0065) \\ &= 0.0058. \end{aligned}$$

This is the value that we originally computed as the probability that an individual has a hearing impairment due to injury,

$$\begin{aligned} P(H) &= \frac{947}{163,157} \\ &= 0.0058. \end{aligned}$$

The more complicated expression

$$P(H) = P(E_1)\, P(H \mid E_1) + P(E_2)\, P(H \mid E_2) + P(E_3)\, P(H \mid E_3)$$

may be useful when we are unable to calculate $P(H)$ directly.

Suppose that we now change our perspective and attempt to find $P(E_1 \mid H)$, the probability that an individual is currently employed given that he or she has a hearing impairment. The multiplicative rule of probability states that

$$P(E_1 \cap H) = P(H)\, P(E_1 \mid H);$$

hence,

$$P(E_1 \mid H) = \frac{P(E_1 \cap H)}{P(H)}.$$

Applying the multiplicative rule to the numerator of the right-hand side of the equation, we have

$$P(E_1 \mid H) = \frac{P(E_1)\, P(H \mid E_1)}{P(H)}.$$

Substituting the identity that was derived above,

$$P(H) = P(E_1)\, P(H \mid E_1) + P(E_2)\, P(H \mid E_2) + P(E_3)\, P(H \mid E_3),$$

results in

$$P(E_1 \mid H) = \frac{P(E_1)\,P(H \mid E_1)}{P(E_1)\,P(H \mid E_1) + P(E_2)\,P(H \mid E_2) + P(E_3)\,P(H \mid E_3)}.$$

This rather daunting expression is known as *Bayes' theorem*. Substituting in the numerical values of all terms,

$$\begin{aligned} P(E_1 \mid H) &= \frac{0.6063 \times 0.0056}{(0.6063 \times 0.0056) + (0.0457 \times 0.0036) + (0.3480 \times 0.0065)} \\ &= 0.583. \end{aligned}$$

The probability that an individual is currently employed given that he or she has a hearing impairment due to injury is approximately 0.583. In this particular example, the result may be checked directly by referring to the original data. Among the 947 persons with hearing impairments, 552 are currently employed. Therefore,

$$\begin{aligned} P(E_1 \mid H) &= \frac{552}{947} \\ &= 0.583. \end{aligned}$$

Bayes' theorem is not restricted to situations in which individuals fall into one of three distinct subgroups. If $A_1, A_2, \ldots,$ and A_n are n mutually exclusive and exhaustive events so that

$$\begin{aligned} P(A_1 \cup A_2 \cup \cdots \cup A_n) &= P(A_1) + P(A_2) + \cdots + P(A_n) \\ &= 1, \end{aligned}$$

then Bayes' theorem states that

$$P(A_i \mid B) = \frac{P(A_i)\,P(B \mid A_i)}{P(A_1)\,P(B \mid A_1) + \cdots + P(A_n)\,P(B \mid A_n)}$$

for each $i, 1 \leq i \leq n$.

Bayes' theorem is valuable because it allows us to recalculate a probability based on some new information. In the National Health Interview Survey example, we know that

$$\begin{aligned} P(E_1) &= \text{P(an individual is currently employed)} \\ &= 0.6063. \end{aligned}$$

If we are then given a new piece of information—the knowledge that a particular individual has a hearing impairment due to injury, for instance—does our assessment of the probability that he or she is currently employed change? We observed that it does. Using Bayes' theorem, we found that

$$\begin{aligned} P(E_1 \mid H) &= \text{P(an individual is currently employed} \mid \\ &\quad \text{he or she has a hearing impairment)} \\ &= 0.5832. \end{aligned}$$

Once we are told that someone has a hearing impairment, the probability that he or she is currently employed decreases slightly.

6.4 Diagnostic Tests

Bayes' theorem is often employed in issues of diagnostic testing, or screening. *Screening* is the application of a test to individuals who have not yet exhibited any clinical symptoms to classify them with respect to their probability of having a particular disease. Those who test positive are considered to be more likely to have the disease and are usually subjected either to further diagnostic procedures or to treatment. Screening is most often employed by health care professionals in situations in which the early detection of disease will contribute to a more favorable prognosis for the individual or the population in general. Bayes' theorem allows us to use probability to evaluate the associated uncertainties.

6.4.1 Sensitivity and Specificity

Suppose that we are interested in two mutually exclusive and exhaustive states of health; D_1 is the event that an individual has a particular disease, and D_2 is the event that he or she does not have the disease. We could use the more succinct notation defined earlier—namely, D and D^c—but we wish to emphasize that the situation may be generalized to include three or more events. Let T^+ represent a positive screening test result. We would like to find $P(D_1 \mid T^+)$, the probability that a person with a positive test result actually does have the disease.

Cervical cancer is a disease for which the chance of containment is high given that it is detected early. The Pap smear is a widely accepted screening procedure that can detect a cancer that is as yet asymptomatic; it has been credited with being primarily responsible for the decreasing death rate due to cervical cancer in recent years. An on-site proficiency test conducted in 1972, 1973, and 1978 assessed the competency of technicians who scan Pap smear slides for abnormalities [1]. Technicians in 306 cytology labs in 44 states were tested. Once again, let us assume that these numbers are large enough to satisfy the frequentist definition of probability.

Overall, 16.25% of the tests performed on women with cancer resulted in false negative outcomes. A *false negative* occurs when the test of a woman who has cancer of the cervix incorrectly indicates that she does not. Therefore, in this study,

$$\text{P(test negative} \mid \text{cancer)} = 0.1625.$$

The other 100 − 16.25 = 83.75% of the women who had cervical cancer did in fact test positive; as a result,

$$\text{P(test positive} \mid \text{cancer)} = 0.8375.$$

The probability of a positive test result given that the individual tested actually has the disease is called the *sensitivity* of a test; in this study, the sensitivity of the Pap smear was 0.8375.

Not all the women who were tested actually suffered from cervical cancer. In fact, 18.64% of the tests were *false positive* outcomes; this implies that

$$\text{P(test positive} \mid \text{no cancer)} = 0.1864.$$

The *specificity* of a test is the probability that its result will be negative given that the individual tested does not have the disease. In this study, the specificity of the Pap smear was

$$\begin{aligned}\text{P(test negative} \mid \text{no cancer)} &= 1 - 0.1864 \\ &= 0.8136.\end{aligned}$$

6.4.2 Applications of Bayes' Theorem

Now that we have examined the accuracy of the Pap smear among women who have cervical cancer and women who do not, we can investigate the question that is of primary concern to the individuals being tested and to the health care professionals involved in the screening: What is the probability that a woman with a Pap smear that is positive for cancer actually does have the disease? Let D_1 represent the event that a woman has cervical cancer and D_2 the event that she does not. Also, let T^+ represent a positive Pap smear. We wish to compute $P(D_1 \mid T^+)$. Using Bayes' theorem, we can write

$$P(D_1 \mid T^+) = \frac{P(D_1)\, P(T^+ \mid D_1)}{P(D_1)\, P(T^+ \mid D_1) + P(D_2)\, P(T^+ \mid D_2)}.$$

We already know that $P(T^+ \mid D_1) = 0.8375$ and $P(T^+ \mid D_2) = 0.1864$. We must now determine $P(D_1)$ and $P(D_2)$.

$P(D_1)$ is the probability that a woman suffers from cervical cancer. It can also be interpreted as the proportion of women who have cancer of the cervix at a given time, or the *prevalence* of the disease. One source reports that the rate of cases of cervical cancer among women studied in 1983–1984 was 8.3 per 100,000 [2]. Using these data,

$$P(D_1) = 0.000083.$$

$P(D_2)$ is the probability that a woman does not have cervical cancer. Since D_2 is the complement of D_1,

$$\begin{aligned} P(D_2) &= 1 - P(D_1) \\ &= 1 - 0.000083 \\ &= 0.999917. \end{aligned}$$

Therefore, substituting these numerical values into Bayes' theorem,

$$\begin{aligned} P(D_1 \mid T^+) &= \frac{0.000083 \times 0.8375}{(0.000083 \times 0.8375) + (0.999917 \times 0.1864)} \\ &= 0.000373. \end{aligned}$$

$P(D_1 \mid T^+)$, the probability of disease given a positive test result, is called the *predictive value* of a positive test. Here, it tells us that for every 1,000,000 positive Pap smears, only 373 represent true cases of cervical cancer.

Bayes' theorem may also be used to calculate the predictive value of a negative test. If T^- represents a negative test result, then the negative predictive value, or the probability of no disease given a negative test result, is equal to

$$\begin{aligned} P(D_2 \mid T^-) &= \frac{P(D_2)\, P(T^- \mid D_2)}{P(D_2)\, P(T^- \mid D_2) + P(D_1)\, P(T^- \mid D_1)} \\ &= \frac{0.999917 \times 0.8136}{(0.999917 \times 0.8136) + (0.000083 \times 0.1625)} \\ &= 0.999983. \end{aligned}$$

Therefore, for every 1,000,000 women with negative Pap smears, 999,983 do not have cervical cancer.

Although the Pap smear is widely accepted as a screening test for cervical cancer, its previously assumed high rate of accuracy is now being questioned. A number of studies estimate the proportion of false negative results to be in the range of 20 to 40%, a fraction considerably greater than the 16.25% noted in the previous example [3]. Some of the lab errors are due to poor cell sampling techniques; others are a consequence of the fatigue suffered by lab technicians who must examine a large number of slides each day.

As a second example of the application of Bayes' theorem in diagnostic testing, consider the following data [4]. Among the 1,820 subjects in a study, 30 suffered from tuberculosis and 1,790 did not. Chest Xrays were administered to all individuals; 73 had a positive Xray–implying that there was significant evidence of inflammatory disease—whereas the results of the other 1,747 were negative. The data for this study are presented below. What is the probability that a randomly selected individual has tuberculosis given that his or her Xray is positive?

	Tuberculosis		
XRay	**No**	**Yes**	**Total**
Negative	1,739	8	1,747
Positive	51	22	73
Total	1,790	30	1,820

Let D_1 represent the event that an individual is suffering from tuberculosis and D_2 the event that he or she is not. These two events are mutually exclusive and exhaustive. In addition, let T^+ represent a positive Xray. We wish to find $\mathrm{P}(D_1 \mid T^+)$, the probability that an individual who tests positive for tuberculosis actually has the disease. This is the positive predictive value of the Xray. Using Bayes' theorem, we can write

$$\mathrm{P}(D_1 \mid T^+) = \frac{\mathrm{P}(D_1)\,\mathrm{P}(T^+ \mid D_1)}{\mathrm{P}(D_1)\,\mathrm{P}(T^+ \mid D_1) + \mathrm{P}(D_2)\,\mathrm{P}(T^+ \mid D_2)}.$$

Therefore, to solve for $\mathrm{P}(D_1 \mid T^+)$ we must first know $\mathrm{P}(D_1), \mathrm{P}(D_2), \mathrm{P}(T^+ \mid D_1)$, and $\mathrm{P}(T^+ \mid D_2)$.

$\mathrm{P}(D_1)$ is the probability that an individual in the general population has tuberculosis; since the 1,820 individuals were not chosen from the population at random, the prevalence of disease cannot be obtained from the preceding table. In 1987, however, there were 9.3 cases of tuberculosis per 100,000 population [5]. With the spread of the human immunodeficiency virus (HIV) this number will increase markedly, but for now we can estimate

$$\mathrm{P}(D_1) = 0.000093.$$

$\mathrm{P}(D_2)$ is the probability that an individual does not have tuberculosis. Since D_2 is the complement of the event D_1,

$$\begin{aligned}\mathrm{P}(D_2) &= 1 - \mathrm{P}(D_1) \\ &= 1 - 0.000093 \\ &= 0.999907.\end{aligned}$$

$\mathrm{P}(T^+ \mid D_1)$ is the probability of a positive Xray given that an individual has tuberculosis—the sensitivity of the test. In this study, the sensitivity of the Xray is

$$\begin{aligned}\mathrm{P}(T^+ \mid D_1) &= \frac{22}{30} \\ &= 0.7333.\end{aligned}$$

$P(T^+ \mid D_2)$, the probability of a positive Xray given that a person does not have tuberculosis, is the complement of the specificity. Therefore,

$$\begin{aligned} P(T^+ \mid D_2) &= 1 - P(T^- \mid D_2) \\ &= 1 - \frac{1,739}{1,790} \\ &= 1 - 0.9715 \\ &= 0.0285. \end{aligned}$$

Using all this information, we can now calculate the probability that an individual suffers from tuberculosis given that he or she has a positive Xray; this probability is

$$\begin{aligned} P(D_1 \mid T^+) &= \frac{P(D_1)\,P(T^+ \mid D_1)}{P(D_1)\,P(T^+ \mid D_1) + P(D_2)\,P(T^+ \mid D_2)} \\ &= \frac{(0.000093)(0.7333)}{(0.000093)(0.7333) + (0.999907)(0.0285)} \\ &= 0.00239. \end{aligned}$$

For every 100,000 positive Xrays, only 239 signal true cases of tuberculosis.

Note that, before an Xray is taken, an individual who is randomly selected from the U. S. population has a

$$9.3/100,000 = 0.000093 = 0.0093\%$$

chance of having tuberculosis. This is called the *prior probability*. After an Xray is taken and the result is positive, the same individual has a

$$239/100,000 = 0.00239 = 0.239\%$$

chance of being afflicted with tuberculosis. This is the *posterior probability*. The posterior probability takes into account a new piece of information–the positive test result. Although $99,761/100,000$ persons with a positive Xray do not actually have the disease, we have greatly increased the chance of properly diagnosing tuberculosis. Since $0.00239/0.000093 = 25.7$, the probability that an individual with a positive Xray has tuberculosis is 25.7 times greater than the probability that a person randomly selected from the population has the disease.

6.4.3 The ROC Curve

Diagnosis is an imperfect process. In theory, it is desirable to have a test that is both highly sensitive and highly specific. In reality, however, such a test is usually not possible. Many tests are based on some clinical measurement that can assume a range of values; in this case, there is an inherent trade-off between sensitivity and specificity.

Consider Table 6.1. This table displays data from a kidney transplant program in which renal allografts were performed [6]. Serum creatinine level in milligrams percent was used as a diagnostic tool for detecting potential transplant rejection. If we use a level greater than 2.9 mg % as an indicator of imminent rejection, then the test has a sensitivity of 0.303 and a specificity of 0.909. To increase the sensitivity, we could lower the arbitrary cutoff point that distinguishes a positive test result from a negative one; if we use 1.2 mg %, for example, a much greater proportion of results are designated positive. In this case, we would rarely fail to identify a patient who will reject the organ. At the same time, however, we would increase the probability of a false positive result, thereby decreasing the specificity. Alternatively, by increasing the specificity we would hardly ever misclassify a person who is not going to reject the organ, and, in turn, we would decrease the sensitivity. In general, a sensitive test is most useful when failing to detect a disease has dangerous consequences; a specific test is important when a false positive result is harmful.

The relationship between sensitivity and specificity may be illustrated using a graph known as a *receiver operator characteristic (ROC) curve*. An ROC curve is a line graph that plots the probability of a true positive result against the probability of a false positive result for a range of cutoff points. As an example, Figure 6.1 displays an ROC curve for the data contained in Table 6.1. When a diagnostic test is being examined, this type of graph may be used to help determine the appropriate

Table 6.1 Sensitivity and specificity of serum creatinine level for predicting transplant rejection

Serum Creatinine (mg %)	Sensitivity	Specificity
>1.2	0.939	0.123
1.3	0.939	0.203
1.4	0.909	0.281
1.5	0.818	0.380
1.6	0.758	0.461
1.7	0.727	0.535
1.8	0.636	0.649
1.9	0.636	0.711
2.0	0.545	0.766
2.1	0.485	0.773
2.2	0.485	0.803
2.3	0.394	0.811
2.4	0.394	0.843
2.5	0.364	0.870
2.6	0.333	0.891
2.7	0.333	0.894
2.8	0.333	0.896
2.9	0.303	0.909

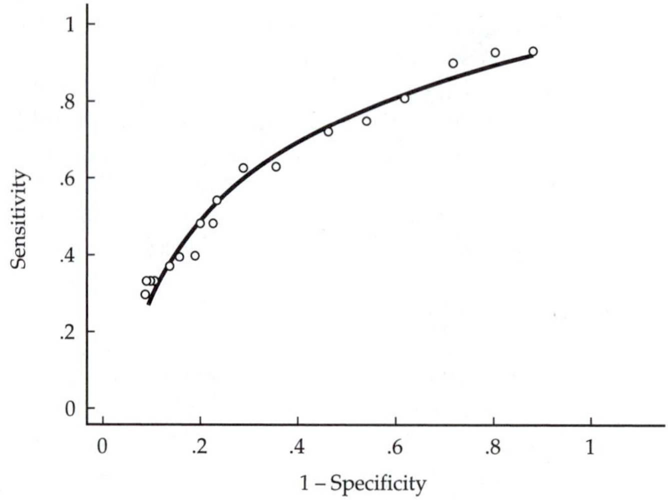

Figure 6.1. ROC curve for serum creatinine level as a predictor of transplant rejection

cutoff point. A new test that simultaneously improves on the sensitivity and specificity of an existing one signals an advance in testing technology.

6.5 Calculation of Prevalence

In addition to being used in applications involving Bayes' theorem, diagnostic testing or screening can also be used to calculate the prevalence of disease in a specified population. For example, the New York State Department of Health initiated a program to screen all infants born over a 28-month period for the human immunodeficiency virus. Since maternal antibodies cross the placenta, the presence of antibodies in an infant signals infection in the mother. Because the tests are performed anonymously, however, no verification of the results is possible. The reported outcomes of the statewide screening are presented in Table 6.2 [7].

Let n^+ represent the number of newborns who test positive and n the total number of infants screened. In each region of New York, HIV seroprevalence–or $P(H)$, where H is the event that a mother is infected with the virus–is calculated as n^+/n. In Manhattan, for example, 50,364 infants were tested, and 799 of the results were positive. In this borough,

$$\begin{aligned} \frac{n^+}{n} &= \frac{799}{50,364} \\ &= 0.0159. \end{aligned}$$

Note, however, that the quantity n^+/n actually represents $P(T^+)$, the probability of a positive test result. If the screening test were perfect, then $P(H)$ and $P(T^+)$

Table 6.2 Percentage of HIV-positive newborns by region for the state of New York, December 1987–March 1990

Region	Number Positive	Total Tested	Percent Positive
New York State exclusive of NYC	601	346,522	0.17
NYC suburban	329	120,422	0.27
Mid-Hudson Valley	71	29,450	0.24
Upstate urban	119	88,088	0.14
Upstate rural	82	108,562	0.08
New York City	3,650	294,062	1.24
Manhattan	799	50,364	1.59
Bronx	998	58,003	1.72
Brooklyn	1,352	104,613	1.29
Queens	424	67,474	0.63
Staten Island	77	13,608	0.57

would be identical. The test is not infallible, however; both false positive and false negative results are possible. In fact, the true probability of a positive test is

$$\begin{aligned} P(T^+) &= P(T^+ \mid H)\,P(H) + P(T^+ \mid H^c)\,P(H^c) \\ &= P(T^+ \mid H)\,P(H) + [1 - P(T^- \mid H^c)]\,[1 - P(H)]. \end{aligned}$$

In addition to the prevalence of infection, this equation incorporates both the sensitivity and the specificity of the diagnostic test.

If n^+/n is the probability of a positive test result, then how do we compute the prevalence of HIV? Using the expression for $P(T^+)$, we are able to solve for the true quantity of interest. After some algebraic manipulation, we find that

$$\begin{aligned} P(H) &= \frac{P(T^+) - P(T^+ \mid H^c)}{P(T^+ \mid H) - P(T^+ \mid H^c)} \\ &= \frac{(n^+/n) - P(T^+ \mid H^c)}{P(T^+ \mid H) - P(T^+ \mid H^c)}. \end{aligned}$$

Since the prevalence of HIV infection is also a probability, its value must lie between 0 and 1. Let us examine the expression for $P(H)$. For any screening test of value,

$$P(T^+ \mid H) > P(T^+ \mid H^c);$$

the probability of a positive test result among individuals who are infected with HIV is higher than the probability among individuals who are not. As a result, the denominator of the ratio is positive. For P(H) to be greater than 0, the numerator is required to be positive as well. Consequently, we must have

$$\begin{aligned} \frac{n^+}{n} &> \mathrm{P}(T^+ \mid H^c) \\ &= 1 - \mathrm{P}(T^- \mid H^c). \end{aligned}$$

The proportion of positive test results in the entire population must be greater than the proportion of positive results among those who are not infected with HIV. Note that the specificity of the screening test plays a critical role in the calculation of prevalence; if the prevalence is very low, it may not be detected by a test with inadequate specificity.

Return to the data in Table 6.2. We do not know the sensitivity and specificity of the diagnostic procedure used, although we can be sure that the test was not perfect. Suppose, however, that the sensitivity of the screening test is 0.99 and that its specificity is 0.998; these values represent the higher end of the range of possible values. Also, recall that the probability of a positive test result in Manhattan is 0.0159. As a result, the prevalence of HIV infection would be calculated as

$$\begin{aligned} \mathrm{P}(H) &= \frac{0.0159 - (1 - 0.998)}{0.99 - (1 - 0.998)} \\ &= 0.0141. \end{aligned}$$

For the upstate urban region of New York, however, $n^+/n = 119/88{,}088 = 0.0014$, and

$$\begin{aligned} \mathrm{P}(H) &= \frac{0.0014 - (1 - 0.998)}{0.99 - (1 - 0.998)} \\ &= -0.0006. \end{aligned}$$

Even with a specificity as high as 0.998, the prevalence is found to be negative. Obviously, this is a nonsensical result; it most likely occurred because the testing procedure was not accurate enough to measure the very low prevalence.

6.6 The Relative Risk and the Odds Ratio

The concept of relative risk is often useful when we wish to compare the probabilities of disease in two different situations or groups. The *relative risk*, abbreviated RR, is the chance that a member of a group receiving some exposure will develop disease relative to the chance that a member of an unexposed group will develop the same disease. It is defined as the probability of disease in the exposed group divided by the probability of disease in the unexposed group, or

$$\text{RR} = \frac{\text{P(disease} \mid \text{exposed)}}{\text{P(disease} \mid \text{unexposed)}}.$$

Consider a study that examines the risk factors for breast cancer among women participating in the first National Health and Nutrition Examination Survey [8]. In a sample of 4,540 women who gave birth to their first child before the age of 25, 65 developed breast cancer. Of the 1,628 women who first gave birth at age 25 or older, 31 were diagnosed with breast cancer. If, in this case, we consider "exposure" to be the condition of having first given birth at age 25 or older—and again assume that the numbers are large enough to satisfy the frequentist definition of probability—then the relative risk of developing breast cancer would be

$$\begin{aligned} \text{RR} &= \frac{\text{P(disease} \mid \text{exposed)}}{\text{P(disease} \mid \text{unexposed)}} \\ &= \frac{31/1,628}{65/4,540} \\ &= 1.33. \end{aligned}$$

A relative risk of 1.33 implies that women who first give birth at 25 years of age or older are 33 % more likely to develop breast cancer than women who give birth at a younger age. In Chapter 15 we explain how to interpret this finding.

In general, a relative risk of 1.0 indicates that the probabilities of disease in the exposed and unexposed groups are identical; consequently, an association between the exposure and the disease does not exist. A relative risk greater than 1.0 implies that there is an increased risk of disease among those with the exposure, whereas a value less than 1.0 suggests that there is a decreased risk of developing disease among the exposed individuals.

Note that the value of the relative risk is independent of the magnitudes of the relevant probabilities; only the ratio of these probabilities is important. This is especially useful when we are concerned with unlikely events. In the United States, for example, the probability that a man over the age of 35 dies of lung cancer is 0.002679 for current smokers and 0.000154 for nonsmokers [9]. The relative risk of death for smokers versus nonsmokers, however, is

$$\begin{aligned} \text{RR} &= \frac{0.002679}{0.000154} \\ &= 17.4. \end{aligned}$$

Similarly, the probability that a woman over the age of 35 dies of lung cancer is 0.001304 for current smokers and 0.000121 for nonsmokers; the relative risk is

$$\begin{aligned} \text{RR} &= \frac{0.001304}{0.000121} \\ &= 10.8. \end{aligned}$$

Even though we are dealing with low probability events, the relative risk allows us to see that smoking has a large effect on the likelihood that a particular individual will die from lung cancer.

Another commonly used measure of the relative probabilities of disease is the *odds ratio*, or *relative odds*. If an event takes place with probability p, then the odds in favor of the event are $p/(1-p)$ to 1. If $p = 1/2$, for instance, then the odds are $(1/2)/(1/2) = 1$ to 1. In this case, the event is equally likely to either occur or not occur. If $p = 2/3$, then the odds of the event are $(2/3)/(1/3) = 2$ to 1; the probability that the event occurs is twice as large as the probability that it does not. If for every 100,000 persons 9.3 new cases of tuberculosis develop each year, then the odds of developing tuberculosis are

$$\frac{(9.3/100,000)}{(99,990.7/100,000)} = 0.00009301 \text{ to } 1.$$

Conversely, we know that if the odds in favor of an event are a to b, then the probability that the event will occur is $a/(a+b)$. The odds ratio is defined as the odds in favor of disease among exposed individuals divided by the odds of disease among the unexposed, or

$$\text{OR} = \frac{\text{P(disease} \mid \text{exposed})/[1 - \text{P(disease} \mid \text{exposed})]}{\text{P(disease} \mid \text{unexposed})/[1 - \text{P(disease} \mid \text{unexposed})]}.$$

Consider the following data taken from a study that examines the association between the use of oral contraceptives and breast cancer [10]. Among the 2,914 women who had previously used oral contraceptives, 273 developed breast cancer and 2,641 did not. Of the 7,976 women who had never used oral contraceptives, 716 developed cancer and 7,260 did not. Assuming that these numbers are large enough to satisfy the frequentist definition of probability,

$$\begin{aligned} \text{OR} &= \frac{(273/2,914)/(1 - 273/2,914)}{(716/7,976)/(1 - 716/7,976)} \\ &= \frac{(273/2,914)/(2,641/2,914)}{(716/7,976)/(7,260/7,976)} \\ &= \frac{273/2,641}{716/7,260} \\ &= 1.05. \end{aligned}$$

These data imply that women who have used oral contraceptives have an odds of developing breast cancer that is only 1.05 times that of nonusers. (We will interpret this result in Chapter 15.) As with the relative risk, however, an odds ratio of 1.0 indicates that exposure does not have an effect on the probability of disease.

The relative risk and the odds ratio are two different measures that attempt to explain the same phenomenon. Although the relative risk might seem more intuitive, the odds ratio has better statistical properties; these properties will be clarified later in the text. In any event, for rare diseases the odds ratio is a close approximation of the relative risk. If

$$\text{P(disease} \mid \text{exposed)} \approx 0$$

and

$$\text{P(disease} \mid \text{unexposed)} \approx 0,$$

then

$$1 - \text{P(disease} \mid \text{exposed)} \approx 1$$

and

$$1 - \text{P(disease} \mid \text{unexposed)} \approx 1.$$

Therefore,

$$\begin{aligned} \text{OR} &= \frac{\text{P(disease} \mid \text{exposed)}/[1 - \text{P(disease} \mid \text{exposed)}]}{\text{P(disease} \mid \text{unexposed)}/[1 - \text{P(disease} \mid \text{unexposed)}]} \\ &\approx \frac{\text{P(disease} \mid \text{exposed)}/1}{\text{P(disease} \mid \text{unexposed)}/1} \\ &= \frac{\text{P(disease} \mid \text{exposed)}}{\text{P(disease} \mid \text{unexposed)}} \\ &= \text{RR}. \end{aligned}$$

6.7 Further Applications

Suppose that we are interested in estimating the probability that a woman who becomes pregnant will give birth to a boy. In 1986, there were 3,756,547 registered births in the United States [11]. Of these infants, 1,924,868 were boys and 1,831,679 were girls. Therefore, if a randomly selected woman were to become pregnant, we could estimate the probability that her child will be a boy as

$$\begin{aligned} P(B) &= P(\text{child will be a boy}) \\ &= \frac{1,924,868}{3,756,547} \\ &= 0.512. \end{aligned}$$

We can discuss the probability that the child will be a boy only before the woman actually becomes pregnant; after conception, the sex of the fetus has been determined, and the concept of probability no longer applies.

The complement of the event that the child is a boy is the event that it is a girl. Consequently,

$$\begin{aligned} P(G) &= P(\text{child will be a girl}) \\ &= 1 - P(B) \\ &= 1 - 0.512 \\ &= 0.488. \end{aligned}$$

Since the child is classified as either a boy or a girl, these two events are mutually exclusive. When events are mutually exclusive, the additive rule of probability states that the probability that either of the events will occur is equal to the sum of the probabilities of the individual events; thus

$$\begin{aligned} P(B \cup G) &= P(B) + P(G) \\ &= 0.512 + 0.488 \\ &= 1.000. \end{aligned}$$

The sum of the probabilities of these two events is 1. This indicates that the events are exhaustive. The child must be classified as either a boy or a girl; there are no other possible outcomes.

Now suppose that we randomly select two women from the population and that they both will become pregnant. What is the probability that both children will be boys? We know that the two events are independent; the sex of the first woman's child has no effect on the sex of the second woman's child. Therefore, using the multiplicative rule of probability and representing the event that both children will be boys by $B_1 \cap B_2$,

$$\begin{aligned} P(B_1 \cap B_2) &= \text{P(both children will be boys)} \\ &= \text{P(first child will be a boy) P(second child will be a boy)} \\ &= (0.512)(0.512) \\ &= 0.262. \end{aligned}$$

There are three other possible events: $B_1 \cap G_2$, the first woman's child will be a boy and the second woman's child a girl; $G_1 \cap B_2$, the first woman will have a girl and the second a boy; and $G_1 \cap G_2$, both children will be girls. The probabilities of these events are

$$\begin{aligned} P(B_1 \cap G_2) &= \text{P(first child will be a boy) P(second child will be a girl)} \\ &= (0.512)(0.488) \\ &= 0.250, \end{aligned}$$

$$\begin{aligned} P(G_1 \cap B_2) &= \text{P(first child will be a girl) P(second child will be a boy)} \\ &= (0.488)(0.512) \\ &= 0.250, \end{aligned}$$

and

$$\begin{aligned} P(G_1 \cap G_2) &= \text{P(first child will be a girl) P(second child will be a girl)} \\ &= (0.488)(0.488) \\ &= 0.238. \end{aligned}$$

What is the probability that both children will be boys given that at least one child is a boy? The chance that a particular event will occur given that another event has already taken place is known as a *conditional probability*. Representing the event that at least one child is a boy by A and applying the formula for a conditional probability,

$$\begin{aligned} P(B_1 \cap B_2 \mid A) &= \text{P(both children will be boys } \mid \text{ at least one boy)} \\ &= \frac{P[(B_1 \cap B_2) \cap A]}{P(A)} \\ &= \frac{P(B_1 \cap B_2)}{P(A)}. \end{aligned}$$

We already know that $P(B_1 \cap B_2) = 0.262$. What is the probability that there is at least one boy? Note that this event can occur in three different ways—either both children will be boys, the first will be a boy and the second a girl, or the first will be a girl and the second a boy. Since these three events are mutually exclusive, we apply the additive rule to find

$$\begin{aligned} P(A) &= P[(B_1 \cap B_2) \cup (B_1 \cap G_2) \cup (G_1 \cap B_2)] \\ &= P(B_1 \cap B_2) + P(B_1 \cap G_2) + P(G_1 \cap B_2) \\ &= 0.262 + 0.250 + 0.250 \\ &= 0.762. \end{aligned}$$

Therefore,

$$\begin{aligned} P(B_1 \cap B_2 \mid A) &= \frac{P(B_1 \cap B_2)}{P(A)} \\ &= \frac{0.262}{0.762} \\ &= 0.344. \end{aligned}$$

If we know that at least one child is a boy, then the probability that both children will be boys increases from 0.262 to 0.344.

At first glance, this result may appear counterintuitive. We are told that one child is a boy; as a result, we might expect the probability that the other child is a boy to be simply $P(B) = 0.512$. Instead, we calculated the probability to be 0.344. The important point to keep in mind is that we did not specify which of the two children was a boy. In this example, order is important. When dealing with probabilities, the seemingly "obvious" answer is not always correct; each problem must be considered carefully.

Conditional probabilities often come into play when we are working with life tables. Suppose we would like to know the probability that a person will live to be 80 years of age given that he or she is now 40. Let A represent the event that the individual is 40 years old and B the event that he or she lives to be 80. Using Table 5.1, the 1988 life table for the U. S. population,

$$\begin{aligned} P(A) &= P(\text{a person lives to be 40 years of age}) \\ &= \frac{95,316}{100,000} \\ &= 0.95316, \end{aligned}$$

and

$$\begin{aligned} P(A \cap B) &= P(\text{a person lives to be 40 and lives to be 80}) \\ &= P(\text{a person lives to be 80 years of age}) \\ &= \frac{46,029}{100,000} \\ &= 0.46029. \end{aligned}$$

Therefore,

$$\begin{aligned}
P(B \mid A) &= P(\text{a person lives to be 80} \mid \text{he or she is now 40}) \\
&= \frac{P(A \cap B)}{P(A)} \\
&= \frac{0.46029}{0.95316} \\
&= 0.4829.
\end{aligned}$$

If A_1 and A_2 are mutually exclusive and exhaustive events so that

$$\begin{aligned}
P(A_1 \cup A_2) &= P(A_1) + P(A_2) \\
&= 1,
\end{aligned}$$

then Bayes' theorem states that

$$P(A_1 \mid B) = \frac{P(A_1)\, P(B \mid A_1)}{P(A_1)\, P(B \mid A_1) + P(A_2)\, P(B \mid A_2)}.$$

Bayes' theorem is important in diagnostic testing. It relates the predictive value of a test to its sensitivity and specificity as well as the prevalence of disease in the population being tested.

Consider the following data taken from a study that investigates the accuracy of three brands of home pregnancy tests [12]. Let A_1 represent the event that a woman is pregnant, A_2 the event that she is not pregnant, and B a positive home pregnancy test result. The average sensitivity for the three brands of test kits is 80%; therefore,

$$P(\text{test positive} \mid \text{pregnant}) = 0.80.$$

Consequently, the probability of a false negative result is

$$\begin{aligned}
P(\text{test negative} \mid \text{pregnant}) &= 1 - 0.80 \\
&= 0.20.
\end{aligned}$$

The specificity of the home pregnancy tests is found to be 68%; therefore,

$$P(\text{test negative} \mid \text{not pregnant}) = 0.68;$$

the probability of a false positive is

$$\begin{aligned}
P(\text{test positive} \mid \text{not pregnant}) &= 1 - 0.68 \\
&= 0.32.
\end{aligned}$$

What is the probability that a woman with a positive home test kit result is pregnant?

Suppose that in the population being tested $P(A_1) = 0.60$; that is, 60% of the women who use the home pregnancy tests are actually pregnant. Since A_2 is the complement of A_1, the probability that a woman is not pregnant is

$$\begin{aligned} P(A_2) &= 1 - P(A_1) \\ &= 1 - 0.60 \\ &= 0.40. \end{aligned}$$

Using Bayes' theorem, the predictive value of a positive test is

$$\begin{aligned} P(A_1 \mid B) &= \frac{P(A_1)\,P(B \mid A_1)}{P(A_1)\,P(B \mid A_1) + P(A_2)\,P(B \mid A_2)} \\ &= \frac{(0.60)(0.80)}{(0.60)(0.80) + (0.40)(0.32)} \\ &= 0.79. \end{aligned}$$

Therefore, a positive home test kit result increases the probability that a woman in this population is pregnant from 0.60 to 0.79.

When we wish to compare the probabilities of a specified event in two different groups, the concept of relative risk is often useful. Consider Figure 6.2. This bar chart illustrates the risks of lung cancer among women who have smoked 21 or more cigarettes a day relative to women who have never smoked [9, 13]. For

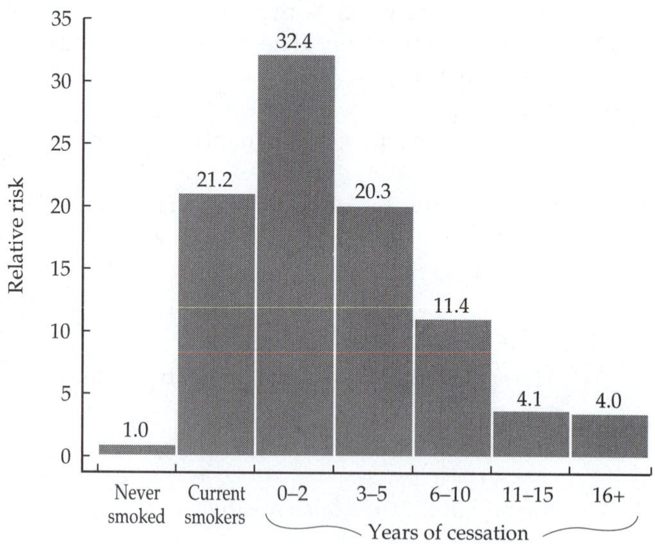

Figure 6.2. Relative risks of lung cancer in female exsmokers, 21 or more cigarettes a day

the group of persons who have stopped smoking within the past two years, for example,

$$\text{RR} = \frac{P(\text{lung cancer} \mid \text{quit within past two years})}{P(\text{lung cancer} \mid \text{nonsmoker})}$$

$$= 32.4.$$

It is rather surprising that this relative risk is higher than the corresponding risk for current smokers; however, this group contains many sick people who were forced to quit because of their illness. Given time, the risk of lung cancer among even heavy smokers gradually decreases once an individual stops smoking.

6.8 Review Exercises

1. Let A represent the event that a particular individual is exposed to high levels of carbon monoxide and B the event that he or she is exposed to high levels of nitrogen dioxide.

a. What is the event $A \cap B$?
b. What is the event $A \cup B$?
c. What is the complement of A?
d. Are the events A and B mutually exclusive?

2. Consider the natality statistics for the U. S. population in 1987 [14]. According to these data, the probabilities that a randomly selected woman who gave birth in 1987 was in each of the following age groups are as follows:

Age	Probability
<15	0.0027
15–19	0.1214
20–24	0.2824
25–29	0.3192
30–34	0.1997
35–39	0.0651
40–44	0.0091
45–49	0.0004
Total	1.0000

a. What is the probability that a woman who gave birth in 1987 was 24 years of age or younger?
b. What is the probability that she was 40 or older?
c. Given that the mother of a particular child was under 30 years of age, what is the probability that she was not yet 20?

3. Recall the previously discussed study regarding the proficiency test of lab technicians who scan Pap smear slides for abnormalities. This study reported a false negative rate of 0.1625,

$$P(\text{test negative} \mid \text{cancer}) = 0.1625,$$

and a specificity of 0.8136,

$$P(\text{test negative} \mid \text{no cancer}) = 0.8136.$$

Assume that consecutive tests are independent.

a. Given these data, what is the probability that a woman who does not have cancer of the cervix will have three negative Pap smears in three consecutive years?
b. What is the probability that a woman who does have cancer will have three negative Pap smears in three consecutive years?
c. What is the probability that a woman who does not have cancer will have at least one positive test in three years?

4. Refer to Table 5.1, the 1988 abridged life table for the United States.

a. What is the probability that an individual who is 60 years old will survive for the next ten years?
b. Consider a man and woman who are married and who are each 60 years of age. What is the probability that both the woman and her husband will be alive on their 70th birthdays? Assume that the two events are independent.
c. What is the probability that either the wife or the husband, but not both, will be alive at age 70?

5. The sensitivity of the mammogram, a screening test for detecting breast cancer, is 0.85; its specificity is 0.80 [15]. In a population in which the probability that a particular woman has breast cancer is 0.0025, what is the probability that she has cancer given that her mammogram is positive?

6. The National Institute for Occupational Safety and Health has developed a case definition of carpal tunnel syndrome that incorporates three criteria: symptoms of median nerve involvement, a history of occupational risk factors, and the presence of physical exam findings [16]. The sensitivity of this definition as a test for carpal tunnel syndrome is 0.67; its specificity is 0.58.

a. In a population in which the prevalence of carpal tunnel syndrome is estimated to be 15%, what is the predictive value of a positive test result?
b. How does this probability change if the prevalence is only 5%?

7. The following data are taken from a study investigating the use of radionuclide ventriculography in detecting coronary artery disease [17].

Test	Disease		Total
	Present	Absent	
Positive	302	80	382
Negative	179	372	551
Total	481	452	933

a. What is the sensitivity of radionuclide ventriculography in this study? What is its specificity?
b. For a population in which the probability of having coronary artery disease is 0.10, estimate the probability that an individual has the disease given that he or she tests positive using radionuclide ventriculography.
c. What is the predictive value of a negative test?

8. For several different methods of contraception, the probabilities that a currently married woman experiences an unplanned pregnancy during the first year of use are provided below [18].

Method of Contraception	Probability of Pregnancy
None	0.431
Diaphragm	0.149
Condom	0.106
IUD	0.071
Pill	0.037

For each method listed, calculate the relative risk of pregnancy for women using this method versus women not using any type of protection.

References

[1] Yobs, A. R., Swanson, R. A., and Lamotte, L. C., "Laboratory Reliability of the Papanicolaou Smear," *Obstetrics and Gynecology*, Volume 65, February 1985, 235–244.

[2] Devesa, S. S., Silverman, D. T., Young, J. L., Pollack, E. S., Brown, C. C., Horm, J. W., Percy, C. L., Myers, M. H., McKay, F. W., and Fraumeni, J. F., "Cancer Incidence and Mortality Trends Among Whites in the United States, 1949–1984," *Journal of the National Cancer Institute*, Volume 79, October 1987, 701–770.

[3] Henig, R. M., "Is the Pap Test Valid?," *The New York Times Magazine*, May 28, 1989, 37–38.

[4] Yerushalmy, J., Harkness, J. T., Cope, J. H., and Kennedy, B. R., "The Role of Dual Reading in Mass Radiography," *American Review of Tuberculosis*, Volume 61, April 1950, 443–464.

[5] Centers for Disease Control, "A Strategic Plan for the Elimination of Tuberculosis in the United States," *Morbidity and Mortality Weekly Report*, Volume 38, Number 16, April 28, 1989.

[6] DeLong, E. R., Vernon, W. B., and Bollinger, R. R., "Sensitivity and Specificity of a Monitoring Test," *Biometrics*, Volume 41, December 1985, 947–958.

[7] Novick, L. F., Glebatis, D. M., Stricof, R. L., MacCubbin, P. A., Lessner, L., and Berns, D. S., "Newborn Seroprevalence Study: Methods and Results," *American Journal of Public Health*, Volume 81, May 1991, 15–21.

[8] Carter, C. L., Jones, D. Y., Schatzkin, A., and Brinton, L. A., "A Prospective Study of Reproductive, Familial, and Socioeconomic Risk Factors for Breast Cancer Using NHANES I Data," *Public Health Reports*, Volume 104, January–February 1989, 45–49.

[9] Garfinkel, L., and Silverberg, E., "Lung Cancer and Smoking Trends in the United States Over the Past 25 Years," *Ca–A Cancer Journal for Clinicians*, Volume 41, May/June 1991, 137–145.

[10] Hennekens, C. H., Speizer, F. E., Lipnick, R. J., Rosner, B., Bain, C., Belanger, C., Stampfer, M. J., Willett, W., and Peto, R., "A Case-Control Study of Oral Contraceptive Use and Breast Cancer," *Journal of the National Cancer Institute*, Volume 72, January 1984, 39–42.

[11] National Center for Health Statistics, *Vital Statistics of the United States, 1986*, Volume I—Natality, 1988.

[12] Doshi, M. L., "Accuracy of Consumer Performed In-home Tests for Early Pregnancy Detection," *American Journal of Public Health*, Volume 76, May 1986, 512–514.

[13] Garfinkel, L., and Stellman, S. D., "Smoking and Lung Cancer in Women: Findings in a Prospective Study," *Cancer Research*, Volume 48, December 1, 1988, 6951–6955.

[14] National Center for Health Statistics, "Advance Report of Final Natality Statistics, 1987," *Monthly Vital Statistics Report*, Volume 38, Number 3, June 29, 1989.

[15] Hulka, B. S., "Cancer Screening: Degrees of Proof and Practical Application," *Cancer*, Volume 62, Supplement to October 15, 1988, 1776–1780.

[16] Katz, J. N., Larson, M. G., Fossel, A. H., and Liang, M. H., "Validation of a Surveillance Case Definition of Carpal Tunnel Syndrome," *American Journal of Public Health*, Volume 81, February 1991, 189–193.

[17] Begg, C. B., and McNeil, B. J., "Assessment of Radiologic Tests: Control of Bias and Other Design Considerations," *Radiology*, Volume 167, May 1988, 565–569.

[18] Grady, W. R., Hayward, M. D., and Yagi, J., "Contraceptive Failure in the United States: Estimates from the 1982 National Survey of Family Growth," *Family Planning Perspectives*, Volume 18, September/October 1986, 200–209.

CHAPTER

7

Theoretical Probability Distributions

Any characteristic that can be measured or categorized is called a *variable*. If a variable is able to assume a number of different values such that any particular outcome is determined by chance, then it is known as a *random variable*. Random variables are typically represented by uppercase letters such as X, Y, and Z. A *discrete random variable* is able to assume only a finite or countable number of outcomes. One example is marital status; an individual can be single, married, divorced, or widowed. A *continuous random variable*, such as weight or height, can take on any value within a specified interval or continuum.

7.1 Probability Distributions

Every random variable has a corresponding probability distribution. A *probability distribution* applies the theory of probability to describe the behavior of a random variable. In the discrete case, it specifies all possible outcomes of the random variable along with the probability that each will occur. In the continuous case, it allows us to determine the probabilities that are associated with specified ranges of values.

For example, let X be a discrete random variable that represents the live birth order of each child born to a woman residing in the United States [1]. If a child is a woman's first-born, for instance, then $X = 1$; if it is her second child, then $X = 2$. To construct a probability distribution for X, we list each of the values x that the random variable is able to assume along with $P(X = x)$ for each one. This has been done in Table 7.1. Note that we use an uppercase X to denote the random variable and a lowercase x to represent the outcome of a particular child.

Table 7.1 resembles the frequency distributions that were introduced in Chapter 2. For a sample of observations, a frequency distribution displays each observed outcome and its corresponding frequency in the set of data; for a discrete random variable, a probability distribution lists each possible outcome and its associated probability. The probabilities represent the relative frequency of occurrence of each outcome x in a large number of trials repeated under essentially identical conditions. They tell us which values are more likely to occur than others. We assume that the numbers involved are large enough to satisfy the frequentist definition of probability. Since all possible values of the random variable are taken into account, the outcomes are exhaustive; therefore, the sum of their associated probabilities must be 1.

In many cases, we are also able to display a probability distribution by means of either a graph or a mathematical formula. Figure 7.1, for example, is a graph of the probability distribution enumerated in Table 7.1. The area of each vertical bar represents $P(X = x)$, the probability associated with that particular outcome of the random variable; the total area is equal to 1.

The probability distribution of X can be used to make statements about the possible outcomes of the random variable. Suppose we wish to know the probability that a randomly chosen newborn is its mother's fourth child. Using the data in Table 7.1, we observe that $P(X = 4) = 0.058$. What is the probability that the newborn is its mother's first or second child? Applying the additive rule of probability,

$$\begin{aligned} P(X = 1 \text{ or } X = 2) &= P(X = 1) + P(X = 2) \\ &= 0.416 + 0.330 \\ &= 0.746. \end{aligned}$$

If a random variable is able to take on a large number of values, then a probability distribution may not be a useful way to summarize its behavior. As with a distribution of grouped data, however, we are able to describe a probability distribution using a measure of central tendency and a measure of dispersion.

Table 7.1 Probability distribution of a random variable X representing the live birth order of children born in the United States

x	$P(X = x)$
1	0.416
2	0.330
3	0.158
4	0.058
5	0.021
6	0.009
7	0.004
8^+	0.004
Total	1.000

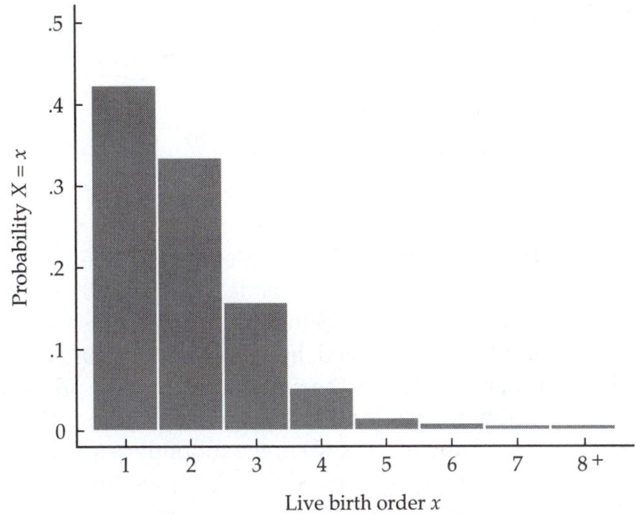

Figure 7.1. Probability distribution of a random variable representing the birth order of children born in the United States

The average value assumed by a random variable is known as the *population mean*; the dispersion of the values relative to this mean is the *population variance*.

The probability distribution of the birth order of children born to women in the United States was generated based on the experience of the U. S. population in 1986; the probability distributions of many other variables of interest, however, can be determined based on theoretical considerations. Distributions of this kind are known as *theoretical probability distributions*.

7.2 The Binomial Distribution

Consider a dichotomous random variable. By definition, the variable must assume one of two possible values; these mutually exclusive outcomes could represent life and death, male and female, or sickness and health. For simplicity, they are often referred to as "failure" and "success." A variable of this type is known as a *Bernoulli random variable*.

For instance, let Y be a random variable that represents smoking status; $Y = 1$ if an adult is currently a smoker and $Y = 0$ if he or she is not. The two outcomes of Y are mutually exclusive and exhaustive. In 1987, 29% of the adults in the United States smoked cigarettes, cigars, or pipes [2]; therefore, we can enumerate the probabilities associated with the respective outcomes of Y as

$$\begin{aligned} P(Y = 1) &= p \\ &= 0.29, \end{aligned}$$

and

$$\begin{aligned} P(Y = 0) &= 1 - p \\ &= 1.00 - 0.29 \\ &= 0.71. \end{aligned}$$

These two equations completely describe the probability distribution of the dichotomous random variable Y; if in 1987 we had traveled around the United States observing whether particular adults were smokers or nonsmokers, then Y would have assumed the value 1 approximately 29% of the time and the value 0 the remaining 71% of the time. Recall that the proportion of times that a dichotomous random variable takes on the value 1 is equal to its population mean.

Suppose that we now randomly select two individuals from the population of adults in the United States. If we introduce a new random variable X that represents the number of persons in the pair who are current smokers, then X can take on three possible values: 0, 1, or 2. Either both of the individuals selected do not smoke, one smokes and the other does not, or both adults smoke. The smoking statuses of the two persons chosen are independent; therefore, we can apply the multiplicative rule to find the probability that X will take on a particular value.

Outcome of Y			
First Person	**Second Person**	**Probability of These Outcomes**	**Number of Smokers**
0	0	$(1-p)(1-p)$	0
1	0	$p(1-p)$	1
0	1	$(1-p)p$	1
1	1	pp	2

Substituting in the value of p, we find that

$$\begin{aligned} P(X = 0) &= (1 - p)^2 \\ &= (0.71)^2 \\ &= 0.5041, \end{aligned}$$

$$\begin{aligned} P(X = 1) &= p(1 - p) + (1 - p)p \\ &= 2p(1 - p) \\ &= 2(0.29)(0.71) \\ &= 0.4118, \end{aligned}$$

and

$$\begin{aligned} P(X=2) &= p^2 \\ &= (0.29)^2 \\ &= 0.0841. \end{aligned}$$

Note that there are two possible situations in which one adult smokes and the other does not; either $Y = 1$ for the first individual and $Y = 0$ for the second, or $Y = 0$ for the first and $Y = 1$ for the second. Since the two outcomes are mutually exclusive, we can apply the additive rule of probability to find $P(X = 1)$. Also note that because all the possible outcomes of X are considered, their probabilities must sum to 1; that is,

$$\begin{aligned} P(X=0) + P(X=1) + P(X=2) &= 0.5041 + 0.4118 + 0.0841 \\ &= 1.0000. \end{aligned}$$

The probability distribution of the discrete random variable X described above is a special case of the *binomial distribution*. In general, if we have a sequence of n independent Bernoulli trials–or equivalently, n independent outcomes of the Bernoulli random variable Y–each with a probability of "success" p, then the total number of successes X is a binomial random variable. The fixed numbers n and p are called the *parameters* of the distribution. Parameters are numerical quantities that summarize the characteristics of a probability distribution. In the previous example, the parameters are $n = 2$, since two individuals are selected, and $p = 0.29$, because the probability that any randomly chosen adult is a current smoker is 0.29. The binomial distribution involves three assumptions:

1. There are a fixed number of trials n, each of which results in one of two mutually exclusive outcomes.
2. The outcomes of the n trials are independent.
3. The probability of success p is constant for each trial.

Suppose that we were to expand the preceding example by randomly selecting three adults instead of two. In this case, X would be a binomial random variable with parameters $n = 3$ and $p = 0.29$.

Outcome of Y First Person	Second Person	Third Person	Probability of These Outcomes	Number of Smokers
0	0	0	$(1-p)(1-p)(1-p)$	0
1	0	0	$p(1-p)(1-p)$	1
0	1	0	$(1-p)p(1-p)$	1
0	0	1	$(1-p)(1-p)p$	1
1	1	0	$pp(1-p)$	2
1	0	1	$p(1-p)p$	2
0	1	1	$(1-p)pp$	2
1	1	1	ppp	3

Substituting in the value of p,

$$\begin{aligned} P(X = 0) &= (1 - p)^3 \\ &= (0.71)^3 \\ &= 0.3579, \end{aligned}$$

$$\begin{aligned} P(X = 1) &= p(1 - p)^2 + p(1 - p)^2 + p(1 - p)^2 \\ &= 3(0.29)(0.71)^2 \\ &= 0.4386, \end{aligned}$$

$$\begin{aligned} P(X = 2) &= p^2(1 - p) + p^2(1 - p) + p^2(1 - p) \\ &= 3(0.29)^2(0.71) \\ &= 0.1791, \end{aligned}$$

and

$$\begin{aligned} P(X = 3) &= p^3 \\ &= (0.29)^3 \\ &= 0.0244. \end{aligned}$$

These equations describe the probability distribution of X. The random variable X is able to take on four possible values, and

$$\begin{aligned} P(X = 0) &+ P(X = 1) + P(X = 2) + P(X = 3) \\ &= 0.3579 + 0.4386 + 0.1791 + 0.0244 \\ &= 1.0000. \end{aligned}$$

Note that $P(X = 1)$ and $P(X = 2)$ each involve the summation of three terms; if we have a total of three individuals, then there are exactly three ways in which one of them can be a smoker and three ways in which any two of them can be smokers.

If we select a total of n adults from the population, then the probability that exactly x of them smoke may be written as

$$\begin{aligned} P(X = x) &= \frac{n!}{x!(n - x)!} p^x (1 - p)^{n-x} \\ &= \binom{n}{x} p^x (1 - p)^{n-x} \\ &= \binom{n}{x} (0.29)^x (0.71)^{n-x} \end{aligned}$$

where $n = 1, 2, 3, \ldots$ and $x = 0, 1, \ldots n$. This is the general expression for the probability distribution of a binomial random variable X where X is the number of smokers in a sample of size n. Given a total of n adults, $n!$—or n *factorial*—allows us to calculate the number of ways in which the n individuals can be ordered; note that we have n choices for the first position, $n - 1$ choices for the second position, and so on. In general,

$$n! = n(n-1)(n-2)\cdots(3)(2)(1).$$

By definition, 0! is equal to 1. The expression

$$\binom{n}{x} = \frac{n!}{x!(n-x)!}$$

is the *combination* of n objects chosen x at a time; it represents the number of ways in which x objects can be selected from a total of n objects without regard to order.

For example, if we were to randomly choose three individuals from the adult population, there would be

$$\binom{3}{0} = \frac{3!}{0!(3-0)!} = \frac{6}{(1)(6)} = 1$$

way in which we could select zero smokers; in particular, all three adults would have to be nonsmokers. Similarly, there would be

$$\binom{3}{1} = \frac{3!}{1!(3-1)!} = \frac{6}{(1)(2)} = 3$$

ways in which we could choose one smoker,

$$\binom{3}{2} = \frac{3!}{2!(3-2)!} = \frac{6}{(2)(1)} = 3$$

ways in which we could select two smokers, and only

$$\binom{3}{3} = \frac{3!}{3!(3-3)!} = \frac{6}{(6)(1)} = 1$$

way for all three adults to be smokers. Therefore, as we found earlier,

$$\begin{aligned} P(X = 0) &= \binom{3}{0} p^0 (1-p)^{3-0} \\ &= 1(0.29)^0 (0.71)^3 \\ &= 0.3579, \end{aligned}$$

$$\begin{aligned} P(X = 1) &= \binom{3}{1} p^1 (1-p)^{3-1} \\ &= 3(0.29)(0.71)^2 \\ &= 0.4386, \end{aligned}$$

$$\begin{aligned} P(X = 2) &= \binom{3}{2} p^2 (1-p)^{3-2} \\ &= 3(0.29)^2(0.71) \\ &= 0.1791, \end{aligned}$$

and

$$\begin{aligned} P(X = 3) &= \binom{3}{3} p^3 (1-p)^{3-3} \\ &= 1(0.29)^3(0.71)^0 \\ &= 0.0244. \end{aligned}$$

In general, if we have a sequence of n independent Bernoulli trials with a constant probability of success p at each trial, then the total number of successes X is a binomial random variable with probability distribution

$$P(X = x) = \binom{n}{x} p^x (1-p)^{n-x}.$$

The random variable X can assume any integer value from 0 to n.

In addition to the probabilities of individual outcomes, we can also compute the numerical summary measures associated with a probability distribution. For example, the mean value of a binomial random variable X—or the average number of successes in samples of size n—is obtained by multiplying the number of independent Bernoulli trials by the probability of success at each trial; hence, the mean value of X is equal to np. The variance of X is $np(1-p)$. These expressions were derived using a method analogous to that for finding the mean and variance of grouped data [3]. Applying these formulas, if we were to select repeated samples of size $n = 10$ from the adult population, the mean number of smokers per sample would be

$$\begin{aligned} np &= 10(0.29) \\ &= 2.9, \end{aligned}$$

and the standard deviation would equal

$$\begin{aligned} \sqrt{np(1-p)} &= \sqrt{10(0.29)(0.71)} \\ &= \sqrt{2.059} \\ &= 1.4. \end{aligned}$$

The expression for the variance of a binomial random variable seems quite reasonable. The quantity $np(1-p)$ is largest when p is equal to 0.5; it decreases as p approaches 0 or 1. When p is very large or very small, nearly all of the outcomes take the same value–for example, almost everyone smokes or almost everyone does not smoke–and the variability among outcomes is small. In contrast, if half of the population takes the value 0 and the other half takes the value 1, then it will be more difficult to predict any particular outcome; in this case, the variability is relatively large.

Figure 7.2 is a graph of the probability distribution of X–the number of smokers–for which $n = 10$ and $p = 0.29$. Because all the possible outcomes of X are considered, the area represented by the vertical bars sums to 1. Figure 7.3 is the probability distribution of another binomial random variable for which $n = 10$ and $p = 0.71$. Note that the distribution is skewed to the right when $p < 0.5$ and skewed to the left when $p > 0.5$. If $p = 0.5$, as is the case in Figure 7.4, then the probability distribution is symmetric.

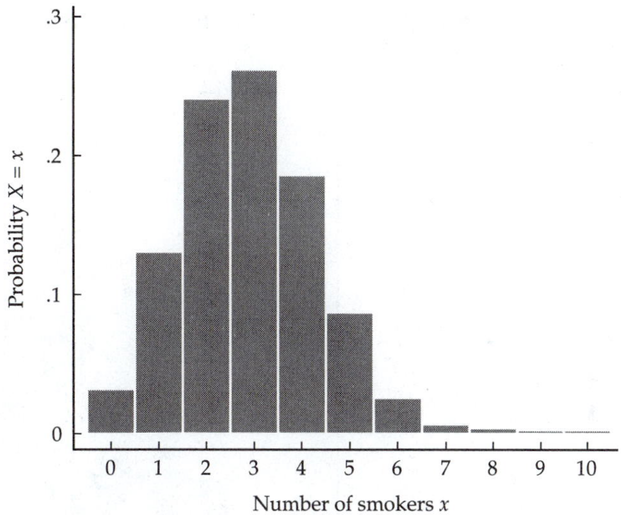

Figure 7.2. Probability distribution of a binomial random variable for which $n = 10$ and $p = 0.29$

7.3 The Poisson Distribution

Suppose that X is a random variable that represents the number of individuals who are involved in a motor vehicle accident each year. In the United States, the probability that a particular individual is involved is 0.00024 [4]. Technically, this is a binomial situation in which there are two distinct outcomes–accident or

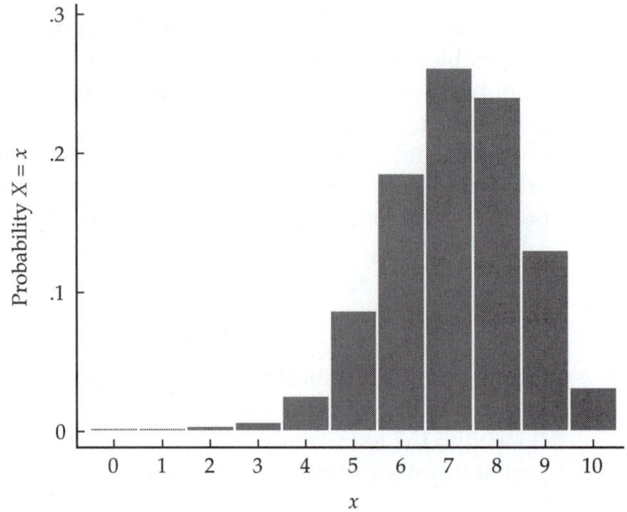

Figure 7.3. Probability distribution of a binomial random variable for which $n = 10$ and $p = 0.71$

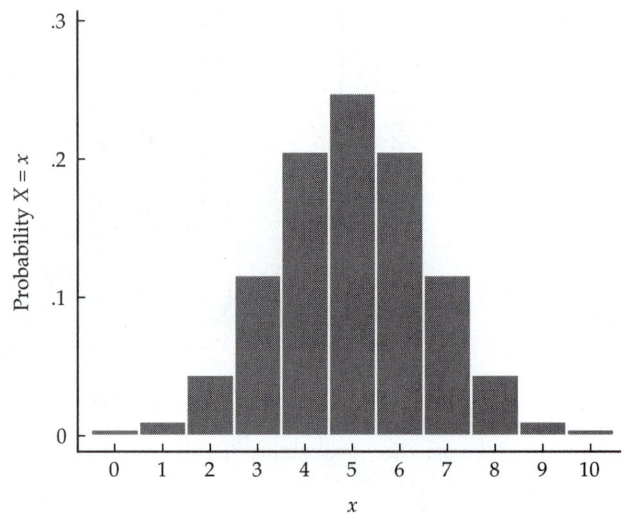

Figure 7.4. Probability distribution of a binomial random variable for which $n = 10$ and $p = 0.50$

no accident. Note, however, that n is very large; we are interested in the entire U. S. population. When n becomes large, the combination of n objects taken x at a time, $n!/x!(n-x)!$, is very tedious to evaluate. As a result, the binomial distribution may be impractical to use as a basis for calculations. However, in situations such as this—when n is very large and p is very small—the binomial distribution is well approximated by the Poisson distribution. The *Poisson distribution* is used to model discrete events that occur infrequently in time or space; thus, it is sometimes called the *distribution of rare events.*

Consider a random variable X that represents the number of occurrences of some event of interest over a given interval. Since X is a count, it is theoretically able to assume any integer value between 0 and infinity. Let λ (lambda) be a constant that denotes the average number of occurrences of the event in an interval. If the probability that X assumes the value x is

$$\mathrm{P}(\mathrm{X} = x) = \frac{e^{-\lambda}\lambda^{x}}{x!},$$

then X is said to have a Poisson distribution with parameter λ. The symbol e represents a constant that is approximated by 2.71828; in fact, e is the base of the natural logarithms. Like the binomial distribution, the Poisson distribution involves a set of underlying assumptions:

1. The probability that a single event occurs within an interval is proportional to the length of the interval.
2. Theoretically, within a single interval an infinite number of occurrences of the event are possible. We are not restricted to a fixed number of trials.
3. The events occur independently both within the same interval and between consecutive intervals.

Recall that the mean of a binomial random variable is equal to np and that its variance is $np(1 - p)$. When p is very small, $1 - p$ is close to 1 and $np(1 - p)$ is approximately equal to np. In this situation, the mean and the variance of the distribution are identical and may be represented by the single parameter λ. The property that the mean is equal to the variance is an identifying characteristic of the Poisson distribution.

Suppose that we are interested in determining the number of people in a population of 10,000 who will be involved in a motor vehicle accident each year. The mean number of persons involved would be

$$\begin{aligned}\lambda &= np \\ &= (10,000)(0.00024) \\ &= 2.4;\end{aligned}$$

this is also the variance. The probability that no one in this population will be involved in an accident in a given year is

$$\begin{aligned}\mathrm{P}(\mathrm{X} = 0) &= \frac{e^{-2.4}(2.4)^{0}}{0!} \\ &= 0.0907.\end{aligned}$$

The probability that exactly one person will be involved is

$$\begin{aligned} P(X = 1) &= \frac{e^{-2.4}(2.4)^1}{1!} \\ &= 0.2177. \end{aligned}$$

Similarly,

$$\begin{aligned} P(X = 2) &= \frac{e^{-2.4}(2.4)^2}{2!} \\ &= 0.2613, \end{aligned}$$

$$\begin{aligned} P(X = 3) &= \frac{e^{-2.4}(2.4)^3}{3!} \\ &= 0.2090, \end{aligned}$$

$$\begin{aligned} P(X = 4) &= \frac{e^{-2.4}(2.4)^4}{4!} \\ &= 0.1254, \end{aligned}$$

$$\begin{aligned} P(X = 5) &= \frac{e^{-2.4}(2.4)^5}{5!} \\ &= 0.0602, \end{aligned}$$

and

$$\begin{aligned} P(X = 6) &= \frac{e^{-2.4}(2.4)^6}{6!} \\ &= 0.0241. \end{aligned}$$

Since the outcomes of X are mutually exclusive and exhaustive,

$$\begin{aligned} P(X \geq 7) &= 1 - P(X < 7) \\ &= 1 - (0.0907 + 0.2177 + 0.2613 + 0.2090 \\ &\quad + 0.1254 + 0.0602 + 0.0241) \\ &= 0.0116. \end{aligned}$$

Figure 7.5 is a graph of the probability distribution of X, the number of individuals in the population involved in a motor vehicle accident each year. Once again, the area represented by the vertical bars sums to 1. As shown in Figure 7.6, the Poisson distribution is highly skewed for small values of λ; as λ increases, the distribution becomes more symmetric.

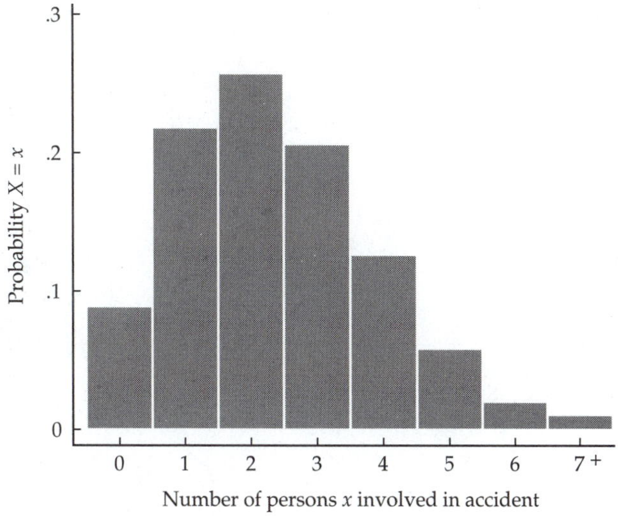

Figure 7.5. Probability distribution of a Poisson random variable for which $\lambda = 2.4$

7.4 The Normal Distribution

When it follows either a binomial or a Poisson distribution, a random variable X is restricted to taking on integer values only. Under different circumstances, however, the outcomes of a random variable may not be limited to integers. Suppose that X represents height. Rarely is an individual exactly 67 inches tall or exactly 68 inches tall; theoretically, X can assume an infinite number of intermediate values, such as 67.04 inches or 67.83 inches. Although one can argue philosophically that we are only able to measure discrete outcomes, treating such a variable as if it were continuous allows us to take advantage of powerful mathematical results.

As we have seen, the probability distribution of a discrete random variable is represented by an equation for $P(X = x)$, the probability that the random variable X will take on the specific value x. For a binomial random variable with parameters n and p, for instance,

$$P(X = x) = \binom{n}{x} p^x (1-p)^{n-x}.$$

These probabilities may be plotted against x, as in Figure 7.4. Suppose that the number of possible outcomes of X were to become very large and the widths of the corresponding intervals very small. In Figure 7.7, for example, $n = 30$ and $p = 0.50$. In general, if the number of possible values of X approaches infinity while the widths of the intervals approach zero, the graph will increasingly resemble a smooth curve.A smooth curve is used to represent the probability

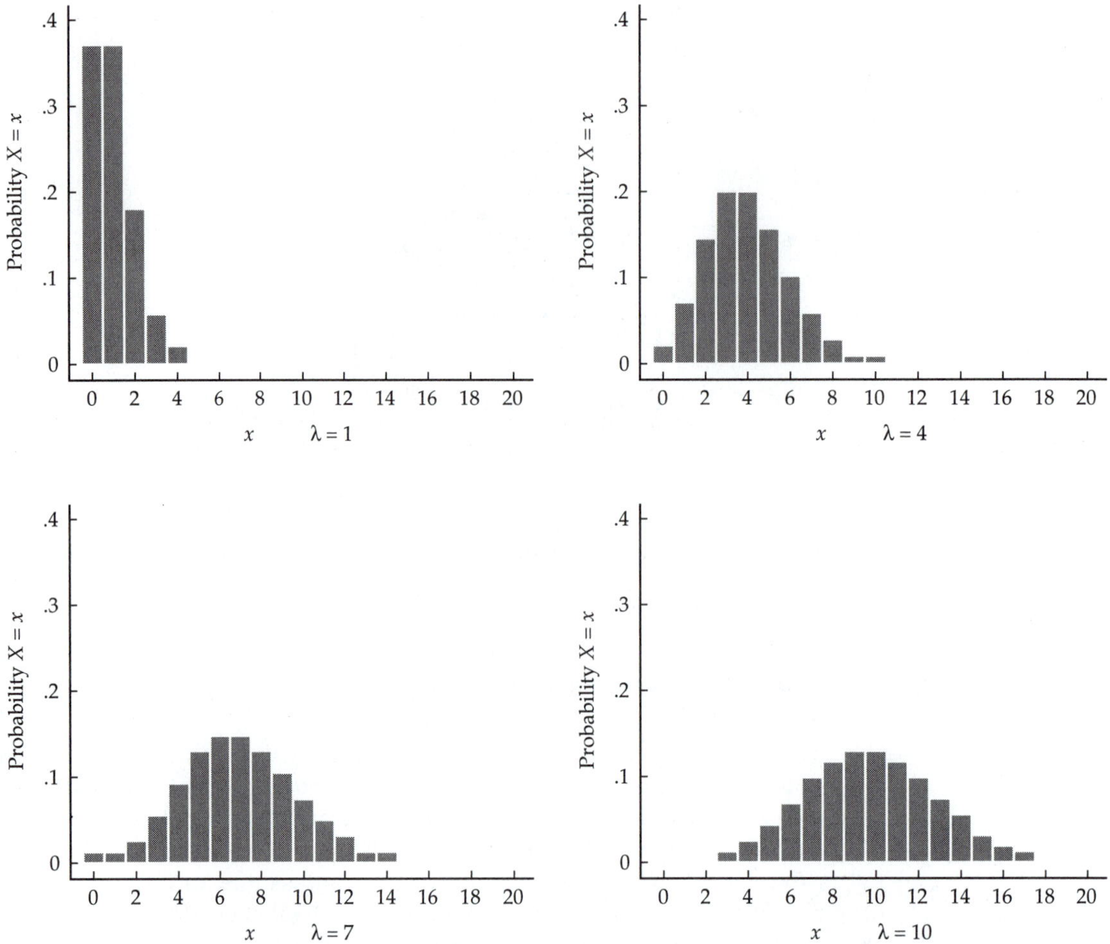

Figure 7.6. Probability distributions of Poisson random variables for various values of λ

distribution of a continuous random variable; the curve is called a *probability density*.

For any graph that illustrates a discrete probability distribution, the area represented by the vertical bars sums to 1. For a probability density, the total area beneath the curve must also be 1. Because a continuous random variable X can take on an uncountably infinite number of values, the probability associated with any specific outcome is equal to 0. The probability that X will assume some value in the interval enclosed by the outcomes x_1 and x_2 is equal to the area beneath the curve that lies between these two values.

The most common continuous distribution is the *normal distribution*, also known as the *Gaussian distribution*, or the *bell-shaped curve*. Its shape is that of a binomial distribution for which p is constant but n approaches infinity, or a

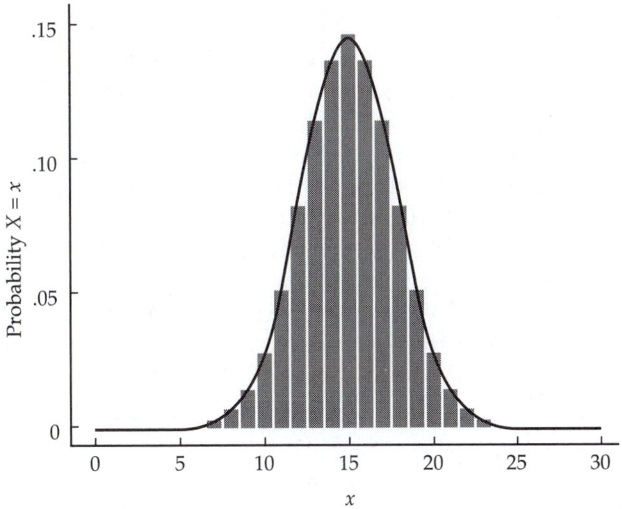

Figure 7.7. Probability distribution of a binomial random variable for which $n = 30$ and $p = 0.50$

Poisson distribution for which λ approaches infinity; its probability density is given by the equation

$$f(x) = \frac{1}{\sqrt{2\pi}\,\sigma} e^{-\frac{1}{2}\left(\frac{x-\mu}{\sigma}\right)^2}$$

where $-\infty < x < \infty$. The symbol π (pi) represents a constant approximated by 3.14159. The normal curve is unimodal and symmetric about its mean μ (mu); in this special case, the mean, median, and mode of the distribution are all identical. The standard deviation, represented by σ (sigma), specifies the amount of dispersion about the mean. Together, the two parameters μ and σ completely define a normal curve.

The value of the normal distribution will become more apparent when we begin to work with the sampling distribution of the mean. For now, however, it is important to note that many random variables of interest–including blood pressure, serum cholesterol level, and weight–are approximately normally distributed. The normal curve may thus be used to estimate probabilities that are associated with these variables. For example, in a population in which serum cholesterol level is normally distributed with mean μ and standard deviation σ, we might wish to find the probability that a randomly chosen individual has a serum cholesterol level greater than 250 mg/100 ml. Perhaps this knowledge will help us to plan for future cardiac services. Since the total area beneath the normal curve is equal to 1, we can estimate the probability in question by determining the proportion of the area under the curve that lies to the right of the point $x = 250$. This may be done using a table of areas calculated for the normal curve.

Since a normal distribution could have an infinite number of values for its mean and standard deviation, it is impossible to tabulate the area associated with each and every normal curve. Instead, only a single curve is tabulated—the special case for which $\mu = 0$ and $\sigma = 1$. This curve is known as the *standard normal distribution.* Figure 7.8 illustrates the standard normal curve, and Table A.1 in Appendix A displays the areas in one tail of the distribution. Outcomes of the random variable are denoted by z; the whole number and tenths decimal place of z are listed in the column to the left of the table, and the hundredths decimal place is shown in the row across the top. For a particular value of z, the entry in the body of the table specifies the area beneath the curve to the right of z. Some sample values of z and their corresponding areas are as follows:

z	**Area in Right Tail**
0.00	0.500
1.65	0.049
1.96	0.025
2.58	0.005
3.00	0.001

Since the standard normal distribution is symmetric about $z = 0$, the area under the curve to the right of z is equal to the area to the left of $-z$.

$-z$	**Area in Left Tail**
0.00	0.500
−1.65	0.049
−1.96	0.025
−2.58	0.005
−3.00	0.001

Suppose that we wish to know the area under the standard normal curve that lies between $z = -1.00$ and $z = 1.00$; since $\mu = 0$ and $\sigma = 1$, this is the area contained in the interval $\mu \pm 1\sigma$, illustrated in Figure 7.9. Looking at Table A.1, we see that the area to the right of $z = 1.00$ is 0.159. Therefore, the area to the left of $z = -1.00$ must be 0.159 as well. Since the total area under the curve is 1.000, the area between $z = -1.00$ and $z = 1.00$ is calculated by subtracting the area in each tail from this value;

$$1.000 - 0.159 - 0.159 = 0.682.$$

Therefore, for the standard normal distribution, approximately 68.2% of the area beneath the curve lies within ±1 standard deviation from the mean.

We might also wish to calculate the area under the standard normal curve that is contained in the interval $\mu \pm 2\sigma$. This area is illustrated in Figure 7.10.

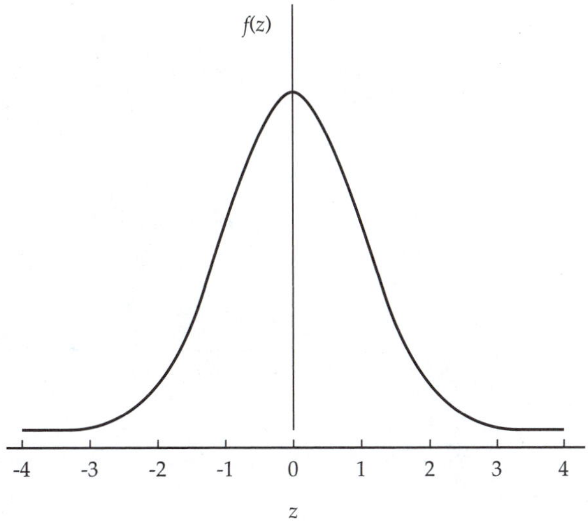

Figure 7.8. The standard normal curve for which $\mu = 0$ and $\sigma = 1$

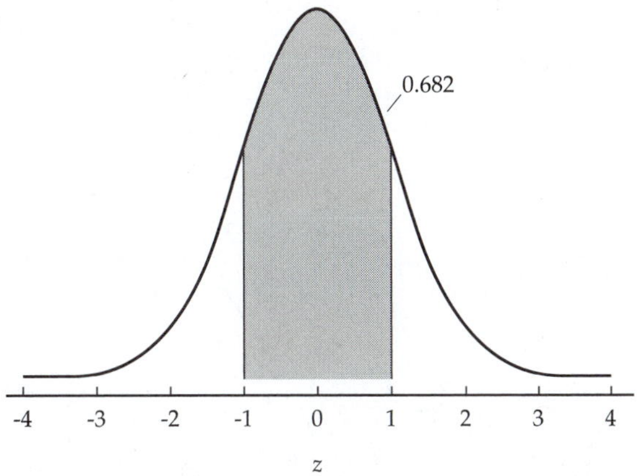

Figure 7.9. The standard normal curve, area between $z = -1.00$ and $z = 1.00$

Table A.1 indicates that the area to the right of $z = 2.00$ is 0.023; the area to the left of $z = -2.00$ is 0.023 as well. Therefore, the area between -2.00 and 2.00 is

$$1.000 - 0.023 - 0.023 \quad = \quad 0.954.$$

Approximately 95.4% of the area under the standard normal curve lies within ± 2 standard deviations from the mean.

Table A.1 can also be used the other way around. For example, we might wish to find the value of z that cuts off the upper 10% of the standard normal

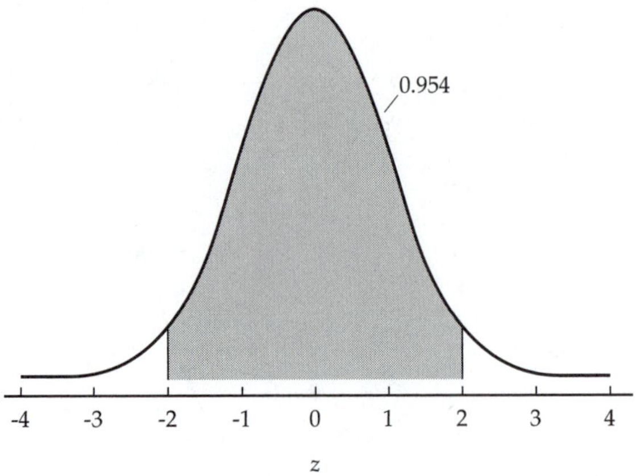

Figure 7.10. The standard normal curve, area between $z = -2.00$ and $z = 2.00$

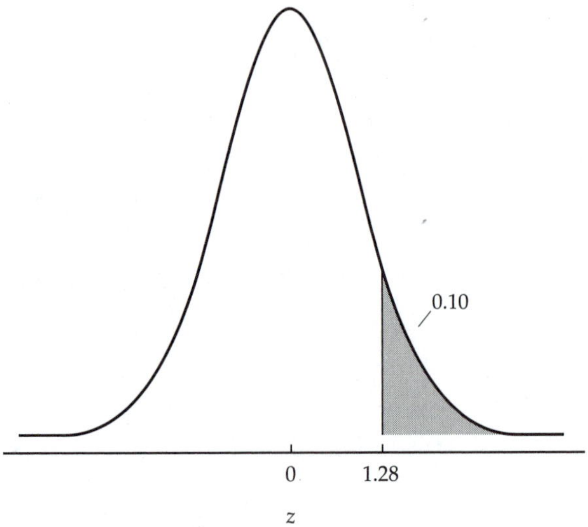

Figure 7.11. The standard normal curve, area to the right of $z = 1.28$

distribution. Locating 0.100 in the body of the table, we observe that the corresponding value of z is 1.28. Therefore, 10% of the area under the standard normal curve lies to the right of $z = 1.28$; this area is illustrated in Figure 7.11.

Now suppose that X is a normal random variable with mean 2 and standard deviation 0.5. Subtracting 2 from X would give us a normal random variable that has mean 0; as shown in Figure 7.12, the whole distribution would be shifted two units to the left. Dividing $(X - 2)$ by 0.5 alters the spread of the distribution so that we have a normal random variable with mean 0 and standard deviation 1.

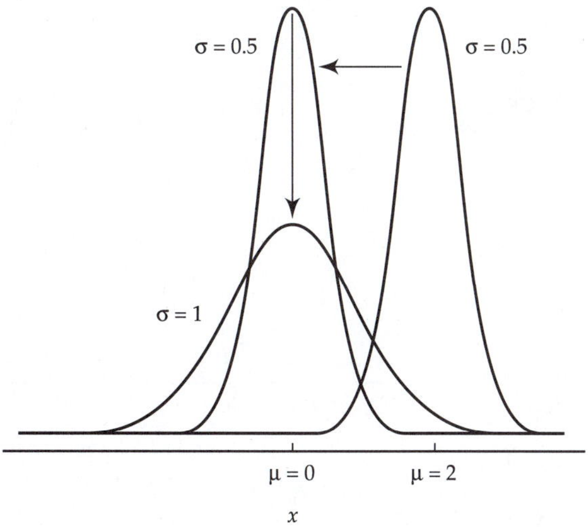

Figure 7.12. Transforming a normal curve with mean 2 and standard deviation 0.5 into the standard normal curve

Therefore, if X is a normal random variable with mean 2 and standard deviation 0.5, then

$$Z = \frac{X - 2}{0.5}$$

is a standard normal random variable. In general, for any arbitrary normal random variable with mean μ and standard deviation σ,

$$Z = \frac{X - \mu}{\sigma}$$

has a standard normal distribution. By transforming X into Z, we can use a table of areas computed for the standard normal curve to estimate probabilities associated with X. An outcome of the random variable Z, denoted z, is known as a *standard normal deviate*.

For example, let X be a random variable that represents systolic blood pressure. For the population of 18- to 74-year-old males in the United States, systolic blood pressure is approximately normally distributed with mean 129 millimeters of mercury and standard deviation 19.8 millimeters of mercury [5]. This distribution is shown in Figure 7.13. Note that

$$Z = \frac{X - 129}{19.8}$$

is normally distributed with mean 0 and standard deviation 1.

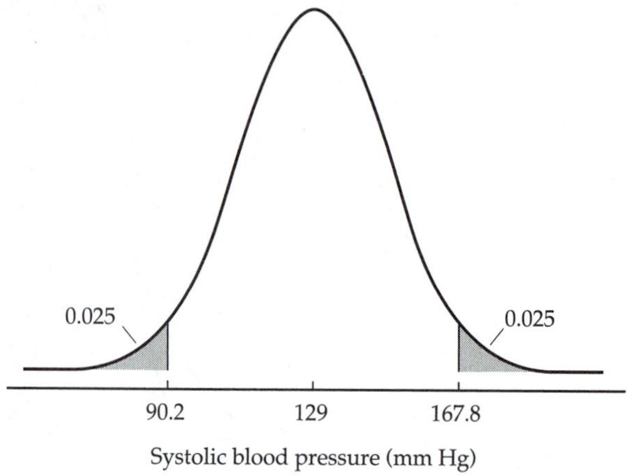

Figure 7.13. Distribution of systolic blood pressure for males 18 to 74 years of age, United States, 1976–1980

Using Table A.1, we see that the area to the right of $z = 1.96$ is 0.025. To obtain the value of x that corresponds to $z = 1.96$, we solve the equation

$$1.96 = \frac{x - 129}{19.8}$$

or

$$\begin{aligned} x &= 129 + (1.96)(19.8) \\ &= 167.8. \end{aligned}$$

Therefore, approximately 2.5% of the men in this population—a minuscule minority—have a systolic blood pressure that is greater than 167.8 mm Hg; 97.5% have a systolic blood pressure less than 167.8 mm Hg. Consequently, if we randomly select an individual from this male population, the probability that his systolic blood pressure is less than 167.8 mm Hg is 0.975.

Because the standard normal curve is symmetric about $z = 0$, we know that the area to the left of $z = -1.96$ is also 0.025. By solving the equation

$$-1.96 = \frac{x - 129}{19.8}$$

or

$$\begin{aligned} x &= 129 + (-1.96)(19.8) \\ &= 90.2, \end{aligned}$$

we find that $x = 90.2$ millimeters of mercury; this means that 2.5% of the men have a systolic blood pressure that is less than 90.2 mm Hg. Equivalently, the probability that a randomly selected male has a systolic blood pressure less than 90.2 mm Hg is 0.025. Since 2.5% of the men in the population have systolic blood pressures greater than 167.8 mm Hg and 2.5% have values less than 90.2 mm Hg, the remaining 95% of the men must have systolic blood pressure readings that lie between 90.2 and 167.8 mm Hg.

We might also be interested in determining the proportion of men in the population who have a systolic blood pressure greater than 150 millimeters of mercury. In this case, we are given the outcome of the random variable X and must solve for the normal deviate z;

$$\begin{aligned} z &= \frac{150 - 129}{19.8} \\ &= 1.06. \end{aligned}$$

We find that the area to the right of $z = 1.06$ is 0.145. Therefore, approximately 14.5% of the men have a systolic blood pressure reading that is greater than 150 mm Hg.

Now consider a situation in which we have two normally distributed random variables. In an Australian national study of risk factor prevalence, two of the populations investigated are men whose blood pressures are within a normal range and who are not taking any corrective medication, and men who have had high blood pressure but who are presently undergoing antihypertensive drug therapy [6].

For the population of men who are not taking corrective medication, diastolic blood pressure is approximately normally distributed with mean $\mu_n =$ 80.7 mm Hg and standard deviation $\sigma_n = 9.2$ mm Hg. For the men who are using antihypertensive drugs, diastolic blood pressure is also approximately normally distributed with mean $\mu_a = 94.9$ mm Hg and standard deviation $\sigma_a = 11.5$ mm Hg. These two distributions are pictured in Figure 7.14. Our goal is to be able to determine whether a man has normal blood pressure or whether he is taking antihypertensive medication solely on the basis of his diastolic blood pressure. This seemingly senseless exercise is valuable in that it provides us with a foundation for hypothesis testing.

If our goal is to identify 90% of the individuals who are currently taking medication, what value of diastolic blood pressure should be designated as the lower cutoff point? Note that this is the equivalent of finding the value of diastolic blood pressure that marks off the lower 10% of the distribution of men currently using antihypertensive drugs. Looking at Table A.1, we find that $z = -1.28$. Therefore, solving for x,

$$-1.28 = \frac{x - 94.9}{11.5}$$

and

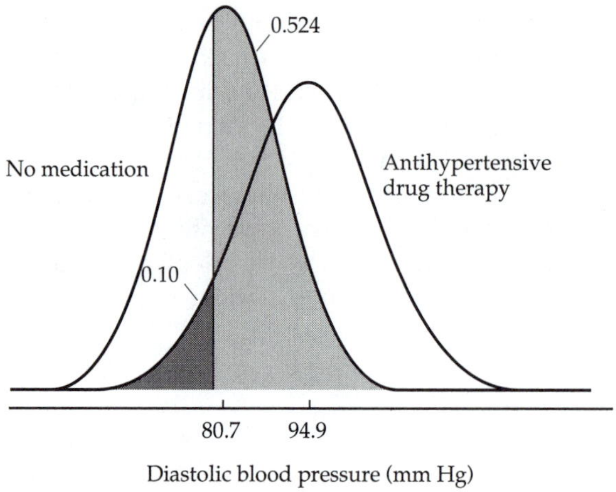

Figure 7.14. Distributions of diastolic blood pressure for two populations of Australian males, 1980

$$
\begin{aligned}
x &= 94.9 + (-1.28)(11.5) \\
&= 80.18.
\end{aligned}
$$

Approximately 90% of the men taking antihypertensive drugs have diastolic blood pressures that are greater than 80.18 mm Hg. If we use this value as our cutoff point, then the other 10% of the men represent false negatives; they are individuals currently using medication but not labeled as such.

What proportion of the men with normal blood pressures will be incorrectly identified as antihypertensive drug users? These are the men in the normal population who have diastolic blood pressure readings greater than 80.18 mm Hg. Solving for z,

$$
\begin{aligned}
z &= \frac{80.18 - 80.7}{9.2} \\
&= -0.06.
\end{aligned}
$$

An area of 0.476 lies to the left of -0.06; therefore, the area to the right of $z = -0.06$ must be

$$1.000 - 0.476 = 0.524.$$

Approximately 52.4% of the men with normal blood pressure would be incorrectly labeled as using medication. Note that these errors are false positives.

To reduce the proportion of false positive errors, the cutoff point for identifying individuals who are currently using antihypertensive drugs could be raised. If the cutoff were 90 mm Hg, for example, then

$$\begin{aligned} z &= \frac{90 - 80.7}{9.2} \\ &= 1.01, \end{aligned}$$

and only 15.6% of the men with normal blood pressures would be classified as taking medication.

When the cutoff is raised, however, the proportion of men correctly labeled as using medication decreases; note that

$$\begin{aligned} z &= \frac{90 - 94.9}{11.5} \\ &= -0.43. \end{aligned}$$

The area to the left of $z = -0.43$ is 0.334, and

$$1.000 - 0.334 = 0.666;$$

therefore, 66.6% of the men using antihypertensive drugs would be identified. The remaining 33.4% of these men would be false negatives.

A tradeoff always exists when we try to manipulate proportions of false negative and false positive outcomes; this is the same phenomenon that we observed when we were investigating the sensitivity and specificity of a diagnostic test. In general, a smaller false positive rate can be achieved only by increasing the false negative error, and the proportion of false negatives can be reduced only by raising the rate of false positives. The relationship between these types of error is determined by the amount of overlap in the two normal populations of interest.

7.5 Further Applications

Suppose that we are interested in investigating the probability that a patient who has been stuck with a needle infected with hepatitis B actually develops the disease. Let Y be a Bernoulli random variable that represents the disease status of a patient who has been exposed to an infected needle; Y takes the value 1 if the individual develops hepatitis and 0 if he or she does not. These two outcomes are mutually exclusive and exhaustive. If 30% of the patients who are exposed to hepatitis B become infected [7], then

$$\begin{aligned} P(Y = 1) &= p \\ &= 0.30, \end{aligned}$$

and

$$\begin{aligned} P(Y = 0) &= 1 - p \\ &= 1 - 0.30 \\ &= 0.70. \end{aligned}$$

If we have n independent observations of a dichotomous random variable such that each observation has a constant probability of "success" p, then the total number of successes X follows a binomial distribution. The random variable X can assume any integer value between 0 and n; the probability that X takes on a particular value x may be expressed as

$$P(X = x) = \binom{n}{x} p^x (1-p)^{n-x}.$$

Now suppose that we select five individuals from the population of patients who have been stuck with a needle infected with hepatitis B. The number of patients in this sample who develop the disease is a binomial random variable with parameters $n = 5$ and $p = 0.30$. Its probability distribution may be represented in the following way:

$$\begin{aligned} P(X = 0) &= \binom{5}{0}(0.30)^0(0.70)^{5-0} \\ &= (1)(1)(0.70)^5 \\ &= 0.168, \end{aligned}$$

$$\begin{aligned} P(X = 1) &= \binom{5}{1}(0.30)^1(0.70)^{5-1} \\ &= (5)(0.30)(0.70)^4 \\ &= 0.360, \end{aligned}$$

$$\begin{aligned} P(X=2) &= \binom{5}{2}(0.30)^2(0.70)^{5-2} \\ &= (10)(0.30)^2(0.70)^3 \\ &= 0.309, \end{aligned}$$

$$\begin{aligned} P(X=3) &= \binom{5}{3}(0.30)^3(0.70)^{5-3} \\ &= (10)(0.30)^3(0.70)^2 \\ &= 0.132, \end{aligned}$$

$$\begin{aligned} P(X=4) &= \binom{5}{4}(0.30)^4(0.70)^{5-4} \\ &= (5)(0.30)^4(0.70) \\ &= 0.028, \end{aligned}$$

and

$$\begin{aligned} P(X=5) &= \binom{5}{5}(0.30)^5(0.70)^{5-5} \\ &= (1)(0.30)^5(1) \\ &= 0.003. \end{aligned}$$

The mean number of persons who develop the disease is $np = 5(0.3) = 1.5$, and the standard deviation is $\sqrt{np(1-p)} = \sqrt{5(0.3)(0.7)} = \sqrt{1.05} = 1.025$. In addition, the probability that at least three individuals among the five develop hepatitis is

$$\begin{aligned} P(X \geq 3) &= P(X=3) + P(X=4) + P(X=5) \\ &= 0.132 + 0.028 + 0.003 \\ &= 0.163; \end{aligned}$$

the probability that at most one patient develops the disease is

$$\begin{aligned} P(X \leq 1) &= P(X=0) + P(X=1) \\ &= 0.168 + 0.360 \\ &= 0.528. \end{aligned}$$

If X represents the number of occurrences of some event in a specified interval of time or space such that both the mean number of occurrences and the population variance are equal to λ, then X has a Poisson distribution with parameter λ. The random variable X can take on any integer value between 0 and ∞; the probability that X assumes a particular value x is

$$P(X = x) = \frac{e^{-\lambda}\lambda^x}{x!}.$$

Suppose that we are concerned with the possible spread of diphtheria and wish to know how many cases we can expect to see in a particular year. Let X represent the number of cases of diphtheria reported in the United States in a given year between 1980 and 1989. The random variable X has a Poisson distribution with parameter $\lambda = 2.5$ [8]; the probability distribution of X may be expressed as

$$P(X = x) = \frac{e^{-2.5}(2.5)^x}{x!}.$$

Therefore, the probability that no cases of diphtheria will be reported during a given year is

$$\begin{aligned} P(X = 0) &= \frac{e^{-2.5}(2.5)^0}{0!} \\ &= 0.0821. \end{aligned}$$

The probability that a single case will be reported is

$$\begin{aligned} P(X = 1) &= \frac{e^{-2.5}(2.5)^1}{1!} \\ &= 0.2052; \end{aligned}$$

similarly,

$$\begin{aligned} P(X = 2) &= \frac{e^{-2.5}(2.5)^2}{2!} \\ &= 0.2565, \end{aligned}$$

$$\begin{aligned} P(X = 3) &= \frac{e^{-2.5}(2.5)^3}{3!} \\ &= 0.2138, \end{aligned}$$

$$\begin{aligned} P(X = 4) &= \frac{e^{-2.5}(2.5)^4}{4!} \\ &= 0.1336, \end{aligned}$$

and

$$
\begin{aligned}
P(X = 5) &= \frac{e^{-2.5}(2.5)^5}{5!} \\
&= 0.0668.
\end{aligned}
$$

Since the outcomes of X are mutually exclusive and exhaustive,

$$
\begin{aligned}
P(X \geq 6) &= 1 - (0.0821 + 0.2052 + 0.2565 + 0.2138 \\
&\qquad + 0.1336 + 0.0668) \\
&= 0.0420.
\end{aligned}
$$

There is only a 4.2% chance that we will observe six or more cases in a given year.

If X is able to assume any real value within a specified interval rather than being restricted to integers only, then X is a continuous random variable. The most common continuous distribution is the normal distribution. The normal distribution is defined by two parameters—its mean μ and standard deviation σ. The mean specifies the center of the distribution; the standard deviation indicates the amount of dispersion about the mean. The shape of the normal distribution suggests that outcomes of the random variable X that are close to the mean are more likely to occur than values that are far from it.

The normal distribution that has mean $\mu = 0$ and standard deviation $\sigma = 1$ is known as the *standard normal distribution*. Because its area has been tabulated, it is used to obtain probabilities associated with normal random variables. For example, suppose that we wish to know the area under the standard normal curve that lies between $z = -3.00$ and $z = 3.00$ or, equivalently, the area in the interval $\mu \pm 3\sigma$. This area is pictured in Figure 7.15. Looking at Table A.1, we find the area to the right of $z = 3.00$ to be 0.001. Since the normal curve is symmetric, the area

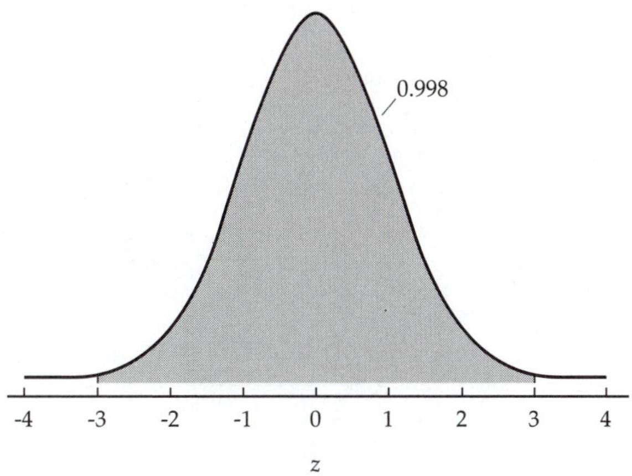

Figure 7.15. The standard normal curve, area between $z = -3.00$ and $z = 3.00$

to the left of $z = -3.00$ must be 0.001 as well. Therefore, the area between -3.00 and 3.00 is

$$1.000 - 0.001 - 0.001 = 0.998;$$

approximately 99.8% of the area under a standard normal curve lies within ± 3 standard deviations from the mean.

If X is an arbitrary normal random variable with mean μ and standard deviation σ, then

$$Z = \frac{X - \mu}{\sigma}$$

is a standard normal random variable. By transforming X into Z, we are able to use a table of areas of the standard normal curve to estimate probabilities associated with X.

For instance, suppose that X is a random variable that represents height. For the population of 18- to 74-year-old females in the United States, height is normally distributed with mean $\mu = 63.9$ inches and standard deviation $\sigma = 2.6$ inches [9]. This distribution is illustrated in Figure 7.16. Observe that

$$Z = \frac{X - 63.9}{2.6}$$

is a standard normal random variable.

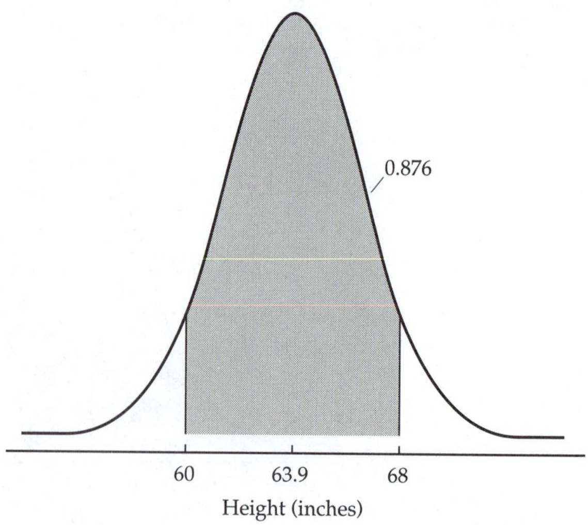

Figure 7.16. Distribution of height for females 18 to 74 years of age, United States, 1976–1980

If we randomly select a woman between the ages of 18 and 74 from this population, what is the probability that she is between 60 and 68 inches tall? For $x = 60$,

$$\begin{aligned} z &= \frac{60 - 63.9}{2.6} \\ &= -1.50, \end{aligned}$$

and, for $x = 68$,

$$\begin{aligned} z &= \frac{68 - 63.9}{2.6} \\ &= 1.58. \end{aligned}$$

As a result, the probability that x—the woman's height—lies between 60 and 68 inches is equal to the probability that z lies between -1.50 and 1.58. The area to the right of $z = 1.58$ is 0.057, and the area to the left of $z = -1.50$ is 0.067. Since the total area under the curve is 1.000, the area between -1.50 and 1.58 must be

$$1.000 - 0.057 - 0.067 = 0.876.$$

The probability that the woman's height is between 60 and 68 inches is 0.876.

We might also wish to know what value of height cuts off the upper 5% of the distribution. We observe that a tail area of 0.050 corresponds to $z = 1.645$. Solving for x,

$$1.645 = \frac{x - 63.9}{2.6}$$

and

$$\begin{aligned} x &= 63.9 + (1.645)(2.6) \\ &= 68.2. \end{aligned}$$

Approximately 5% of the women in this population are taller than 68.2 inches.

7.6 Review Exercises

1. Consider a group of seven individuals selected from the population of 65- to 74-year-olds in the United States. The number of persons in this sample who suffer from diabetes is a binomial random variable with parameters $n = 7$ and $p = 0.125$ [10].

a. If you wish to make a list of the seven persons chosen, in how many ways can they be ordered?
b. Without regard to order, in how many ways can you select four individuals from this group of seven?
c. What is the probability that exactly two of the individuals in the sample suffer from diabetes?
d. What is the probability that four of them have diabetes?

2. According to the National Health Survey, 9.8% of the population of 18- to 24-year-olds in the United States are left-handed [9].

a. Suppose that you select ten individuals from this population. In how many ways can the ten persons be ordered?
b. Without regard to order, in how many ways can you select four individuals from this group of ten?
c. What is the probability that exactly three of the ten persons are left-handed?
d. What is the probability that at least six of the ten are left-handed?
e. What is the probability that at most two individuals are left-handed?

3. According to the Behavioral Risk Factor Surveillance System, 58% of all Americans adhere to a sedentary lifestyle [11]. Suppose that you select a sample of twelve individuals and find that ten of them do not exercise regularly. Assuming that the Surveillance System is correct, what is the chance that you would obtain results as bad as or worse than those you have observed?

4. According to the Massachusetts Department of Health, 224 women who gave birth in the state of Massachusetts in 1988 tested positive for the HIV antibody. Assume that, in time, 25% of the babies born to such mothers will also become HIV positive.

a. If samples of size 224 were repeatedly selected from the population of children born to mothers with the HIV antibody, what would be the mean number of infected children per sample?
b. What would be the standard deviation?
c. Use Chebychev's inequality to describe this distribution.

5. The number of cases of tetanus reported in the United States during a single month in 1989 has a Poisson distribution with parameter $\lambda = 4.5$ [8].

a. What is the probability that exactly one case of tetanus will be reported during a given month?
b. What is the probability that at most two cases of tetanus will be reported?
c. What is the probability that five or more cases will be reported?

6. In a particular county, the average number of suicides reported each month is 2.75 [12]. Assume that the number of suicides follows a Poisson distribution.

a. What is the probability that no suicides will be reported during a given month?

b. What is the probability that at most four suicides will be reported?
c. What is the probability that six or more suicides will be reported?

7. Consider the standard normal distribution with mean $\mu = 0$ and standard deviation $\sigma = 1$.

a. What is the probability that an outcome z is greater than 2.6?
b. What is the probability that z is less than 1.3?
c. What is the probability that z is between -1.7 and 3.1?
d. What value of z cuts off the upper 15% of the standard normal distribution?
e. What value of z marks off the lower 20% of the distribution?

8. Among females in the United States between 18 and 74 years of age, diastolic blood pressure is normally distributed with mean $\mu = 77$ mm Hg and standard deviation $\sigma = 11.6$ mm Hg [5].

a. What is the probability that a randomly selected woman has a diastolic blood pressure less than 60 mm Hg?
b. What is the probability that she has a diastolic blood pressure greater than 90 mm Hg?
c. What is the probability that the woman has a diastolic blood pressure between 60 and 90 mm Hg?

9. The distribution of weights for the population of males in the United States is approximately normal with mean $\mu = 172.2$ pounds and standard deviation $\sigma = 29.8$ pounds [9].

a. What is the probability that a randomly selected man weighs less than 130 pounds?
b. What is the probability that he weighs more than 210 pounds?
c. What is the probability that among five males selected at random from the population, at least one will have a weight outside the range 130 to 210 pounds?

10. In the Framingham Study, serum cholesterol levels were measured for a large number of healthy males. The population was then followed for 16 years. At the end of this time, the men were divided into two groups—those who had developed coronary heart disease and those who had not. The distributions of the initial serum cholesterol levels for each group were found to be approximately normal. Among individuals who eventually developed coronary heart disease, the mean serum cholesterol level was $\mu_d = 244$ mg/100 ml, and the standard deviation was $\sigma_d = 51$ mg/100 ml; for those who did not develop the disease, the mean serum cholesterol level was $\mu_{nd} = 219$ mg/100 ml, and the standard deviation was $\sigma_{nd} = 41$ mg/100 ml [13].

a. Suppose that an initial serum cholesterol level of 260 mg/100 ml or higher is used to predict future coronary heart disease. What is the probability of predicting the disease for a man who will never develop it?

b. What is the probability of failing to predict coronary heart disease for a man who will develop it?
c. What would happen to the probabilities of false positive and false negative errors if the cutoff point for predicting future disease was lowered to 250 mg/100 ml?
d. In this population, does initial serum cholesterol level appear to be useful for predicting future coronary heart disease?

References

[1] National Center for Health Statistics, "Supplements to the Monthly Vital Statistics Reports: Advance Reports, 1986," *Vital and Health Statistics*, Series 24, Number 3, March 1990.

[2] Centers for Disease Control, "The Surgeon General's 1989 Report on Reducing the Health Consequences of Smoking: 25 Years of Progress," *Morbidity and Mortality Weekly Report Supplement*, Volume 38, March 24, 1989.

[3] Ross, S. M., *Introduction to Probability Models*, Orlando, Florida: Academic Press, 1985.

[4] Wilson, R., and Crouch, E. A. C., "Risk Assessment and Comparisons: An Introduction," *Science*, Volume 236, April 17, 1987, 267–270.

[5] National Center for Health Statistics, Drizd, T., Dannenberg, A. L., and Engel, A., "Blood Pressure Levels in Persons 18–74 Years of Age in 1976–1980, and Trends in Blood Pressure From 1960 to 1980 in the United States," *Vital and Health Statistics*, Series 11, Number 234, July 1986.

[6] Castelli, W. P., and Anderson, K., "Antihypertensive Treatment and Plasma Lipoprotein Levels: The Associations in Data from a Population Study," *American Journal of Medicine Supplement*, Volume 80, February 14, 1986, 23–32.

[7] Tye, L., "Many States Tackling Issue of AIDS-Infected Health Care Workers," *The Boston Globe*, May 27, 1991, 29–30.

[8] Centers for Disease Control, "Summary of Notifiable Diseases, United States, 1989," *Morbidity and Mortality Weekly Report*, Volume 39, October 5, 1990.

[9] National Center for Health Statistics, Najjar, M. F., and Rowland, M., "Anthropometric Reference Data and Prevalence of Overweight: United States, 1976–1980," *Vital and Health Statistics*, Series 11, Number 238, October 1987.

[10] Centers for Disease Control, "Regional Variation in Diabetes Mellitus Prevalence—United States, 1988 and 1989," *Morbidity and Mortality Weekly Report*, Volume 39, November 16, 1990.

[11] Centers for Disease Control, "Coronary Heart Disease Attributable to Sedentary Lifestyle—Selected States, 1988," *Morbidity and Mortality Weekly Report*, Volume 39, August 17, 1990.

[12] Gibbons, R. D., Clark, D. C., and Fawcett, J., "A Statistical Method for Evaluating Suicide Clusters and Implementing Cluster Surveillance," *American Journal of Epidemiology Supplement*, Volume 132, July 1990, 183–191.

[13] MacMahon, S. W., and MacDonald, G. J., "A Population at Risk: Prevalence of High Cholesterol Levels in Hypertensive Patients in the Framingham Study," *American Journal of Medicine Supplement*, Volume 80, February 14, 1986, 40–47.

CHAPTER 8

Sampling Distribution of the Mean

In the previous chapter we examined several theoretical probability distributions. In all cases the relevant population parameters were assumed to be known; this allowed us to describe the distributions completely. In most practical applications, however, we are not given the values of these parameters. Instead, we attempt to describe some characteristic of a population using the information contained in a sample of observations. The process of drawing conclusions about an entire population based on such a sample is known as *statistical inference*.

8.1 Sampling Distributions

Suppose that our focus is on estimating the mean value of some continuous random variable of interest. For example, we might wish to make a statement about the mean serum cholesterol level of all men residing in the United States, based on a sample drawn from this population. The obvious approach would be to use the mean of the sample as an estimate of the unknown population mean. The quantity $\overline{X}$ is called an *estimator* of the parameter μ. Philosophically, there are many different approaches to the process of estimation; because the population is assumed to be normally distributed, the sample mean $\overline{X}$ is a *maximum likelihood estimator* [1]. The method of maximum likelihood generates the value of the parameter that is most likely to have produced the observed sample. This method can be relied on to yield reasonable estimators. Keep in mind, however, that two different samples are likely to result in different sample means; consequently, there is some degree of uncertainty involved. Before we apply this procedure, we first investigate some of the properties of the sample mean.

The population under investigation can be any group we choose. In general, we are able to estimate a population mean μ with greater precision when the group is relatively homogeneous. If there is only a small amount of variation among individuals, then we can be more certain that the observations in any given sample are representative of the entire population.

A sample must provide an accurate representation of the population from which it is selected. If it does not, then the conclusions that we draw about the population could be distorted or biased. For instance, if we intend to make a statement about the mean serum cholesterol level of all 20- to 74-year-old males in the United States but sample only men over the age of 60, then our estimate of the population mean is likely to be too high. It is crucial that the sample drawn be *random*; each individual should have an equal chance of being selected. This point is discussed further in Chapter 22. In addition, we would expect that the larger the sample, the more reliable our estimate of the population mean will be.

Consider the distribution of the continuous random variable serum cholesterol level. In a specified population, the mean serum cholesterol level is μ, and the standard deviation is σ. Suppose that we randomly select a sample of n observations from the population and compute the mean of this sample. Call the sample mean $\bar{x}_1$. We then obtain a second random sample of n observations and calculate the mean of the new sample. Label this second sample mean $\bar{x}_2$. Unless everyone in the population has the same serum cholesterol level, it is very unlikely that $\bar{x}_1$ will equal $\bar{x}_2$. If we were to continue this procedure indefinitely—selecting all possible samples of size n and computing their means—we would end up with a set of values consisting entirely of sample means.

If each mean in the series is treated as a unique observation, their collective distribution is known as a *sampling distribution* of means of samples of size n. For example, if we were to select repeated samples of size 25 from the population of males residing in the United States and calculate the mean serum cholesterol level for each sample, we would end up with the sampling distribution of mean serum cholesterol levels of samples of size 25. In practice it is not common to select repeated samples of size n from a population; understanding the properties of the theoretical distribution of their means, however, allows us to make inference based upon a **single** sample of size n.

8.2 The Central Limit Theorem

Given that the distribution of serum cholesterol levels in the underlying population has mean μ and standard deviation σ, the distribution of sample means computed for samples of size n has three important properties:

1. The mean of the sampling distribution is identical to the population mean μ.
2. The standard deviation of the distribution of sample means is equal to $\sigma/\sqrt{n}$. This quantity is known as the *standard error* of the mean.
3. Provided that n is large enough, the shape of the sampling distribution is approximately normal.

Intuitively, we would expect the means of our samples to cluster around the mean of the population. Although the standard deviation of the sampling distribution is related to the population standard deviation σ, there is less variability among the sample means than there is among the individual observations. Even if a particular sample contains one or two extreme values, it is likely that they would be offset by the other measurements in the group. In addition, as n increases, the amount of sampling variation decreases. Finally, if n is large enough, the distribution of means is approximately normal. This remarkable result is known as the *central limit theorem;* it applies to any population with a finite standard deviation regardless of the shape of the underlying distribution [2]. The farther the underlying population departs from being normally distributed, however, the larger the value of n necessary to ensure normality of the sampling distribution. If the underlying population is itself normal, then samples of size 1 are large enough.

The central limit theorem is very powerful. It holds true not only for serum cholesterol levels but for almost any other type of measurement as well. The central limit theorem allows us to quantify the uncertainty inherent in statistical inference without having to make a great many assumptions that cannot be verified. Because the distribution of the sample means is normal with mean μ and standard deviation $\sigma/\sqrt{n}$, we know that

$$Z = \frac{\overline{X} - \mu}{\sigma/\sqrt{n}}$$

is normally distributed with mean 0 and standard deviation 1. As a result, we can use tables of the standard normal distribution—such as Table A.1 in Appendix A—to make inference about the value of a population mean.

8.3 Applications

Consider the distribution of serum cholesterol levels for all 20- to 74-year-old males living in the United States. The mean of this population is $\mu = 211$ mg/100 ml, and the standard deviation is $\sigma = 46$ mg/100 ml [3]. If we select repeated samples of size 25 from the population, what proportion of the samples will have a mean value of 230 mg/100 ml or above? The central limit theorem states that the distribution of means of samples of size 25 is approximately normal with mean $\mu = 211$ mg/100 ml and standard deviation $\sigma/\sqrt{n} = 46/\sqrt{25} = 9.2$ mg/100 ml. This sampling distribution and the underlying population distribution are illustrated in Figure 8.1. Note that

$$Z = \frac{\overline{X} - 211}{9.2}$$

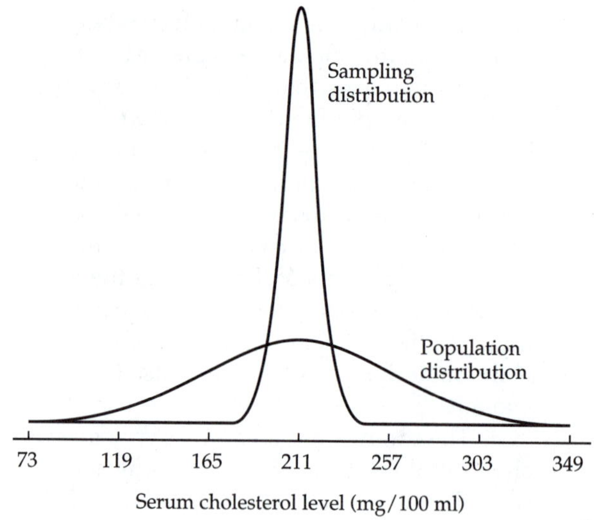

Figure 8.1. Distributions of individual values and means of samples of size 25 for the serum cholesterol levels of 20- to 74-year-old males, United States, 1976–1980

is a standard normal random variable. If $\bar{x} = 230$, then

$$\begin{aligned} z &= \frac{230 - 211}{9.2} \\ &= 2.07. \end{aligned}$$

Referring to Table A.1, we see that the area to the right of $z = 2.07$ is 0.019. Only about 1.9% of the samples will have a mean greater than 230 mg/100 ml.

What mean value of serum cholesterol level cuts off the lower 10% of the sampling distribution? Locating 0.100 in the body of Table A.1, we see that it corresponds to the value $z = -1.28$. Solving for $\bar{x}$,

$$-1.28 = \frac{\bar{x} - 211}{9.2}$$

and

$$\begin{aligned} \bar{x} &= 211 + (-1.28)(9.2) \\ &= 199.2. \end{aligned}$$

Approximately 10% of the samples of size 25 have means that are less than or equal to 199.2 mg/100 ml.

Let us now calculate the upper and lower limits that enclose 95% of the means of samples of size 25 drawn from the population. Since 2.5% of the area under the standard normal curve lies above $z = 1.96$ and another 2.5% lies below $z = -1.96$,

$$P(-1.96 \leq Z \leq 1.96) = 0.95.$$

As a result, we are interested in outcomes z of Z for which

$$-1.96 \leq z \leq 1.96.$$

We would like to transform this into a statement about $\bar{x}$. Substituting $(\bar{x}-211)/9.2$ for z,

$$-1.96 \leq \frac{\bar{x}-211}{9.2} \leq 1.96.$$

Multiplying all three terms of the inequality by 9.2 and adding 211 results in

$$211 - 1.96(9.2) \leq \bar{x} \leq 211 + 1.96(9.2)$$

or

$$193.0 \leq \bar{x} \leq 229.0.$$

Approximately 95% of the means of samples of size 25 lie between 193.0 and 229.0 mg/100 ml. Consequently, if we select a random sample of size 25 that is presumed to be from the population of serum cholesterol levels for all 20- to 74-year-old males, and the sample has a mean that is either greater than 229.0 or less than 193.0 mg/100 ml, we should be suspicious of this claim. Either the random sample was drawn from a different population, or a rare event has taken place. Under these circumstances, a "rare event" is defined as an outcome that occurs less than 5% of the time.

If we had selected samples of size 10 from the population rather than samples of size 25, we would have constructed the inequality

$$-1.96 \leq \frac{\bar{x}-211}{46/\sqrt{10}} \leq 1.96$$

and would have found the upper and lower limits that enclose 95% of the means to be

$$182.5 \leq \bar{x} \leq 239.5.$$

Note that this interval is wider than the one calculated for samples of size 25. Drawing samples of size 50 would result in upper and lower limits

$$198.2 \leq \bar{x} \leq 223.8;$$

not surprisingly, this interval is more narrow than the one constructed for samples of size 25. Samples of size 100 produce the limits

$$202.0 \leq \bar{x} \leq 220.0.$$

In summary, if we include the case for which $n = 1$, we have the following results:

n	$\sigma/\sqrt{n}$	**Interval Enclosing 95% of the Means**	**Length of Interval**
1	46.0	$120.8 \leq \bar{x} \leq 301.2$	180.4
10	14.5	$182.5 \leq \bar{x} \leq 239.5$	57.0
25	9.2	$193.0 \leq \bar{x} \leq 229.0$	36.0
50	6.5	$198.2 \leq \bar{x} \leq 223.8$	25.6
100	4.6	$202.0 \leq \bar{x} \leq 220.0$	18.0

As the sizes of the samples increase, the amount of sampling variability among the means decreases; consequently, the limits encompassing 95% of these means move closer together.

Note that all the intervals that we have constructed have been symmetric about the population mean 211 mg/100 ml. Clearly, there are other intervals that would also capture the appropriate proportion of the sample means. Suppose that we wish to construct an interval that contains 95% of the means of samples of size 25. Since 1% of the area under the standard normal curve lies above $z = 2.32$ and 4% lies below $z = -1.75$,

$$P(-1.75 \leq Z \leq 2.32) = 0.95.$$

In this case, we are interested in the outcomes z for which

$$-1.75 \leq z \leq 2.32.$$

Again substituting $(\bar{x} - 211)/9.2$ for z, we would find the interval to be

$$194.9 \leq \bar{x} \leq 232.3.$$

Therefore, we are able to say that approximately 95% of the means of samples of size 25 lie between 194.9 and 232.3 mg/100 ml. It is usually preferable to construct a symmetric interval primarily because it is the shortest interval that captures the appropriate proportion of the means. In this case, the asymmetrical interval has length $232.3 - 194.9 = 37.4$ mg/100 ml; the length of the symmetric interval is $229.0 - 193.0 = 36.0$ mg/100 ml.

Now we move on to a slightly more complicated question. How large would the samples need to be so that 95% of their means lie within ±5 mg/100 ml of the population mean? To answer this, we must find the sample size n for which

$$P(\mu - 5 \leq \overline{X} \leq \mu + 5) = 0.95$$

or

$$P(-5 \leq \overline{X} - \mu \leq 5) = 0.95.$$

To begin, we divide all three terms of the inequality by $\sigma/\sqrt{n} = 46/\sqrt{n}$; this results in

$$P\left(\frac{-5}{46/\sqrt{n}} \leq \frac{\overline{X} - \mu}{46/\sqrt{n}} \leq \frac{5}{46/\sqrt{n}}\right) = 0.95.$$

Since Z is equal to $(\overline{X} - \mu)/(46/\sqrt{n})$,

$$P\left(\frac{-5}{46/\sqrt{n}} \leq Z \leq \frac{5}{46/\sqrt{n}}\right) = 0.95.$$

Recall that 95% of the area under the standard normal curve lies between $z = -1.96$ and $z = 1.96$. Therefore, to find the sample size n we could use the upper bound of the interval and solve the equation

$$1.96 = \frac{5}{46/\sqrt{n}};$$

equivalently, we could use the lower bound and solve

$$-1.96 = \frac{-5}{46/\sqrt{n}}.$$

Taking

$$1.96 = \frac{5\sqrt{n}}{46}$$

and multiplying both sides of the equation by 46/5, we find that

$$\sqrt{n} = \frac{1.96(46)}{5}$$

and

$$\begin{aligned} n &= \left[\frac{1.96(46)}{5}\right]^2 \\ &= 325.2. \end{aligned}$$

When dealing with sample sizes it is conventional to round up. Therefore, samples of size 326 would be required for 95% of the sample means to lie within ±5 mg/100 ml of the population mean. Equivalently, if we select a sample of size 326 from the population and calculate its mean, the probability that the sample mean is within ±5 mg/100 ml of the true population mean μ is 0.95.

Up to this point we have focused on two-sided intervals; we have found the upper and lower limits that enclose a specified proportion of sample means. In some situations, however, we are interested in a one-sided interval instead. For example, we might wish to find the upper bound for 95% of the mean serum cholesterol levels of samples of size 25. Since 95% of the area under the standard normal curve lies below $z = 1.645$,

$$P(Z \leq 1.645) = 0.95.$$

As a result, we are interested in outcomes z for which

$$z \leq 1.645.$$

Substituting $(\bar{x} - 211)/9.2$ for z produces

$$\frac{\bar{x} - 211}{9.2} \leq 1.645$$

or

$$\bar{x} \leq 226.1.$$

Therefore, approximately 95% of the means of samples of size 25 lie below 226.1 mg/100 ml.

If we are interested in constructing a lower bound for 95% of the mean serum cholesterol levels, we focus on values of z that lie above -1.645; in this case, we find that

$$\bar{x} \geq 195.9.$$

Approximately 95% of the means of samples of size 25 lie above 195.9 mg/100 ml.

We must be cautious when we make multiple statements about the sampling distribution. For samples of serum cholesterol levels of size 25, we found that the probability is 0.95 that a sample mean lies within the interval

$$(193.0\,,\ 229.0).$$

We also said that the probability is 0.95 that the mean lies below 226.1 mg/100 ml and 0.95 that it is above 195.9 mg/100 ml. Although these three statements are correct individually, they are not true simultaneously. We must keep in mind that the three events are not independent. For them to occur at the same time, the sample mean would have to lie in the interval

$$(195.9\,,\ 226.1),$$

and the probability that this happens is not 0.95.

8.4 Further Applications

Consider the distribution of age at the time of death for the U. S. population in 1979–1981. This distribution is pictured in Figure 8.2; it has mean $\mu = 73.9$ years and standard deviation $\sigma = 18.1$ years [4]. What happens when we sample from this population of ages?

Rather than draw samples from the population physically, we can use the computer to perform a simulation. In a *simulation*, a computer program is used to model an experiment or procedure according to a specified probability distribution; in our example, the procedure would consist of selecting an individual observation from the distribution pictured in Figure 8.2. The computer is then instructed to repeat the process a given number of times while keeping track of the results.

To illustrate this technique, we can use the computer to simulate the selection of four random samples of size 25 from the population of ages at time of death. Histograms of these samples are shown in Figure 8.3; their means and standard deviations are as follows:

Sample of Size 25	$\bar{x}$	s
1	71.3	18.1
2	69.2	25.6
3	74.0	14.0
4	76.8	15.0

Note that the four random samples are not identical. Each time that we select a set of 25 measurements from the population, the observations included in the

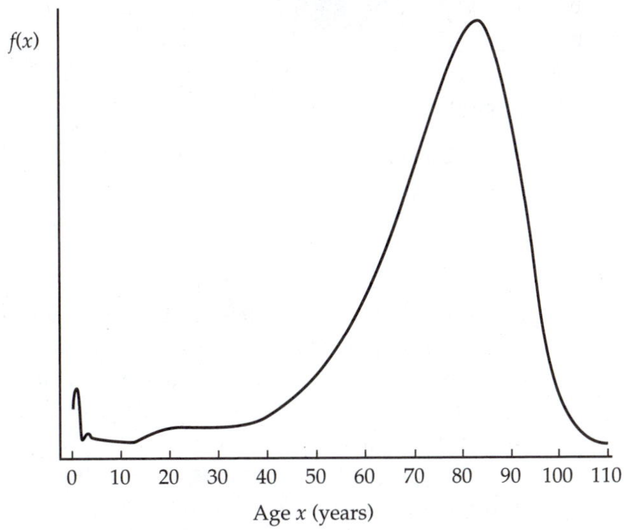

Figure 8.2. Distribution of age at time of death, United States, 1979–1981

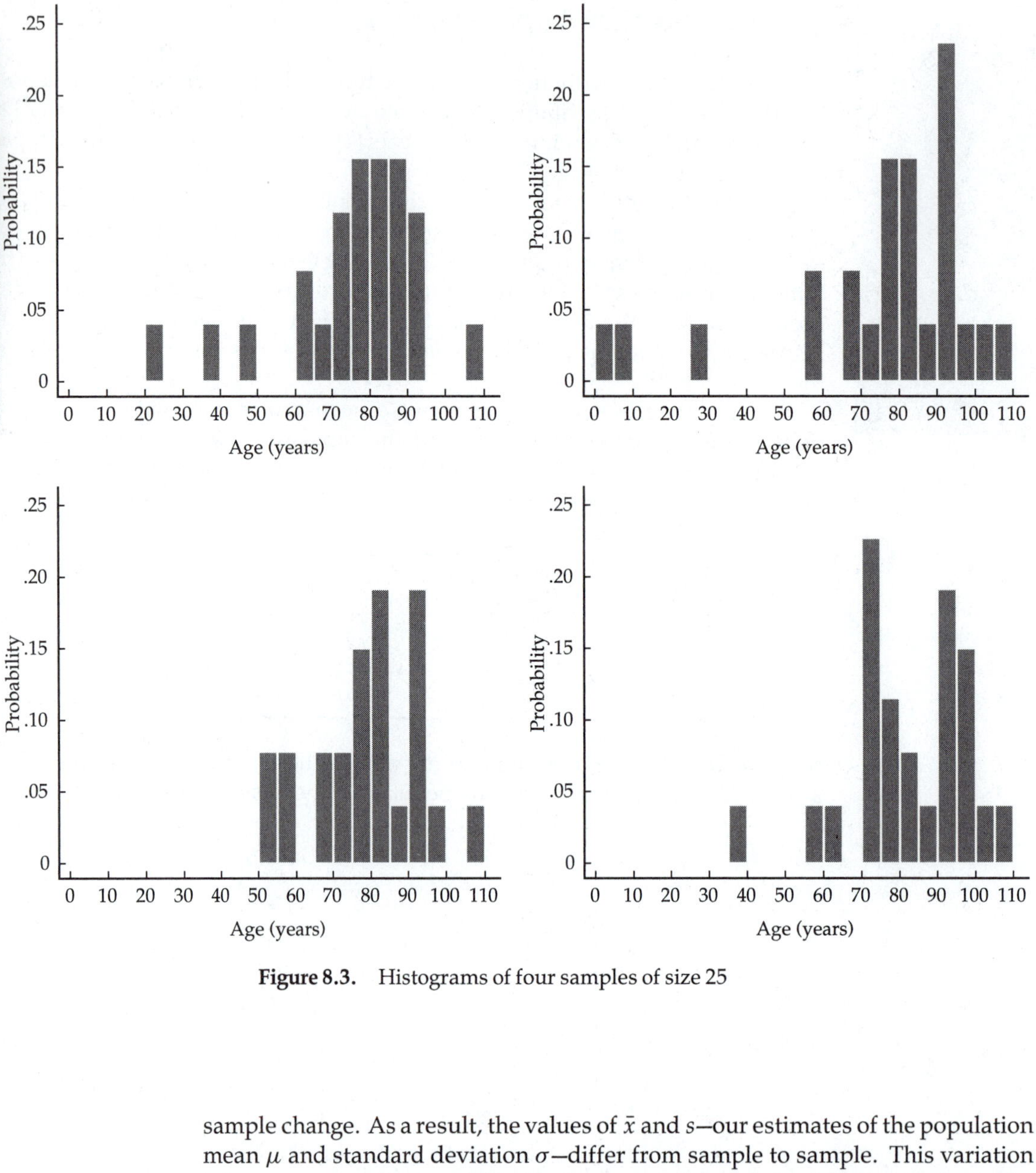

Figure 8.3. Histograms of four samples of size 25

sample change. As a result, the values of $\bar{x}$ and s—our estimates of the population mean μ and standard deviation σ—differ from sample to sample. This variation is known as *sampling variability*. In the four samples of size 25 selected above, the estimates of μ range from 69.2 years to 76.8 years. Similarly, the estimates of σ range from 14.0 to 25.6 years.

Suppose that now, instead of selecting samples of size 25, we choose four random samples of size 100 from the population of ages at the time of death. Histograms of these samples are displayed in Figure 8.4, and their means and standard deviations are provided on the next page.

Sample of Size 100	$\bar{x}$	s
1	75.4	16.5
2	75.0	19.9
3	73.5	18.1
4	72.1	20.2

For these samples, estimates of μ range from 72.1 to 75.4 years and estimates of σ from 16.5 to 20.2 years; these ranges are smaller than the corresponding intervals for samples of size 25.

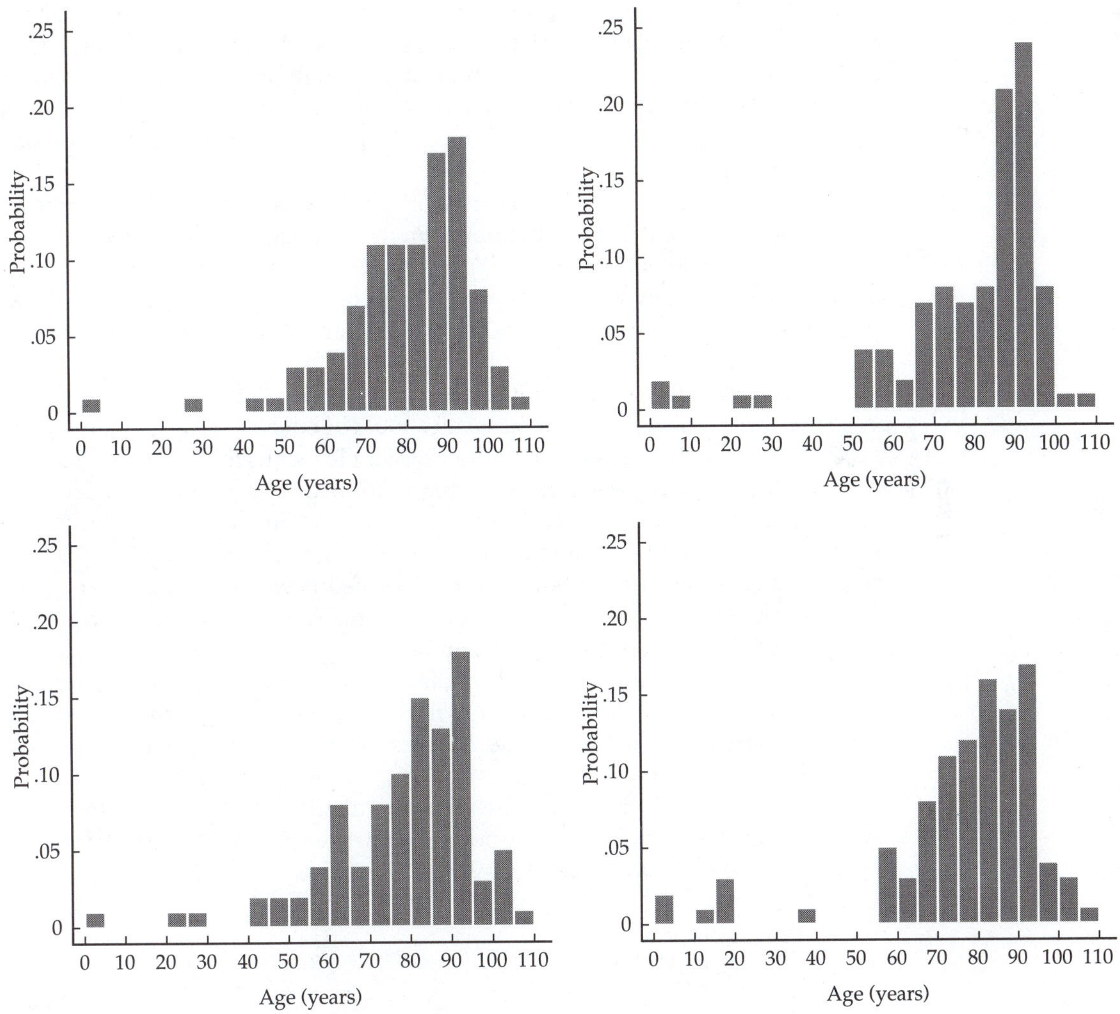

Figure 8.4. Histograms of four samples of size 100

We next select four random samples of size 500 from the population of ages; histograms appear in Figure 8.5, and the means and standard deviations are listed below.

Sample of Size 500	$\bar{x}$	s
1	74.3	17.1
2	73.4	18.1
3	73.5	18.6
4	74.2	17.8

Again, the ranges of estimates for μ and σ decrease.

As the sizes of the samples increase, their distributions approach the shape of the population distribution pictured in Figure 8.2. Although there is still fluctuation among the samples, the amount of variability in the estimates $\bar{x}$ and s decreases. This property is known as *consistency*; as the samples that we select become larger and larger, the estimates of the population parameters approach their target values.

The population of ages at time of death can also be used to demonstrate an application of the central limit theorem. First, we must select repeated samples of size n from the population and examine the distribution of the means of these samples. Theoretically we must enumerate all possible samples; for now, however, we select 100 samples of size 25. A histogram of the 100 sample means is displayed in Figure 8.6.

According to the central limit theorem, the distribution of the sample means possesses three properties. First, its mean should be equal to the population mean $\mu = 73.9$ years. In fact, the mean of the 100 sample means is 74.1 years. Second, we expect the standard deviation of the distribution to be $\sigma/\sqrt{n} = 18.1/\sqrt{25} = 3.6$ years. In reality, the standard deviation is 3.7 years. Finally, the distribution of sample means should be approximately normal. The shape of Figure 8.6—and the theoretical normal distribution superimposed over the histogram—suggest that this third property holds true. Note that this is a large departure from the population distribution illustrated in Figure 8.2.

Now refer back to the population distribution of ages at the time of death. Among samples of size 100 that are drawn from this population, what proportion have a mean that lies between 70 and 78 years? To answer this question, we must find $P(70 \leq \overline{X} \leq 78)$.

The central limit theorem states that the distribution of sample means of samples of size 100 is approximately normal with mean μ=73.9 years and standard deviation $\sigma/\sqrt{n} = 18.1/\sqrt{100} = 1.81$ years. Therefore,

$$
\begin{aligned}
Z &= \frac{\overline{X} - \mu}{\sigma/\sqrt{n}} \\
&= \frac{\overline{X} - 73.9}{1.81}
\end{aligned}
$$

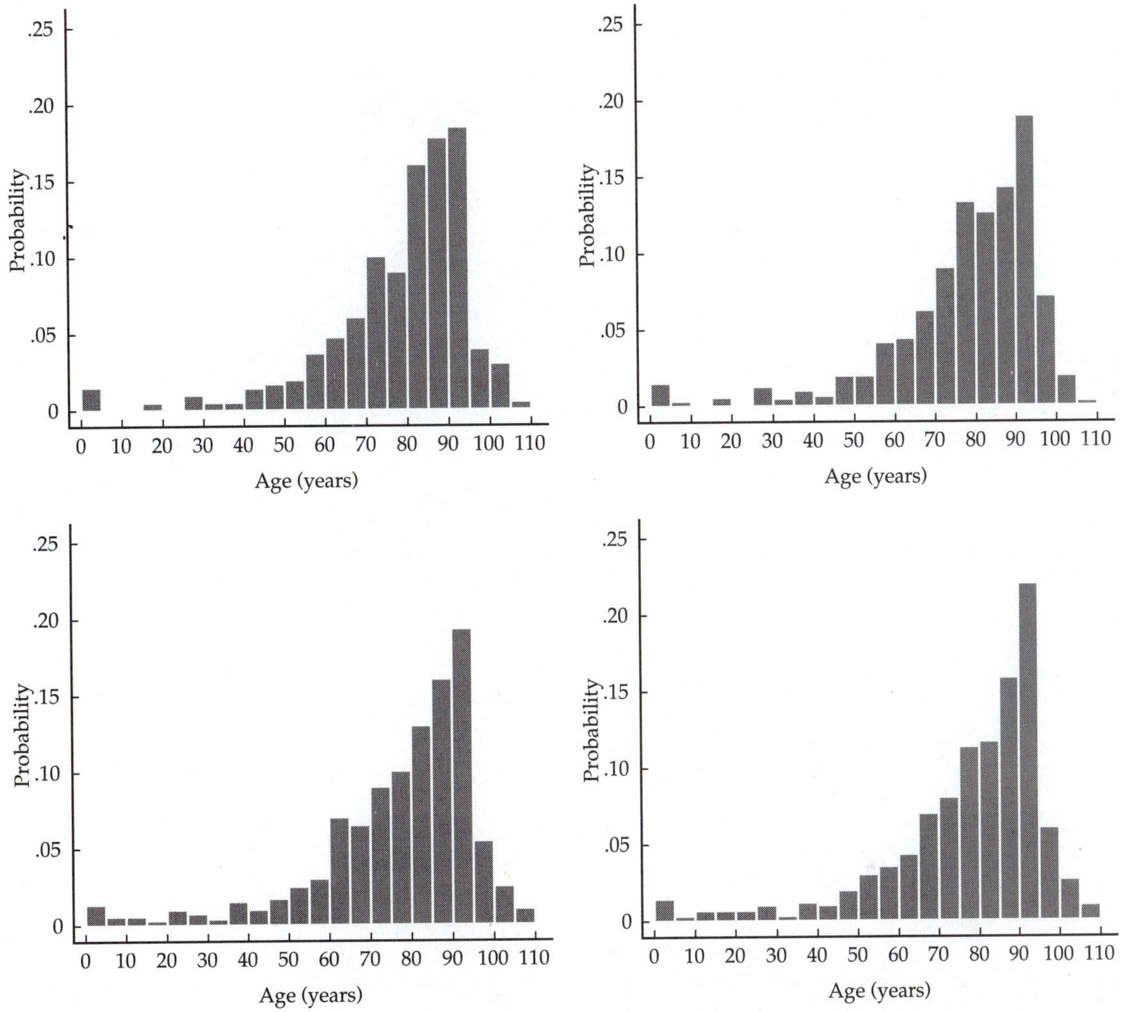

Figure 8.5. Histograms of four samples of size 500

is a standard normal random variable. If we represent the inequality in the expression

$$P(70 \leq \overline{X} \leq 78)$$

in terms of Z rather than $\overline{X}$, we can use Table A.1 to find the proportion of samples that have a mean value in this range.

We begin by subtracting 73.9 from each term in the inequality and dividing by 1.81; thus, we can express

$$P(70 \leq \overline{X} \leq 78)$$

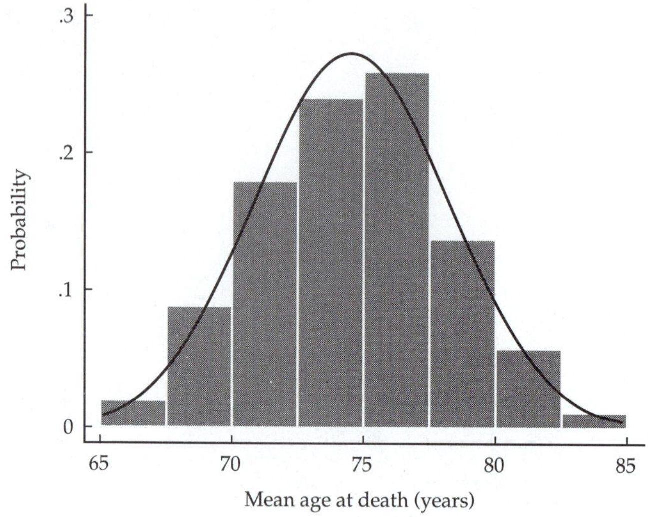

Figure 8.6. Histogram of 100 sample means from samples of size 25

as

$$P\left(\frac{70 - 73.9}{1.81} \leq \frac{\overline{X} - 73.9}{1.81} \leq \frac{78 - 73.9}{1.81}\right)$$

or

$$P(-2.15 \leq Z \leq 2.27).$$

We know that the total area under the standard normal curve is equal to 1.000. According to Table A.1, the area to the right of $z = 2.27$ is 0.012, and the area to the left of $z = -2.15$ is 0.016. Therefore,

$$\begin{aligned} P(-2.15 \leq Z \leq 2.27) &= 1.000 - 0.012 - 0.016 \\ &= 0.972. \end{aligned}$$

Approximately 97.2% of the samples of size 100 have a mean that lies between 70 and 78 years.

What are the upper and lower limits that enclose 80% of the means of samples of size 100? Referring to Table A.1, we find that 10% of the area under a standard normal curve lies above $z = 1.28$, and another 10% lies below $z = -1.28$. Since 80% of the area lies between -1.28 and 1.28, we are interested in values of z for which

$$-1.28 \leq z \leq 1.28$$

and values of $\bar{x}$ for which

$$-1.28 \leq \frac{\bar{x} - 73.9}{1.81} \leq 1.28.$$

Multiplying all three terms of the inequality by 1.81 and adding 73.9 results in

$$73.9 + (-1.28)(1.81) \leq \bar{x} \leq 73.9 + (1.28)(1.81)$$

or, equivalently,

$$71.6 \leq \bar{x} \leq 76.2.$$

Therefore, 80% of the means of samples of size 100 lie between 71.6 and 76.2 years.

8.5 Review Exercises

1. Consider the standard normal distribution with mean $\mu = 0$ and standard deviation $\sigma = 1$.

   a. What can you say about the distribution of means of samples of size 10 that are drawn from this population? List three properties.
   b. Construct an interval that encloses 95% of the means of samples of size 10.
   c. What proportion of the means lie above $z = 1.60$?
   d. What proportion of the means lie below $z = -0.85$?
   e. What value of z cuts off the lower 20% of the distribution of means?

2. Among adults in the United States, the distribution of albumin levels in cerebrospinal fluid has mean $\mu = 29.5$ mg/100 ml and standard deviation $\sigma = 9.25$ mg/100 ml [5].

   a. What proportion of the means of samples of size 20 lie above 35 mg/100 ml?
   b. What proportion of the means lie below 22 mg/100 ml?
   c. What value of $\bar{x}$ cuts off the upper 5% of the distribution of means of samples of size 20?

3. For the population of females between the ages of 3 and 74 who participated in the National Health Interview Survey, the distribution of hemoglobin levels has mean $\mu = 13.3$ g/100 ml and standard deviation $\sigma = 1.12$ g/100 ml [6].

   a. If repeated samples of size 15 are selected from this population, what proportion of the samples will have a mean hemoglobin level that is between 13.0 and 13.6 g/100 ml?

b. If the samples are of size 30, what proportion will have a mean between 13.0 and 13.6 g/100 ml?
c. How large must the samples be so that 95% of their means lie within ± 0.1 g/100 ml of the population mean?

4. In Denver, Colorado, the distribution of daily measures of ambient nitric acid is skewed to the right; it has mean $\mu = 1.81\ \mu g/m^3$ and standard deviation $\sigma = 2.25\ \mu g/m^3$ [7]. Describe the distribution of means of samples of size 40.

5. In the Netherlands, healthy males between the ages of 65 and 79 have a distribution of serum uric acid levels that is approximately normal with mean $\mu = 341\ \mu mol/l$ and standard deviation $\sigma = 79\ \mu mol/l$ [8].

a. What proportion of the males have a serum uric acid level between 300 and 400 μmol/l?
b. What proportion of samples of size 5 have a mean serum uric acid level between 300 and 400 μmol/l?

References

[1] Lindgren, B. W., *Statistical Theory*, New York: Macmillan, 1976.

[2] Snedecor, G. W., and Cochran, W. G., *Statistical Methods*, Ames, Iowa: Iowa State University Press, 1980.

[3] National Center for Health Statistics, Fulwood, R., Kalsbeek, W., Rifkind, B., Russell-Briefel, R., Muesing, R., LaRosa, J., and Lippel, K., "Total Serum Cholesterol Levels of Adults 20–74 Years of Age: United States, 1976–1980," *Vital and Health Statistics*, Series 11, Number 236, May 1986.

[4] National Center for Health Statistics, *United States Decennial Life Tables for 1979–1981*, Volume I, Number 1, August 1985.

[5] Scully, R. E., McNeely, B. U., and Mark, E. J., "Case Record of the Massachusetts General Hospital: Weekly Clinicopathological Exercises," *The New England Journal of Medicine*, Volume 314, January 2, 1986, 39–49.

[6] National Center for Health Statistics, Fulwood, R., Johnson, C. L., Bryner, J. D., Gunter, E. W., and McGrath, C. R., "Hematological and Nutritional Biochemistry Reference Data for Persons 6 Months – 74 Years of Age: United States, 1976–1980," *Vital and Health Statistics*, Series 11, Number 232, December 1982.

[7] Ostro, B. D., Lipsett, M. J., Wiener, M. B., and Selner, J. C., "Asthmatic Responses to Airborne Acid Aerosols," *American Journal of Public Health*, Volume 81, June 1991, 694–702.

[8] Loenen, H. M. J. A., Eshuis, H., Lowik, M. R. H., Schouten, E. G., Hulshof, K. F. A. M., Odink, J., and Kok, F. J., "Serum Uric Acid Correlates in Elderly Men and Women with Special Reference to Body Composition and Dietary Intake (Dutch Nutrition Surveillance System)," *Journal of Clinical Epidemiology*, Volume 43, Number 12, 1990, 1297–1303.

CHAPTER 9

Confidence Intervals

In the previous chapter, we investigated the properties of a distribution of sample means. In all cases, the mean and the standard deviation of the underlying population were assumed to be known. We now apply these results to the situation in which the population mean is not known; our goal is to estimate this parameter using the observations contained in a single sample.

Two methods of estimation are commonly used. The first is called *point estimation*; it involves calculating a single number to estimate the parameter of interest. For instance, we might use the sample mean $\bar{x}$ to estimate the population mean μ. A point estimate, however, does not provide any information about the inherent variability of the estimator; we do not know how close $\bar{x}$ is to μ. As a result, a second method known as *interval estimation* is often preferred. This technique provides a range of reasonable values that are intended to contain the parameter of interest—the population mean μ, in this case—with a certain degree of confidence. This range of values is called a *confidence interval*.

9.1 Two-Sided Confidence Intervals

To construct a confidence interval for μ, we draw on our knowledge of the sampling distribution of the mean. Given a random variable X that has mean μ and standard deviation σ, the central limit theorem states that

$$Z = \frac{\overline{X} - \mu}{\sigma/\sqrt{n}}$$

is a standard normal random variable if n is sufficiently large. For a standard normal distribution, 95% of the observations lie between −1.96 and 1.96. In other words, the probability that Z assumes a value between −1.96 and 1.96 is

$$P(-1.96 \leq Z \leq 1.96) = 0.95.$$

Equivalently, we could substitute the quantity $(\overline{X} - \mu)/(\sigma/\sqrt{n})$ for Z and write

$$P\left(-1.96 \leq \frac{\overline{X} - \mu}{\sigma/\sqrt{n}} \leq 1.96\right) = 0.95.$$

We are able to manipulate the inequality inside the parentheses without altering the probability statement. We begin by multiplying all three terms of the inequality by the standard error $\sigma/\sqrt{n}$; therefore,

$$P\left(-1.96\frac{\sigma}{\sqrt{n}} \leq \overline{X} - \mu \leq 1.96\frac{\sigma}{\sqrt{n}}\right) = 0.95.$$

We next subtract $\overline{X}$ from each term so that

$$P\left(-1.96\frac{\sigma}{\sqrt{n}} - \overline{X} \leq -\mu \leq 1.96\frac{\sigma}{\sqrt{n}} - \overline{X}\right) = 0.95.$$

Finally, we multiply through by −1. Bear in mind that multiplying an inequality by a negative number reverses the direction of the inequality. Consequently,

$$P\left(1.96\frac{\sigma}{\sqrt{n}} + \overline{X} \geq \mu \geq -1.96\frac{\sigma}{\sqrt{n}} + \overline{X}\right) = 0.95$$

and, rearranging the terms,

$$P\left(\overline{X} - 1.96\frac{\sigma}{\sqrt{n}} \leq \mu \leq \overline{X} + 1.96\frac{\sigma}{\sqrt{n}}\right) = 0.95.$$

The quantities $\overline{X} - 1.96(\sigma/\sqrt{n})$ and $\overline{X} + 1.96(\sigma/\sqrt{n})$ are 95% confidence limits for the population mean μ; we are 95% confident that the interval

$$\left(\overline{X} - 1.96\frac{\sigma}{\sqrt{n}}, \overline{X} + 1.96\frac{\sigma}{\sqrt{n}}\right)$$

will cover μ. This statement does not imply that μ is a random variable that assumes a value within the interval 95% of the time, nor that 95% of the population values lie between these limits; rather, it means that if we were to select 100 random samples from the population and use these samples to calculate 100 different confidence intervals for μ, approximately 95 of the intervals would cover the true population mean and 5 would not.

Keep in mind that $\overline{X}$ is a random variable, whereas the parameter μ is a constant. Therefore, the interval

$$\left(\overline{X} - 1.96\frac{\sigma}{\sqrt{n}}, \overline{X} + 1.96\frac{\sigma}{\sqrt{n}}\right)$$

has a 95% chance of covering μ **before** a sample is selected. Since μ has a fixed value, once a sample has been drawn and the confidence limits

$$\left(\bar{x} - 1.96\frac{\sigma}{\sqrt{n}}, \bar{x} + 1.96\frac{\sigma}{\sqrt{n}}\right)$$

have been calculated, μ is either within the interval or it is not.

Although a 95% confidence interval is used most often in practice, we are not restricted to this choice. We might prefer to have a greater degree of certainty regarding the value of the population mean; in this case, we could choose to construct a 99% confidence interval instead. Since 99% of the observations in a standard normal distribution lie between −2.58 and 2.58, a 99% confidence interval for μ is

$$\left(\overline{X} - 2.58\frac{\sigma}{\sqrt{n}}, \overline{X} + 2.58\frac{\sigma}{\sqrt{n}}\right).$$

Approximately 99 out of 100 confidence intervals obtained from 100 independent random samples would cover the true population mean μ. As we would expect, the 99% confidence interval is wider than the 95% interval; the smaller the range of values we consider, the less confident we are that the interval covers μ.

If we want to make an interval tighter without reducing the level of confidence, we need more information about the population mean μ; therefore, we must select a larger sample. As the sample size n increases, the standard error $\sigma/\sqrt{n}$ decreases; this results in a more narrow confidence interval. Consider the 95% confidence limits $\bar{x} \pm 1.96(\sigma/\sqrt{n})$. If we choose a sample of size 10, the confidence limits are $\bar{x} \pm 1.96(\sigma/\sqrt{10})$. If the selected sample is of size 100, then the limits are $\bar{x} \pm 1.96(\sigma/\sqrt{100})$. For an even larger sample of size 1,000, the confidence limits would be $\bar{x} \pm 1.96(\sigma/\sqrt{1,000})$. Summarizing these calculations, we have:

n	95% Confidence Limits for μ	Length of Interval
10	$\bar{x} \pm 0.620\,\sigma$	$1.240\,\sigma$
100	$\bar{x} \pm 0.196\,\sigma$	$0.392\,\sigma$
1,000	$\bar{x} \pm 0.062\,\sigma$	$0.124\,\sigma$

As we select larger and larger random samples, the variability of $\overline{X}$—our estimator of the population mean μ—becomes smaller. The inherent variability of the underlying population, measured by σ, is always present.

Consider the distribution of serum cholesterol levels for all males in the United States who are hypertensive and who smoke. This distribution has an unknown mean μ and a standard deviation of 46 mg/100 ml. Before we select a sample, the probability that the interval

$$\left(\overline{X} - 1.96\frac{46}{\sqrt{n}}\,,\ \overline{X} + 1.96\frac{46}{\sqrt{n}}\right)$$

covers the true population mean μ is 0.95.

Suppose that we draw a random sample of size 12 from the population of hypertensive smokers and that these men have a mean serum cholesterol level of $\bar{x} = 217$ mg/100 ml [1]. Based on this sample, a 95% confidence interval for the population mean μ is

$$\left(217 - 1.96\frac{46}{\sqrt{12}}\,,\ 217 + 1.96\frac{46}{\sqrt{12}}\right)$$

or

$$(191\,,\ 243).$$

We are 95% confident that these limits cover the true mean serum cholesterol level of the population of male hypertensive smokers. We do **not** say that there is a 95% probability that μ lies between 191 and 243 mg/100 ml; the value of μ is fixed and it is either between 191 and 243 or it is not.

As we previously mentioned, this confidence interval also has a frequency interpretation. Suppose that the true mean serum cholesterol level of the population of male hypertensive smokers is equal to 211 mg/100 ml, the mean level of all 20- to 74-year-old males in the United States [2]. If we were to draw 100 random samples of size 12 from this population and use each one to construct a 95% confidence interval, we would expect that, on average, 95 of the intervals cover the true population mean μ and 5 do not. This procedure was simulated and the results illustrated in Figure 9.1. Although the locations of the intervals differ, they all have the same length. The confidence limits that do not contain the true value of μ are marked by a dot; note that exactly five intervals fall into this category.

Instead of calculating a 95% confidence interval, we might prefer to calculate a 99% confidence interval for the parameter μ. Using the same sample of 12 hypertensive smokers, we find the limits to be

$$\left(217 - 2.58\frac{46}{\sqrt{12}}\,,\ 217 + 2.58\frac{46}{\sqrt{12}}\right)$$

or

$$(183\,,\ 251).$$

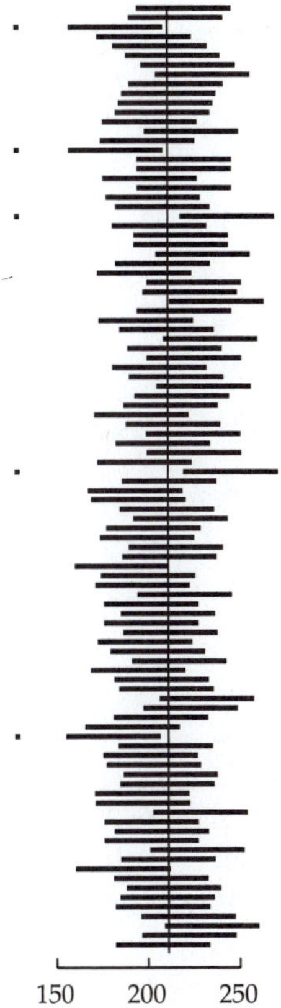

Figure 9.1. Set of 95% confidence intervals constructed from samples of size 12 drawn from a normal population with mean 211 (marked by the vertical line) and standard deviation 46

We are 99% confident that this interval covers the true mean serum cholesterol level of the population. As we previously noted, this interval is wider than the 95% confidence interval.

In the example above, the length of the 99% confidence interval is $251-183 = 68$ mg/100 ml. How large a sample would we need to reduce its length to 20 mg/100 ml? Since the interval is centered on the sample mean $\bar{x} = 217$ mg/100 ml, we are interested in the sample size necessary to produce the interval

$$(217 - 10\,,\ 217 + 10)$$

or

$$(207\,,\ 227).$$

Recall that the 99% confidence interval is of the form

$$\left(217 - 2.58\frac{46}{\sqrt{n}},\ 217 + 2.58\frac{46}{\sqrt{n}}\right).$$

Therefore, to find the sample size n, we must solve the equation

$$10 = \frac{2.58(46)}{\sqrt{n}}.$$

Multiplying both sides of the equation by $\sqrt{n}$ and dividing by 10, we find that

$$\sqrt{n} = \frac{2.58(46)}{10}$$

and

$$n = 140.8.$$

We would need to select a sample of 141 men to reduce the length of the 99% confidence interval to 20 mg/100 ml. Although the sample mean 217 mg/100 ml lies at the center of the interval, it does not play any part in determining its length; the length is a function of σ, n, and the level of confidence.

9.2 One-Sided Confidence Intervals

In some situations we are concerned with either an upper limit or a lower limit for the population mean μ, but not both. Consider the distribution of hemoglobin levels for the population of children under the age of 6 years who have been exposed to high levels of lead. This distribution has an unknown mean μ and standard deviation 0.85 g/100 ml [3]. We know that children who have lead poisoning tend to have much lower levels of hemoglobin than children who do not. Therefore, we are interested in an upper bound for μ.

To construct a one-sided confidence interval, we consider the area in only one tail of a standard normal distribution. Referring to Table A.1, we find that 95% of the observations lie above $z = -1.645$. Therefore,

$$P(Z \geq -1.645) = 0.95,$$

or, substituting $(\overline{X} - \mu)/(\sigma/\sqrt{n})$ for Z,

$$P\left(\frac{\overline{X} - \mu}{\sigma/\sqrt{n}} \geq -1.645\right) = 0.95.$$

Multiplying both sides of the inequality inside the probability statement by $\sigma/\sqrt{n}$ and then subtracting $\overline{X}$, we find that

$$P\left(-\mu \geq -\overline{X} - 1.645\frac{\sigma}{\sqrt{n}}\right) = 0.95$$

and

$$P\left(\mu \leq \overline{X} + 1.645\frac{\sigma}{\sqrt{n}}\right) = 0.95.$$

Therefore, $\overline{X} + 1.645(\sigma/\sqrt{n})$ is an upper 95% confidence bound for μ.

Suppose that we select a sample of 74 children who have been exposed to high levels of lead; these children have a mean hemoglobin level of $\bar{x} = 10.6$ g/100 ml [4]. Based on this sample, a one-sided 95% confidence interval for μ is

$$\begin{aligned} \mu &\leq 10.6 + 1.645\frac{0.85}{\sqrt{74}} \\ &\leq 10.8. \end{aligned}$$

We are 95% confident that the true mean hemoglobin level for this population of children is at most 10.8 g/100 ml.

9.3 Student's t Distribution

When computing confidence intervals for an unknown population mean μ, we have always assumed that σ, the population standard deviation, is known. In reality, this is unlikely to be the case; if μ is unknown, then σ is probably unknown as well. In this situation, confidence intervals are calculated in much the same way as we have already seen. Instead of involving the normal distribution, however, the analysis depends on a new probability distribution known as *Student's t distribution*.

To construct a two-sided confidence interval for the population mean μ, we began by noting that

$$Z = \frac{\overline{X} - \mu}{\sigma/\sqrt{n}}$$

has a standard normal distribution if n is sufficiently large. When the population standard deviation is unknown, it might seem logical to substitute s, the standard deviation of a sample drawn from the population, for σ. This is, in fact, what is done. However, the ratio

$$t = \frac{\overline{X} - \mu}{s/\sqrt{n}}$$

does not have a standard normal distribution. In addition to the sampling variation inherent in $\overline{X}$—which we are using as an estimate of the population mean μ—there is also variability in s. We must account for the fact that s may not be a reliable estimate of σ, especially when the sample with which we are working is small.

For a sample of size n randomly chosen from an underlying normal population, the distribution of the random variable

$$t = \frac{\overline{X} - \mu}{s/\sqrt{n}}$$

is known as Student's t distribution with $n - 1$ degrees of freedom. We represent this using the notation t_{n-1}. Like the normal distribution, the t distribution is symmetrical and the total area under the curve is equal to 1. It is, however, somewhat flatter than the normal distribution and has thicker tails. This is illustrated in Figure 9.2; the shape reflects the extra variability introduced by the estimate s. In addition, the t distribution has a property called the *degrees of freedom*, abbreviated df. The degrees of freedom measure the amount of information available in the data that can be used to estimate σ^2; hence, they measure the reliability of s^2 as an estimate of σ^2. Recall that df $= n - 1$ is the quantity by which we divided the sum of the squared deviations about the mean, $\sum_{i=1}^{n}(x_i - \bar{x})^2$, to obtain the sample variance.

For each possible value of the degrees of freedom, there is a different t distribution. The distributions with small degrees of freedom are more spread out; as df increases, the t distribution approaches the normal. This occurs because, as the sample size increases, s becomes a more reliable estimate of σ; if n is very large, knowing the value of s is nearly equivalent to knowing the value of σ.

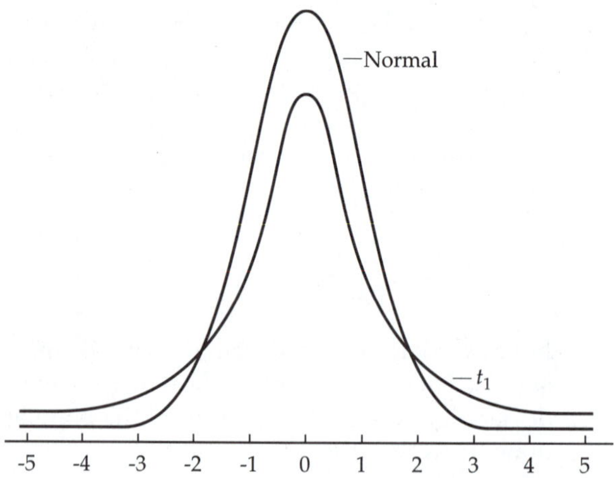

Figure 9.2. The standard normal distribution and Student's t distribution with 1 degree of freedom

Since there is a different t distribution for every value of the degrees of freedom, it would be quite cumbersome to have a complete table of areas corresponding to each possible value. As a result, we typically use a condensed table that lists the areas under the curve for selected percentiles only; for example, it might contain the upper 5.0, 2.5, 1.0, 0.5, and 0.05% of the distributions. Condensed tables are sufficient for most applications involving the construction of confidence intervals.

Table A.2 in Appendix A is a condensed table of areas computed for the family of t distributions. For a particular value of df, the entry in the table represents the value of t_{n-1} that cuts off the specified area in the tail of the distribution. Given a t distribution with 10 degrees of freedom, for instance, $t_{10} = 2.228$ cuts off the upper 2.5% of the area under the curve. Since the distribution is symmetric, $t_{10} = -2.228$ marks off the lower 2.5%. The values of t_{n-1} that cut off the upper 2.5% of the distributions with various degrees of freedom are listed below.

df $(n-1)$	t_{n-1}
1	12.706
2	4.303
5	2.571
10	2.228
20	2.086
30	2.042
40	2.021
60	2.000
120	1.980

For the standard normal curve, $z = 1.96$ marks the upper 2.5% of the distribution. Observe that, as n increases, t_{n-1} approaches this value. In fact, when we have more than 30 degrees of freedom, we are able to substitute the normal distribution for the t; if we do this, our calculations will be off by less than 5%.

Consider a random sample of ten children selected from the population of infants receiving antacids that contain aluminum. These antacids are often used to treat peptic disorders. The mean and standard deviation of the distribution of plasma aluminum levels for this population are unknown. However, we do know that the mean aluminum level for the sample of ten infants is $\bar{x} = 37.2\ \mu$g/l and that the sample standard deviation is $s = 7.13\ \mu$g/l [5].

Since the population standard deviation σ is not known, we use the t distribution to find 95% confidence limits for the mean of the population from which the sample was drawn. For a t distribution with $10 - 1 = 9$ degrees of freedom, 95% of the observations lie between -2.262 and 2.262. Therefore, replacing σ with s, a 95% confidence interval for the population mean μ is

$$\left(\bar{x} - 2.262\frac{s}{\sqrt{10}} , \bar{x} + 2.262\frac{s}{\sqrt{10}} \right).$$

Substituting in the values of $\bar{x}$ and s, the interval becomes

$$\left(37.2 - 2.262\frac{7.13}{\sqrt{10}}\,,\ 37.2 + 2.262\frac{7.13}{\sqrt{10}}\right)$$

or

$$(32.1\,,\ 42.3).$$

We are 95% confident that these limits cover the true mean plasma aluminum level for the population of infants receiving antacids containing aluminum. If the population standard deviation σ had been known and had been equal to the sample value of 7.13 μg/l, then the 95% confidence interval for μ would have been

$$\left(37.2 - 1.96\frac{7.13}{\sqrt{10}}\,,\ 37.2 + 1.96\frac{7.13}{\sqrt{10}}\right)$$

or

$$(32.8\,,\ 41.6).$$

In this case, the confidence interval is slightly shorter. In general, confidence intervals based on the t distribution are longer than the corresponding intervals based on the normal distribution. This principle does not always apply, however; because of sampling variability, it is possible that the value of the estimate s will be considerably smaller than σ for a given sample.

In a previous example, we examined the distribution of serum cholesterol levels for all males in the United States who are hypertensive and who smoke. Recall that the standard deviation of this population is 46 mg/100 ml. On the left-hand side, Figure 9.3 contains the 95% confidence intervals for μ that were calculated from 100 random samples and previously displayed in Figure 9.1. The right-hand side of the figure contains 100 additional intervals that were computed using the same samples; in each case, however, the standard deviation was assumed to be unknown. Once again, 95 of the intervals contain the true mean μ, and the other 5 do not. Note that this time, however, the intervals vary in length.

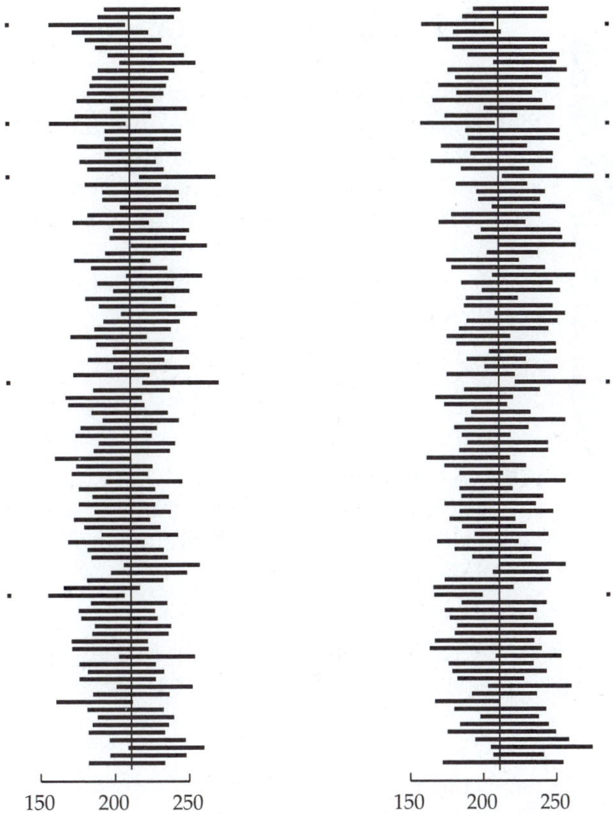

Figure 9.3. Two sets of 95% confidence intervals constructed from samples of size 12 drawn from normal populations with mean 211 (marked by the vertical lines), one with standard deviation 46 and the other with standard deviation unknown

9.4 Further Applications

Confidence intervals are often constructed to provide a range of values that may be used to illustrate the uncertainty inherent in parameter estimation. Consider the distribution of height for the population of individuals between the ages of 12 and 40 who suffer from fetal alcohol syndrome. Fetal alcohol syndrome refers to the severe end of the spectrum of disabilities caused by maternal alcohol use during pregnancy. The distribution of height has an unknown mean μ and standard deviation $\sigma = 6$ centimeters [6]. Before a sample is drawn from the population,

$$\left(\overline{X} - 1.96\frac{6}{\sqrt{n}}, \overline{X} + 1.96\frac{6}{\sqrt{n}}\right)$$

is a 95% confidence interval for μ.

A random sample of 31 patients is selected; the mean height for these individuals is $\bar{x} = 147.4$ cm. Therefore, a 95% confidence interval based on this sample is

$$\left(147.4 - 1.96\frac{6}{\sqrt{31}}, 147.4 + 1.96\frac{6}{\sqrt{31}}\right)$$

or

$$(145.3, 149.5).$$

We are 95% confident that these limits cover the true mean height for this population. In reality, however, the fixed value of μ is either between 145.3 cm and 149.5 cm or it is not.

Now consider the distribution of gestational age for the population of infants who were subjected to fetal surgery for congenital anomalies. The mean and standard deviation of this underlying normal distribution are unknown; therefore, we apply the t distribution when constructing a confidence interval.

Before a random sample of 14 infants is selected, a 95% confidence interval for μ is

$$\left(\overline{X} - 2.160\frac{s}{\sqrt{14}}, \overline{X} + 2.160\frac{s}{\sqrt{14}}\right).$$

This expression makes use of the fact that, given a t distribution with $14 - 1 = 13$ degrees of freedom, 95% of the observations lie between -2.160 and 2.160. If the mean gestational age for the sample of infants is $\bar{x} = 29.6$ weeks, and the standard deviation is $s = 3.6$ weeks [7], then a 95% confidence interval for μ would be

$$\left(29.6 - 2.160\frac{3.6}{\sqrt{14}}, 29.6 + 2.160\frac{3.6}{\sqrt{14}}\right)$$

or

$$(27.5\,,\,31.7).$$

We are 95% confident that these limits cover the true mean gestational age for this population of infants.

We might prefer to calculate a 99% confidence interval rather than a 95% interval. For a t distribution with 13 degrees of freedom, 99% of the observations lie between −3.012 and 3.012. Therefore, using the same sample of 14 infants, we find the confidence limits to be

$$\left(29.6 - 3.012\frac{3.6}{\sqrt{14}}\,,\, 29.6 + 3.012\frac{3.6}{\sqrt{14}}\right)$$

or

$$(26.7\,,\,32.5).$$

We are 99% confident that this interval encloses the true mean gestational age. As we would expect, this interval is wider than the corresponding 95% confidence interval.

9.5 Review Exercises

1. The distributions of systolic and diastolic blood pressure for female diabetics between the ages of 30 and 34 have unknown means. However, their standard deviations are $\sigma_s = 11.8$ mm Hg and $\sigma_d = 9.1$ mm Hg, respectively [8].

   a. A random sample of ten women is selected from this population. The mean systolic blood pressure for the sample is $\bar{x}_s = 130$ mm Hg. Calculate a two-sided 95% confidence interval for μ_s, the true mean systolic blood pressure.
   b. Interpret this confidence interval.
   c. The mean diastolic blood pressure for the sample of size 10 is $\bar{x}_d = 84$ mm Hg. Find a two-sided 90% confidence interval for μ_d, the true mean diastolic blood pressure of the population.
   d. Find a two-sided 99% confidence interval for μ_d.
   e. How does the 99% confidence interval differ from the 90% confidence interval?

2. Consider the t distribution with 5 degrees of freedom.

   a. What proportion of the area under the curve lies to the right of $t = 2.015$?
   b. What proportion of the area lies to the left of $t = -3.365$?
   c. What proportion of the area lies between $t = -4.032$ and $t = 4.032$?

d. What value of t cuts off the upper 2.5% of the distribution?

3. For the population of expectant mothers whose fetuses undergo surgical treatment, the distribution of lengths of stay in the hospital has an unknown mean and standard deviation. A random sample of 17 such mothers has mean length of stay $\bar{x} = 9.5$ days and standard deviation $s = 3.6$ days [7].

a. Construct a 95% confidence interval for the true population mean μ.
b. What is the length of this interval?
c. How large a sample would be required so that the 95% confidence interval has length 2 days? Assume that the value of s does not change.

4. The percentage of ideal body weight was determined for 18 randomly selected insulin-dependent diabetics [9]. The outcomes are recorded below.

107	119	99	114	120	104	88	114	124	
116	101	121	152	100	125	114	95	117	(%)

a. Compute a two-sided 95% confidence interval for the mean percentage of ideal body weight of the population.
b. Does this confidence interval contain the value 100%? What does this tell you?

References

[1] Kaplan, N. M., "Strategies to Reduce Risk Factors in Hypertensive Patients Who Smoke," *American Heart Journal*, Volume 115, January 1988, 288–294.

[2] National Center for Health Statistics, Fulwood, R., Kalsbeek, W., Rifkind, B., Russell-Briefel, R., Muesing, R., LaRosa, J., and Lippel, K., "Total Serum Cholesterol Levels of Adults 20–74 Years of Age: United States, 1976–1980," *Vital and Health Statistics*, Series 11, Number 236, May 1986.

[3] National Center for Health Statistics, Fulwood, R., Johnson, C. L., Bryner, J. D., Gunter, E. W., and McGrath, C. R., "Hematological and Nutritional Biochemistry Reference Data for Persons 6 Months–74 Years of Age: United States, 1976–1980," *Vital and Health Statistics*, Series 11, Number 232, December 1982.

[4] Clark, M., Royal, J., and Seeler, R., "Interaction of Iron Deficiency and Lead and the Hematologic Findings in Children with Severe Lead Poisoning," *Pediatrics*, Volume 81, February 1988, 247–253.

[5] Tsou, V. M., Young, R. M., Hart, M. H., and Vanderhoof, J. A., "Elevated Plasma Aluminum Levels in Normal Infants Receiving Antacids Containing Aluminum," *Pediatrics*, Volume 87, February 1991, 148–151.

[6] Streissguth, A. P., Aase, J. M., Clarren, S. K., Randels, S. P., LaDue, R. A., and Smith, D. F., "Fetal Alcohol Syndrome in Adolescents and Adults," *Journal of the American Medical Association*, Volume 265, April 17, 1991, 1961–1967.

[7] Longaker, M. T., Golbus, M. S., Filly, R. A., Rosen, M. A., Chang, S. W., and Harrison, M. R., "Maternal Outcome After Open Fetal Surgery," *Journal of the American Medical Association*, Volume 265, February 13, 1991, 737–741.

[8] Klein, B. E. K., Klein, R., and Moss, S. E., "Blood Pressure in a Population of Diabetic Persons Diagnosed After 30 Years of Age," *American Journal of Public Health*, Volume 74, April 1984, 336–339.

[9] Saudek, C. D., Selam, J. L., Pitt, H. A., Waxman, K., Rubio, M., Jeandidier, N., Turner, D., Fischell, R. E., and Charles, M. A., "A Preliminary Trial of the Programmable Implantable Medication System for Insulin Delivery," *The New England Journal of Medicine*, Volume 321, August 31, 1989, 574–579.

CHAPTER

10

Hypothesis Testing

In our study of confidence intervals, we encountered the distribution of serum cholesterol levels for the population of males who are hypertensive and who smoke. The mean of this distribution is unknown. We do know, however, that the mean serum cholesterol level for the population of all 20- to 74-year-old males in the United States is 211 mg/100 ml [1]. Therefore, we might wonder whether the mean cholesterol level of the subpopulation of men who are hypertensive smokers is 211 mg/100 ml as well. If we select a random sample of 25 men from this population and their mean serum cholesterol level is 220 mg/100 ml, is this sample mean compatible with a hypothesized mean of 211 mg/100 ml? What if the sample mean is 230 mg/100 ml or 250 mg/100 ml?

10.1 General Concepts

We again concentrate on drawing some conclusion about a population parameter–the mean, in this case–using the information contained in a sample of observations. One approach is to construct a confidence interval for μ; another is to conduct a *statistical hypothesis test.*

We begin by claiming that the mean of the population is equal to some postulated value μ_0. This statement about the value of the population parameter is called the *null hypothesis*, or H_0. If we want to test whether the mean serum cholesterol level of the subpopulation of male hypertensive smokers is equal to the mean of the general population of 20- to 74-year-olds, for instance, the null hypothesis would be

$$H_0: \mu = \mu_0 = 211 \text{ mg/100 ml}.$$

The *alternative hypothesis* is a second statement that contradicts H_0; in this case, we have

$$H_A: \mu \neq 211 \text{ mg/100 ml}.$$

Together the null and the alternative hypotheses cover all possible values of μ; consequently, one of the two statements must be true.

We next draw a random sample of size n from the population of interest. In the case of the hypertensive smokers, we selected a sample of size 12. We compare the mean of this sample, $\bar{x}$, to the postulated mean μ_0; specifically, we wish to know whether the difference between the sample mean and the hypothesized mean is too large to be attributed to chance alone.

If there is evidence that the sample could **not** have come from a population with mean μ_0, then we reject the null hypothesis. This occurs when, given that H_0 is true, the probability of obtaining a sample mean as extreme or more extreme than the observed value $\bar{x}$ is sufficiently small. In this case, the data are not compatible with the null hypothesis; instead, they are more supportive of the alternative. We therefore conclude that the population mean could not be μ_0. In keeping with the vernacular, such a test result is said to be *statistically significant*. Note that statistical significance does not imply clinical significance; the test result could actually have little practical consequence.

If there is not sufficient evidence to doubt the validity of the null hypothesis, then we cannot reject this claim. Instead, we concede that the population mean may be equal to μ_0. However, we do not say that we accept H_0; the test does not prove the null hypothesis. It is still possible that the population mean is some value other than μ_0 but that the random sample selected does not confirm this. Such an event could occur, for instance, when the sample chosen is too small. This point is discussed further later in the chapter.

We stated above that if the probability of obtaining a sample mean as extreme or more extreme than the observed $\bar{x}$ is sufficiently small, then we reject the null hypothesis. But what is a "sufficiently small" probability? In most applications, 0.05 is chosen. Thus we reject H_0 when the chance that the sample could have come from a population with mean μ_0 is less than or equal to 5%. This implies that we reject incorrectly 5% of the time; given many repeated tests of significance, five times out of one hundred we erroneously reject the null hypothesis when it is true. To be more conservative, a probability of 0.01 is sometimes chosen. In this case we mistakenly reject H_0 when it is true only 1% of the time. The probability value that we choose–whether 0.05, 0.01, or some other value–is known as the *significance level* of the hypothesis test. The significance level is denoted by α and must be specified **before** the test is actually carried out.

A test of hypothesis can be compared to a criminal trial by jury in the United States. The individual on trial is either innocent or guilty but is assumed innocent by law. After evidence pertaining to the case has been presented, the jury finds the defendant either guilty or not guilty. If the defendant is innocent and the decision of the jury is that he or she is not guilty, then the right verdict has been reached. The verdict is also correct if the defendant is guilty and is convicted of the crime.

Verdict of Jury	Defendant	
	Innocent	**Guilty**
Not guilty	Correct	Incorrect
Guilty	Incorrect	Correct

Analogously, the true population mean is either μ_0 or it is not μ_0. We begin by assuming that the null hypothesis

$$H_0: \mu = \mu_0$$

is correct, and we consider the "evidence" that is presented in the form of a sample of size n. Based on our findings, the null hypothesis is either rejected or it is not rejected. Again there are two situations in which the conclusion drawn is correct—when the population mean is μ_0 and the null hypothesis is not rejected, and when the population mean is not μ_0 and H_0 is rejected.

Result of Test	Population	
	$\mu = \mu_0$	$\mu \neq \mu_0$
Do not reject	Correct	Incorrect
Reject	Incorrect	Correct

Like our legal system, the process of hypothesis testing is not perfect; there are two kinds of errors that can be made. In particular, we could either reject the null hypothesis when μ is equal to μ_0 or fail to reject it when μ is not equal to μ_0. These two types of error are discussed in more detail in a later section.

The probability of obtaining a mean as extreme or more extreme than the observed sample mean $\bar{x}$—given that the null hypothesis

$$H_0: \mu = \mu_0$$

is true—is called the *p-value* of the test, or simply p. The p-value is compared to the predetermined significance level α to decide whether the null hypothesis should be rejected. If p is less than or equal to α, we reject H_0. If p is greater than α, we do not reject H_0. In addition to the conclusion of the test, the p-value itself is often reported in the literature.

10.2 Two-Sided Tests of Hypotheses

To conduct a test of hypothesis, we again draw on our knowledge of the sampling distribution of the mean. Assume that the population standard deviation σ is known and that μ_0 is the true population mean. If the value of n is large enough, then the sampling distribution of means of samples of size n is approximately normal and

$$Z = \frac{\overline{X} - \mu_0}{\sigma/\sqrt{n}}$$

is a standard normal random variable. For a given sample of size n with mean $\bar{x}$, we can calculate the outcome z and use a table of the standard normal curve to determine the probability of obtaining a value of Z that is as extreme or more extreme than that observed. By "more extreme," we mean farther away from the center of the distribution in the direction of the alternative hypothesis. Because it relies on the normal distribution, a test of this kind is often called a *z-test*.

When the population standard deviation is not known, we substitute the sample value s for σ. If the underlying population is normally distributed, then the random variable

$$t = \frac{\overline{X} - \mu_0}{s/\sqrt{n}}$$

has a t distribution with $n - 1$ degrees of freedom. In this case, we can calculate the outcome of t and consult Table A.2 to find the probability of obtaining a sample mean that is more extreme than the one observed.

Again consider the distribution of serum cholesterol levels for all males in the United States who are hypertensive and who smoke. The standard deviation of the distribution is assumed to be 46 mg/100 ml; the null hypothesis is

$$H_0: \mu = 211 \text{ mg/100 ml},$$

where 211 mg/100 ml is the mean serum cholesterol level for all 20- to 74-year-old males. Since the mean of the subpopulation of hypertensive smokers could be either smaller or larger than μ_0, we are concerned with deviations from the population mean that occur in either direction. As a result, we conduct a two-sided test at the $\alpha = 0.05$ level of significance. The alternative hypothesis is

$$H_A: \mu \neq 211 \text{ mg/100 ml}.$$

The previously mentioned random sample of 12 hypertensive smokers has a mean serum cholesterol level $\bar{x} = 217$ mg/100 ml [2]. Is it likely that this sample comes from a population with mean 211 mg/100 ml? To answer this question, we compute the test statistic

$$\begin{aligned} z &= \frac{\bar{x} - \mu_0}{\sigma/\sqrt{n}} \\ &= \frac{217 - 211}{46/\sqrt{12}} \\ &= 0.45. \end{aligned}$$

According to Table A.1, the area to the right of $z = 0.45$—which is the probability of observing $Z = 0.45$ or anything larger, given that H_0 is true—is 0.326. The area to the left of $z = -0.45$ is 0.326 as well. The area in the two tails of the standard normal distribution sums to 0.652; this is the p-value of the test. Since $p > 0.05$, we do not reject the null hypothesis. Based on this sample, the evidence is insufficient to conclude that the mean serum cholesterol level of the population of hypertensive smokers is different from 211 mg/100 ml.

Since we conducted a two-sided test, any value of z that is between -1.96 and 1.96 would result in a p-value greater than 0.05. In all such cases, the null hypothesis would not be rejected. The numbers -1.96 and 1.96 are the *critical values* of the test statistic. Another way to look at this is that the null hypothesis would not be rejected when μ_0 is any value that lies within the 95% confidence interval for μ. Recall that in Chapter 9 we found a 95% confidence interval for the mean serum cholesterol level of hypertensive smokers to be

$$(191\,,\,243).$$

Therefore, if the null hypothesis had been

$$H_0: \mu = 240 \text{ mg/100 ml},$$

H_0 would not have been rejected. In contrast, any value of μ_0 that lies outside of the 95% confidence interval for μ—such as $\mu_0 = 260$ mg/100 ml—would cause the null hypothesis to be rejected at the $\alpha = 0.05$ level.

The value $\mu_0 = 211$ mg/100 ml was selected because it is the mean serum cholesterol level of the population of all 20- to 74-year-old males. As a result, the null hypothesis claims that the mean serum cholesterol level of the population of males who are hypertensive smokers is identical to the mean cholesterol level of the general population of males. The hypothesis was established in the interest of obtaining evidence that would cause it to be rejected; a rejection would have implied that the mean serum cholesterol level of male hypertensive smokers is **not** equal to the mean of the population as a whole.

As a second example, consider the random sample of ten children selected from the population of infants receiving antacids containing aluminum. The mean and standard deviation of the distribution of plasma aluminum levels for this population are unknown. However, we do know that the mean plasma aluminum level for the sample of ten infants is $\bar{x} = 37.20$ μg/l and that its standard deviation is $s = 7.13$ μg/l [3]. The mean plasma aluminum level for the population of infants not receiving antacids is 4.13 μg/l. Is it likely that the data in our sample could have come from a population with mean $\mu_0 = 4.13$ μg/l? To find out, we conduct a test of hypothesis; the null hypothesis is

$$H_0: \mu = 4.13\ \mu\text{g/l},$$

and the alternative is

$$H_A: \mu \neq 4.13\ \mu\text{g/l}.$$

We are interested in deviations from the mean that could occur in either direction. Therefore, we conduct a two-sided test at the $\alpha = 0.05$ level of significance.

Because we do not know the population standard deviation σ, we use a t-test rather than a z-test. The test statistic is

$$t = \frac{\bar{x} - \mu_0}{s/\sqrt{n}}$$

or

$$\begin{aligned} t &= \frac{37.20 - 4.13}{7.13/\sqrt{10}} \\ &= 14.67. \end{aligned}$$

Consulting Table A.2, we observe that, for a t distribution with 9 degrees of freedom, the total area to the right of $t_9 = 14.67$ and to the left of $t_9 = -14.67$ is less than 0.001. Therefore, $p < 0.05$, and we reject the null hypothesis

$$H_0: \mu = 4.13\ \mu\text{g/l}.$$

This sample of infants provides evidence that the mean plasma aluminum level of children receiving antacids is higher than the mean aluminum level of children not receiving them.

10.3 One-Sided Tests of Hypotheses

Before we conduct a test of hypothesis, we must decide whether we are concerned with deviations from μ_0 that could occur in both directions or in one direction only. This choice determines whether we consider the area in two tails of the appropriate curve or the area in only one. The decision must be made **before** a random sample is selected; it should not be influenced by the outcome of the sample. If prior knowledge indicates that μ cannot be less than μ_0, then the only values of $\bar{x}$ that will provide evidence against the null hypothesis

$$H_0: \mu = \mu_0$$

are those that are much larger than μ_0. In a situation such as this, the null hypothesis is more properly stated as

$$H_0: \mu \leq \mu_0$$

and the alternative hypothesis as

$$H_A: \mu > \mu_0.$$

For example, it is unreasonable to believe that exposure to a toxic substance—such as ambient carbon monoxide or sulphur dioxide—could be beneficial to humans. Therefore, we anticipate only harmful effects and conduct a one-sided test. A two-sided test is the more conservative choice; in general, the p-value of a two-sided test is twice as large as that of a one-sided test.

Consider the distribution of hemoglobin levels for the population of children under the age of 6 who have been exposed to high levels of lead. This distribution has an unknown mean μ; its standard deviation is assumed to be 0.85 g/100 ml [4]. We might wish to know whether the mean hemoglobin level of this population is equal to the mean hemoglobin level of the general population of children under the age of 6, $\mu = 12.29$ g/100 ml. However, since we believe that the hemoglobin levels of exposed children must on the average be lower than the levels of unexposed children, we are concerned only with deviations from the mean that are below μ_0. As a result, the null hypothesis is

$$H_0: \mu \geq 12.29 \text{ g/100 ml},$$

and the alternative is

$$H_A: \mu < 12.29 \text{ g/100 ml}.$$

We conduct a one-sided test at the $\alpha = 0.05$ level of significance; since σ is known, we use the normal distribution rather than the t.

A random sample of 74 children who have been exposed to high levels of lead has a mean hemoglobin level of $\bar{x} = 10.6$ g/100 ml [5]. Therefore, the appropriate test statistic is

$$\begin{aligned} z &= \frac{\bar{x} - \mu_0}{\sigma/\sqrt{n}} \\ &= \frac{10.6 - 12.29}{0.85/\sqrt{74}} \\ &= -17.10. \end{aligned}$$

According to Table A.1, the area to the left of $z = -17.10$ is less than 0.001. Since this p-value is smaller than α, we reject the null hypothesis

$$H_0: \mu \geq 12.29 \text{ g/100 ml}$$

at the 0.05 level. The sample provides evidence that the mean hemoglobin level for the population of children who have been exposed to high levels of lead is smaller than 12.29 g/100 ml. Since this is a one-sided test, any value of z less than or equal to the critical value -1.645 would have led us to reject the null hypothesis.

In this example, the null hypothesis was chosen to test the statement that the mean hemoglobin level of the population of children who have been exposed to lead is the same as that of the general population. By rejecting H_0, we conclude that this is not the case; the mean hemoglobin level among children who have been exposed to lead is in fact lower than the mean level among children who have not been exposed.

The choice between a one- and a two-sided test can be extremely controversial. Not infrequently, a one-sided test achieves significance when a two-sided test does not. Consequently, the decision is often made on nonscientific grounds. In response to this, some journal editors are reluctant to publish studies that employ one-sided tests. This may be an overreaction to something that the intelligent reader can discern. In any event, we avoid further discussion of this debate.

10.4 Types of Error

In the application of the technique of hypothesis testing, there are two kinds of errors that can be made. The first is called a *type I error*; it is also known as a *rejection error*, or an *α error*. A type I error is made if we reject the null hypothesis

$$H_0: \mu = \mu_0$$

when H_0 is true. The probability of committing a type I error is determined by the significance level of the test; recall that

$$\alpha = \text{P(reject } H_0 \mid H_0 \text{ is true).}$$

If we were to conduct repeated tests of hypotheses setting the significance level at 0.05, we would erroneously reject a true null hypothesis 5% of the time.

Consider the case of a drug that has been deemed effective for reducing high blood pressure. After being treated with this drug for a given period of time, a population of individuals suffering from hypertension has mean diastolic blood pressure μ_d, a value that is clinically lower than the mean diastolic blood pressure of the untreated hypertensives. Now suppose that another company produces a generic version of the same drug. We would like to know whether the generic drug is as effective at reducing high blood pressure as the brand name version. To determine this, we examine the distribution of diastolic blood pressures for a sample of individuals who have been treated with the generic drug; if μ is the mean of this population, we use the sample to test the null hypothesis

$$H_0: \mu = \mu_d.$$

What if the manufacturer of the generic drug actually submits the brand name product for testing in place of its own version? Vitarine Pharmaceuticals, a New York based drug company, has reportedly done just that [6]. The company has made similar substitutions on four different occasions. In a situation such as this, we know that the null hypothesis must be true—we are testing the drug that itself set the standard. Therefore, if the test of hypothesis leads us to reject H_0 and pronounce the "generic" drug to be either more or less efficacious than the brand name version, a type I error has been made.

The second kind of error that can be committed during a hypothesis test is a *type II error*, also known as an *acceptance error*, or a β *error*. A type II error is made if we fail to reject the null hypothesis

$$H_0: \mu = \mu_0$$

when H_0 is false. The probability of committing a type II error is represented by

$$\beta \quad = \quad \text{P(do not reject } H_0 \mid H_0 \text{ is false).}$$

If $\beta = 0.10$, for instance, then the probability that we do not reject the null hypothesis when $\mu \neq \mu_0$ is 0.10, or 10%. The two types of errors that can be made are summarized below:

	Population	
Result of Test	$\mu = \mu_0$	$\mu \neq \mu_0$
Do not reject	Correct	Type II error
Reject	Type I error	Correct

Recall the distribution of serum cholesterol levels for all 20- to 74-year-old males in the United States. The mean of this population is $\mu = 211$ mg/100 ml, and the standard deviation is $\sigma = 46$ mg/100 ml. Suppose that we do not know the true mean of this population; however, we do know that the mean serum cholesterol level of the subpopulation of 20- to 24-year-old males is 180 mg/100 ml. Since older men tend to have higher cholesterol levels than do younger men on average, we would expect the mean cholesterol level of the population of 20- to 74-year-olds to be higher than 180 mg/100 ml. Therefore, if we were to test the null hypothesis

$$H_0: \mu \leq 180 \text{ mg/100 ml,}$$

we would expect H_0 to be rejected. It is possible, however, that it would not be. The chance of reaching this incorrect conclusion is β, the probability of making a type II error.

What is the value of β associated with a test of the null hypothesis

$$H_0: \mu \leq 180 \text{ mg/100 ml,}$$

assuming that we select a sample of size 25? To determine this, we must first find the mean value of serum cholesterol level that our sample must have to cause the null hypothesis to be rejected. We believe that the mean serum cholesterol level of the population of 20- to 74-year-olds must be greater than the mean of the population of 20- to 24-year-olds. If we conduct a one-sided test at the $\alpha = 0.05$ level of significance, H_0 would be rejected for $z \geq 1.645$. Since

$$z = \frac{\bar{x} - \mu_0}{\sigma/\sqrt{n}},$$

we have

$$1.645 = \frac{\bar{x} - 180}{46/\sqrt{25}},$$

and

$$\begin{aligned} \bar{x} &= 180 + \frac{1.645(46)}{\sqrt{25}} \\ &= 195.1. \end{aligned}$$

As can be seen in Figure 10.1, the area to the right of $\bar{x} = 195.1$ corresponds to the upper 5% of the sampling distribution of means of samples of size 25 when $\mu = 180$. Therefore, the null hypothesis

$$H_0: \mu \leq 180 \text{ mg/100 ml}$$

would be rejected if our sample has a mean $\bar{x}$ which is greater than or equal to 195.1 mg/100 ml. Any sample with a smaller mean would not provide sufficient evidence to reject H_0 at the 0.05 level.

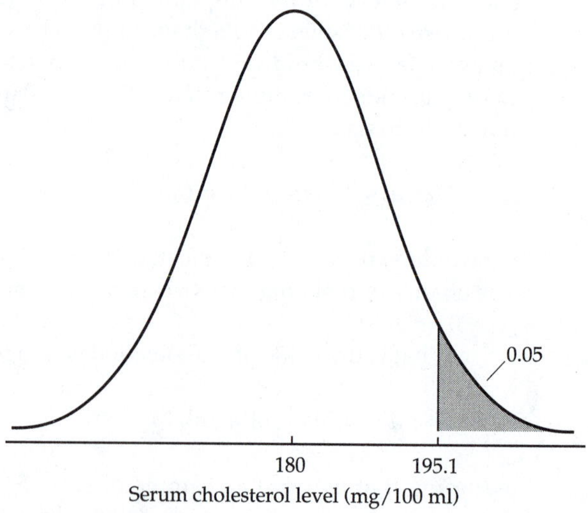

Figure 10.1. Distribution of means of samples of size 25 for the serum cholesterol levels of males 20 to 74 years of age, $\mu = 180$ mg/100 ml

The probability of making a type II error, or β, is the probability of not rejecting the null hypothesis given that H_0 is false and the true population mean is $\mu_1 = 211$ mg/100 ml. To find the value of β, we again consider the sampling distribution of means of samples of size 25; this time, however, we let $\mu = 211$. This distribution is pictured in Figure 10.2. Since a sample mean less than $\bar{x} = 195.1$ mg/100 ml will not lead us to reject H_0, we would like to know what proportion of this new distribution centered at 211 mg/100 ml lies below 195.1. Observe that

$$\begin{aligned} z &= \frac{195.1 - 211.0}{46/\sqrt{25}} \\ &= -1.73. \end{aligned}$$

The area under the standard normal curve that lies to the left of $z = -1.73$ is 0.042. Therefore, β—the probability of failing to reject

$$H_0: \mu \leq 180 \text{ mg/100 ml}$$

when the true population mean is $\mu_1 = 211$ mg/100 ml—is equal to 0.042.

Whereas α, the probability of committing a type I error, is determined by looking at the case in which H_0 is true and μ is equal to μ_0, β is found by considering the situation in which H_0 is false and μ does not equal μ_0. If μ is not equal to μ_0, however, there are an infinite number of possible values that μ could assume. The type II error is calculated for a **single** such value, μ_1; in the previous example, μ_1 was equal to 211 mg/100 ml. If we had chosen a different alternative population mean, then we would have computed a different value for β. The closer the true mean is to the hypothesized mean, the more difficult it is to reject the null hypothesis.

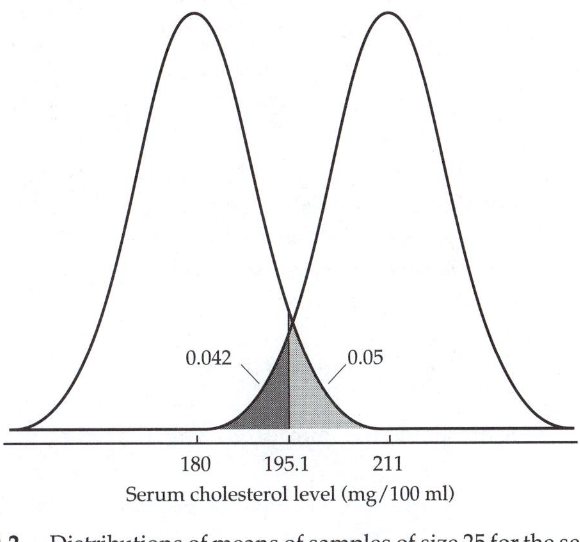

Figure 10.2. Distributions of means of samples of size 25 for the serum cholesterol levels of males 20 to 74 years of age, $\mu = 180$ mg/100 ml versus $\mu = 211$ mg/100 ml

10.5 Power

If β is the probability of committing a type II error, then $1 - \beta$ is called the *power* of the test of hypothesis. The power is the probability of rejecting H_0 when the null hypothesis is false. In other words, it is the probability of avoiding a type II error;

$$\text{power} = P(\text{reject } H_0 \mid H_0 \text{ is false}).$$

The power may also be thought of as the likelihood that a particular study will detect a deviation from the null hypothesis given that one exists. Like β, the power must be computed for a particular alternative population mean μ_1.

In the serum cholesterol example above, the power of the test is

$$\begin{aligned} 1 - \beta &= 1 - 0.042 \\ &= 0.958. \end{aligned}$$

Therefore, there is a 95.8% chance of rejecting the null hypothesis

$$H_0: \mu \leq 180 \text{ mg/100 ml}$$

given that H_0 is false and the true population mean is $\mu_1 = 211$ mg/100 ml. Note that this could also have been expressed in the following way:

$$\begin{aligned} \text{power} &= P(\text{reject } \mu \leq 180 \mid \mu = 211) \\ &= P(\overline{X} \geq 195.1 \mid \mu = 211) \\ &= P(Z \geq -1.73) \\ &= 1 - 0.042 \\ &= 0.958. \end{aligned}$$

The quantity $1 - \beta$ would have assumed a different value if we had set μ_1 equal to 200 mg/100 ml and yet another value if we had let μ_1 be 220 mg/100 ml. If we were to plot the values of $1 - \beta$ against all possible alternative population means, we would have what is known as a *power curve*. A power curve for the test of the null hypothesis

$$H_0: \mu \leq 180 \text{ mg/100 ml}$$

is presented in Figure 10.3. Note that when $\mu_1 = 180$,

$$\begin{aligned} \text{power} &= P(\text{reject } \mu \leq 180 \mid \mu = 180) \\ &= \alpha \\ &= 0.05. \end{aligned}$$

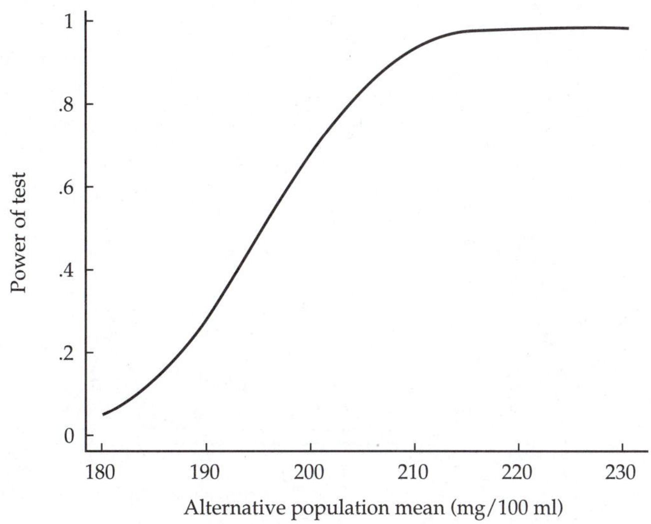

Figure 10.3. Power curve for $\mu_0 = 180, \alpha = 0.05$, and $n = 25$

The power of the test approaches 1 as the alternative mean moves farther and farther away from the null value of 180 mg/100 ml.

Investigators generally try to design tests of hypotheses that have high power. It is not enough to know that we have a small probability of rejecting H_0 when it is true; we would also like there to be a large probability of rejecting the null hypothesis when it is false. One way to increase the power of a test is to raise the significance level, α. If we increase α, then we cut off a smaller portion of the tail of the distribution centered at μ_1. Correspondingly, β becomes smaller, and the power, $1 - \beta$, increases. If α had been equal to 0.10 for the test of the null hypothesis

$$H_0: \mu \leq 180 \text{ mg/100 ml},$$

for example, then β would have been 0.018 and the power 0.982. This situation is illustrated in Figure 10.4; compare it to Figure 10.2, wherein α was equal to 0.05. Bear in mind, however, that by raising α we are increasing the probability of making a type I error.

This tradeoff between α and β is similar to that observed to exist between the sensitivity and the specificity of a diagnostic test. The balance between the two types of error is a delicate one, and their relative importance varies depending on the time and the situation. In the 18th century, Benjamin Franklin said:

> It is better that 100 guilty persons should escape than that one innocent person should suffer.

More recently, however, an editorial on child abuse claimed that it is "equally important" to identify and punish child molesters and to exonerate those who are falsely accused [7]. The more information that we have–meaning the larger the

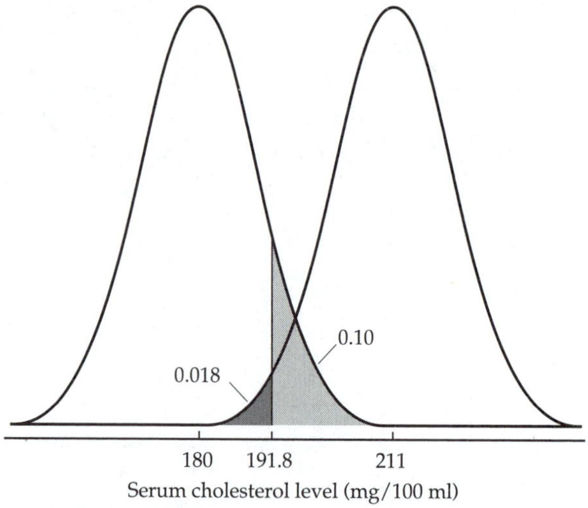

Figure 10.4. Distributions of means of samples of size 25 for the serum cholesterol levels of males 20 to 74 years of age, $\mu = 180$ mg/100 ml versus $\mu = 211$ mg/100 ml

sample–the less likely we are to make an error of either type. Regardless of our decision, however, the possibility that we have committed an error always exists.

The only way to diminish α and β simultaneously is to reduce the amount of overlap in the two normal distributions. One way that this can be done is by considering only large deviations from μ_0. The farther apart the values of μ_0 and μ_1 are, the greater the power of the test. An alternative is to increase the sample size, n. By increasing n, we decrease the standard error $\sigma/\sqrt{n}$; this causes the two sampling distributions to become more narrow, which, in turn, lessens the amount of overlap. Another possibility that we have not discussed is to find a "more powerful" test statistic. This topic is discussed further in Chapter 13.

10.6 Sample Size

When designing an experiment, investigators must determine the sample size n that is necessary to provide a specified power. For example, suppose that we wish to test the null hypothesis

$$H_0: \mu \leq 180 \text{ mg/100 ml}$$

at the $\alpha = 0.01$ level of significance. Once again, μ is the mean serum cholesterol level of the population of 20- to 74-year-old males in the United States. If the true population mean is as large as 211 mg/100 ml, then we want to risk only a 5% chance of failing to reject the null hypothesis; consequently, we set β equal to 0.05 and the power of the test to 0.95. Under these circumstances, how large a sample would we require?

Since $\alpha = 0.01$, we begin by noting that H_0 would be rejected for $z \geq 2.32$. Substituting $(\bar{x} - 180)/(46/\sqrt{n})$ for the normal deviate z, we let

$$2.32 = \frac{\bar{x} - 180}{46/\sqrt{n}}.$$

Solving for $\bar{x}$,

$$\bar{x} = 180 + 2.32\frac{46}{\sqrt{n}}.$$

Therefore, we would reject the null hypothesis if $\bar{x}$ is greater than this value.

Now consider the desired power of the test. If the true mean serum cholesterol level were actually $\mu_1 = 211$ mg/100 ml—so that the normal deviate z could be expressed as $(\bar{x} - 211)/(46/\sqrt{n})$—then we would want to reject the null hypothesis with probability $1 - \beta = 0.95$. The value of z that corresponds to $\beta = 0.05$ is $z = -1.645$; therefore,

$$-1.645 = \frac{\bar{x} - 211}{46/\sqrt{n}}$$

and

$$\bar{x} = 211 - 1.645\frac{46}{\sqrt{n}}.$$

Setting the two expressions for the sample mean $\bar{x}$ equal to each other,

$$180 + 2.32\frac{46}{\sqrt{n}} = 211 - 1.645\frac{46}{\sqrt{n}}.$$

Multiplying both sides of the equation by $\sqrt{n}$ and collecting terms,

$$\sqrt{n}\,(211 - 180) = (2.32 - (-1.645))(46)$$

and

$$n = \left[\frac{(2.32 - (-1.645))(46)}{(211 - 180)}\right]^2$$
$$= 34.6.$$

Therefore, a sample of size 35 would be required.

When $\alpha = 0.01$ and $\beta = 0.05$, the general formula for the sample size n can be written

$$n = \left[\frac{(2.32 - (-1.645))(\sigma)}{(\mu_1 - \mu_0)}\right]^2.$$

Several factors influence the size of n. If we reduce the type I error α, for instance, then the cutoff point for rejecting H_0 would be some value greater than $z = 2.32$. This would result in a larger value of n. Similarly, if we lower the type II error β, then the corresponding value of z must lie to the left of -1.645. Again this would produce a larger value of n. If we consider an alternative population mean that is closer to the hypothesized value, then the difference $\mu_1 - \mu_0$ would decrease and the sample size would increase. Finally, the larger the underlying variability σ, the larger is the sample size required.

In the serum cholesterol level example, we knew that the hypothesized population mean μ_0 had to be smaller than the alternative μ_1; consequently, we conducted a one-sided test. If it is not known whether μ_0 is larger or smaller than μ_1, then a two-sided test is appropriate. In this case, we must modify the critical value of z that would cause the null hypothesis to be rejected. When $\alpha = 0.01$, for instance,

$$H_0: \mu = 180 \text{ mg/100 ml}$$

would be rejected for $z \geq 2.58$. Substituting this value into the equation above,

$$\begin{aligned} n &= \left[\frac{(2.58 - (-1.645))(46)}{(211 - 180)} \right]^2 \\ &= 39.3, \end{aligned}$$

and a sample of size 40 would be required. Note that the sample size for a two-sided test is always larger than the corresponding sample size for a one-sided test.

10.7 Further Applications

Suppose that we are interested in investigating the physical characteristics of adolescents and adults who suffer from fetal alcohol syndrome. As an example, consider the distribution of heights for the population of 12- to 40-year-olds who have this condition. The distribution has an unknown mean μ; its standard deviation is $\sigma = 6$ centimeters [8]. Is the mean height of this population equal to the mean height of the population of individuals who do not have fetal alcohol syndrome?

The first step in conducting a test is to make a claim about the value of μ_0. Since the mean height of persons in this age group who do not suffer from fetal alcohol syndrome is 160.0 centimeters, the null hypothesis would be

$$H_0 : \mu = 160.0 \text{ cm}.$$

We are concerned with deviations from μ_0 that could occur in either direction, thus we conduct a two-sided test at the $\alpha = 0.05$ level of significance. The alternative hypothesis is

$$H_A : \mu \neq 160.0 \text{ cm}.$$

For a random sample of 31 individuals selected from the population of 12- to 40-year-olds with fetal alcohol syndrome, the mean height is $\bar{x} = 147.4$ cm. If the true mean height is $\mu = 160.0$ cm, then what is the probability of selecting a random sample with a mean as low as 147.4? To answer this question, we calculate the test statistic

$$\begin{aligned} z &= \frac{\bar{x} - \mu_0}{\sigma/\sqrt{n}} \\ &= \frac{147.4 - 160.0}{6/\sqrt{31}} \\ &= -11.69. \end{aligned}$$

The area to the left of $z = -11.69$ and to the right of $z = 11.69$ is much less than 0.001. Since $p < 0.05$, we reject the null hypothesis at the 0.05 level of significance. The random sample provides evidence that the mean height of the population of individuals who have fetal alcohol syndrome is different from the mean height of those who do not; the persons who suffer from fetal alcohol syndrome tend to be shorter.

When the standard deviation of a population is not known, we use the sample standard deviation s instead. Consider the distribution of the concentration of benzene in a specified brand of cigar. This distribution has an unknown mean μ and standard deviation σ. We are told that the mean concentration of benzene in a brand of cigarette that is being used as the standard is 81 μg/g tobacco [9]. We would like to know whether the mean concentration of benzene in the cigars is equal to that in the cigarettes. To determine this, we test the null hypothesis

$$H_0: \mu = 81 \ \mu g/g.$$

We are interested in deviations from the mean that could occur in either direction and thus conduct a two-sided test at the $\alpha = 0.05$ level of significance. The alternative hypothesis is

$$H_A: \mu \neq 81 \ \mu g/g.$$

A random sample of seven cigars has mean benzene concentration $\bar{x} = 151 \ \mu g/g$ and standard deviation $s = 9 \ \mu g/g$. Is it possible that these observations could have come from a population with mean $\mu = 81 \ \mu g/g$? To determine this, we calculate the statistic

$$t = \frac{\bar{x} - \mu_0}{s/\sqrt{n}}$$

or

$$\begin{aligned} t &= \frac{151 - 81}{9/\sqrt{7}} \\ &= 20.6. \end{aligned}$$

The statistic t is the outcome of a random variable that has a t distribution with $7 - 1 = 6$ degrees of freedom. Looking at Table A.2, we observe that the total area under the curve to the left of -20.6 and to the right of 20.6 is less than 0.001. Since $p < 0.001$, we reject the null hypothesis. The random sample of size seven suggests that the cigars contain a higher concentration of benzene than do the cigarettes.

Rather than work through the calculations ourselves, we could have let a computer package perform the test of hypothesis for us. Suppose that the seven measurements of benzene concentration for the sample of cigars are saved under the variable name `benzene`. Table 10.1 provides the appropriate output from Stata. The upper portion of the table includes the sample mean and the standard deviation; the lower part contains the null hypothesis, the test statistic t and its associated degrees of freedom, and the p-value of the test.

Now recall the distribution of hemoglobin levels for the population of children under the age of 6 years who have been exposed to high levels of lead. The mean of this population is $\mu = 10.60$ g/100 ml, and its standard deviation is $\sigma = 0.85$ g/100 ml. Suppose that we do not know the true population mean μ; however, we do know that the mean hemoglobin level of the general population of children under the age of 6 is 12.29 g/100 ml. If we were to conduct a test of the null hypothesis

$$H_0: \mu = 12.29 \text{ g/100 ml},$$

Table 10.1 Stata output for the t-test

Variable	Obs	Mean	Std. Dev.
benzene	7	151	9

Ho: mean	=	81
t	=	20.6 with 6 df
Prob > | t |	=	0.0001

we would expect that this false hypothesis would be rejected. Assuming that we select a random sample of size 5 from the population of children who have been exposed to lead, what is the probability that we will make a type II error—or fail to reject H_0 when it is false—given that the true population mean is $\mu_1 = 10.60$ g/100 ml?

To answer this question, we begin by finding the mean hemoglobin level that the sample must have for H_0 to be rejected. We believe that the mean hemoglobin level of children who have been exposed to lead must be less than that of unexposed children. If we conduct a one-sided test at the $\alpha = 0.05$ level of significance, the null hypothesis would be rejected for $z \leq -1.645$. Since

$$z = \frac{\bar{x} - \mu_0}{\sigma/\sqrt{n}},$$

we have

$$-1.645 = \frac{\bar{x} - 12.29}{0.85/\sqrt{5}}$$

and

$$\begin{aligned} \bar{x} &= 12.29 - \frac{1.645(0.85)}{\sqrt{5}} \\ &= 11.66. \end{aligned}$$

Therefore, the null hypothesis

$$H_0: \mu \geq 12.29 \text{ g/100 ml}$$

would be rejected if the sample of size 5 has a mean $\bar{x}$ that is less than or equal to 11.66 g/100 ml. This area corresponds to the lower 5% of the sampling distribution of means of samples of size 5 when $\mu = 12.29$ g/100 ml; it is illustrated in Figure 10.5.

The quantity β is the probability of failing to reject H_0 given that the true population mean is $\mu_1 = 10.60$ g/100 ml. To find β, we consider the sampling distribution of means of samples of size 5 when $\mu = 10.60$ g/100 ml. Since a sample mean greater than 11.66 g/100 ml will not lead us to reject H_0, we must

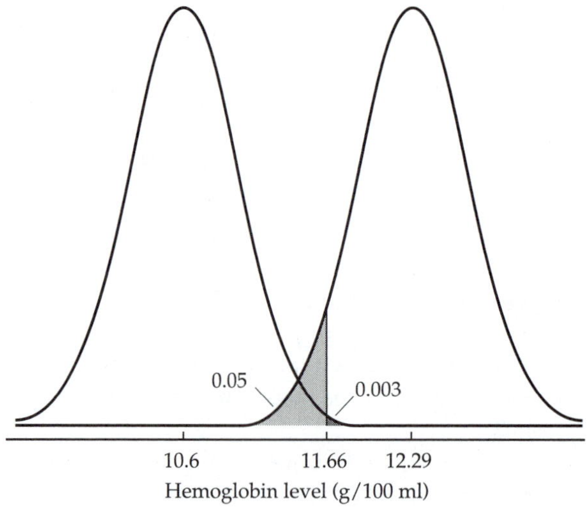

Figure 10.5. Distribution of means of samples of size 5 for the hemoglobin levels of children under the age of 6, $\mu = 10.60$ g/100 ml versus $\mu = 12.29$ g/100 ml

determine what proportion of the distribution centered at 10.60 g/100 ml lies to the right of 11.66. Observe that

$$\begin{aligned} z &= \frac{11.66 - 10.60}{0.85/\sqrt{5}} \\ &= 2.79. \end{aligned}$$

Referring to Table A.1, we find that the area under the standard normal curve that lies to the right of $z = 2.79$ is 0.003. Therefore, β is equal to 0.003.

The power of the test–or the probability of rejecting the null hypothesis given that H_0 is false and the true population mean is $\mu_1 = 10.60$ g/100 ml–is

$$\begin{aligned} 1 - \beta &= 1 - 0.003 \\ &= 0.997. \end{aligned}$$

Even with a sample size as small as 5, we are almost certain to reject H_0. This is in part because the standard deviation of the underlying population is so small.

10.8 Review Exercises

1. The distribution of diastolic blood pressures for the population of female diabetics between the ages of 30 and 34 has an unknown mean μ_d and a standard deviation $\sigma_d = 9.1$ mm Hg. It may be useful to physicians to know whether the

mean of this population is equal to the mean diastolic blood pressure of the general population of females in this age group, 74.4 mm Hg [10].

a. What is the null hypothesis of the test?
b. What is the alternative hypothesis?
c. A sample of ten diabetic women is selected; their mean diastolic blood pressure is $\bar{x}_d = 84$ mm Hg. Using this information, conduct a two-sided test at the $\alpha = 0.05$ level of significance. What is the p-value of the test?
d. What conclusion do you draw from the results of the test?
e. Would your conclusion have been different if you had chosen $\alpha = 0.01$ instead of $\alpha = 0.05$?

2. *E. canis* infection is a tick-borne disease of dogs that is sometimes contracted by humans. Among infected humans, the distribution of white blood cell counts has an unknown mean μ and a standard deviation σ. In the general population, the mean white blood cell count is 7250/mm^3 [11]. It is believed that persons infected with *E. canis* must on average have a lower white blood cell count.

a. What are the null and the alternative hypotheses for a one-sided test?
b. For a sample of 15 infected persons, the mean white blood cell count is $\bar{x} = 4767$/mm^3, and the standard deviation is $s = 3204$/mm^3 [12]. Conduct the test at the $\alpha = 0.05$ level.
c. What do you conclude?

3. The body mass index is calculated by dividing a person's weight by the square of his or her height; it is a measure of the extent to which the individual is overweight. For the population of middle-aged men who later develop diabetes mellitus, the distribution of baseline body mass indices has an unknown mean μ and a standard deviation σ. A sample of 58 men selected from this group has mean $\bar{x} = 25.0$ kg/m^2 and standard deviation $s = 2.7$ kg/m^2 [13].

a. Construct a 95% confidence interval for the population mean μ.
b. At the 0.05 level of significance, test whether the mean baseline body mass index for the population of middle-aged men who do develop diabetes is equal to 24.0 kg/m^2, the mean for the population of men who do not.
c. What do you conclude?

4. Data from the Framingham Study allow us to compare the distributions of initial serum cholesterol levels for two populations of males: those who go on to develop coronary heart disease and those who do not. The mean serum cholesterol level of the population of men who do not develop heart disease is $\mu = 219$ mg/100 ml, and the standard deviation is $\sigma = 41$ mg/100 ml [14]. Suppose, however, that you do not know the true population mean; instead, you hypothesize that μ is equal to 244 mg/100 ml. This is the mean initial serum cholesterol level of men who eventually develop disease. Since it is believed that the mean serum cholesterol level of the men who do not develop heart disease must be less than or equal to the mean level of men who do, a one-sided test conducted at the $\alpha = 0.05$ level of significance is appropriate.

a. What is the probability of making a type I error?
b. If a sample of size 25 is selected from the population of men who do not go on to develop coronary heart disease, what is the probability of committing a type II error?
c. What is the power of the test?
d. How could you increase the power?
e. You wish to test the null hypothesis

$$H_0: \mu \geq 244 \text{ mg/100 ml}$$

at the $\alpha = 0.05$ level of significance. If the true population mean is as low as 219 mg/100 ml, then you want to risk only a 5% chance of failing to reject H_0. How large a sample would be required?
f. How would the sample size change if you were willing to risk a 10% chance of failing to reject a false null hypothesis?

References

[1] National Center for Health Statistics, Fulwood, R., Kalsbeek, W., Rifkind, B., Russell-Briefel, R., Muesing, R., LaRosa, J., and Lippel, K., "Total Serum Cholesterol Levels of Adults 20–74 Years of Age: United States, 1976–1980," *Vital and Health Statistics*, Series 11, Number 236, May 1986.
[2] Kaplan, N. M., "Strategies to Reduce Risk Factors in Hypertensive Patients Who Smoke," *American Heart Journal*, Volume 115, January 1988, 288–294.
[3] Tsou, V. M., Young, R. M., Hart, M. H., and Vanderhoof, J. A., "Elevated Plasma Aluminum Levels in Normal Infants Receiving Antacids Containing Aluminum," *Pediatrics*, Volume 87, February 1991, 148–151.
[4] National Center for Health Statistics, Fulwood, R., Johnson, C. L., Bryner, J. D., Gunter, E. W., and McGrath, C. R., "Hematological and Nutritional Biochemistry Reference Data for Persons 6 Months–74 Years of Age: United States, 1976–1980," *Vital and Health Statistics*, Series 11, Number 232, December 1982.
[5] Clark, M., Royal, J., and Seeler, R., "Interaction of Iron Deficiency and Lead and the Hematologic Findings in Children with Severe Lead Poisoning," *Pediatrics*, Volume 81, February 1988, 247–253.
[6] "Firm Admits Using Rival's Drug in Tests," *The Boston Globe*, July 1, 1989, 41.
[7] "Child Abuse–and Trial Abuse," *The New York Times*, January 20, 1990, 24.
[8] Streissguth, A. P., Aase, J. M., Clarren, S. K., Randels, S. P., LaDue, R. A., and Smith, D. F., "Fetal Alcohol Syndrome in Adolescents and Adults," *Journal of the American Medical Association*, Volume 265, April 17, 1991, 1961–1967.
[9] Appel, B. R., Guirguis, G., Kim, I., Garbin, O., Fracchia, M., Flessel, C. P., Kizer, K. W., Book, S. A., and Warriner, T. E., "Benzene, Benzo(a)Pyrene, and Lead in Smoke from Tobacco Products Other Than Cigarettes," *American Journal of Public Health*, Volume 80, May 1990, 560–564.

[10] Klein, B. E. K., Klein, R., and Moss, S. E., "Blood Pressure in a Population of Diabetic Persons Diagnosed After 30 Years of Age," *American Journal of Public Health*, Volume 74, April 1984, 336–339.

[11] National Center for Health Statistics, Fulwood, R., Johnson, C. L., Bryner, J. D., Gunter, E. W., and McGrath, C. R., "Hematological and Nutritional Biochemistry Reference Data for Persons 6 Months–74 Years of Age: United States, 1976–1980," *Vital and Health Statistics*, Series 11, Number 232, December 1982.

[12] Rohrbach, B. W., Harkess, J. R., Ewing, S. A., Kudlac, J., McKee, G. L., and Istre, G. R., "Epidemiologic and Clinical Characteristics of Persons with Serologic Evidence of *E. canis* Infection," *American Journal of Public Health*, Volume 80, April 1990, 442–445.

[13] Feskens, E. J. M., and Kromhout, D., "Cardiovascular Risk Factors and the 25 Year Incidence of Diabetes Mellitus in Middle-Aged Men," *American Journal of Epidemiology*, Volume 130, December 1989, 1101–1108.

[14] MacMahon, S. W., and MacDonald, G. J., "A Population at Risk: Prevalence of High Cholesterol Levels in Hypertensive Patients in the Framingham Study," *American Journal of Medicine Supplement*, Volume 80, February 14, 1986, 40–47.

CHAPTER

11

Comparison of Two Means

In the preceding chapter we used a statistical test of hypothesis to compare the mean of a single population to some known value μ_0. In practical applications, however, the more common situation centers on comparing two different population means, both of which are unknown.

The idea of comparing populations to draw some conclusion has been around for hundreds of years. For example, in the 16th century it was believed that gunshot wounds were susceptible to infection and therefore needed to be cauterized. One of the earliest large scale uses of gunpowder occurred during the French invasion of Italy in 1537. This expedition was the first for a French army surgeon named Ambroise Paré; he reports on an attack on Turin [1]:

> Now all the said soldiers at the Château, seeing our men coming with great fury, did all they could to defend themselves and killed and wounded a great number of our soldiers with pikes, arquebuses, and stones, where the surgeons had much work cut out for them. Now I was at that time a freshwater soldier, I had not yet seen wounds made by gunshot at the first dressing. It is true that I had read Jean de Vigo, first book, "Of Wounds in General," chapter eight, that wounds made by firearms participate of venenosity, because of the powder, and for their cure he commands to cauterize them with oil of elder, scalding hot, in which should be mixed a little theriac, and in order not to err before using the said oil, knowing that such a thing could bring great pain to the patient, I wished to know first, how the other surgeons did for the first dressing which was to apply the said oil as hot as possible, into the wound with tents and setons, of whom I took courage to do as they did. At last my oil lacked and I was constrained to apply in its place a digestive made of the yolks of eggs, oil of roses and turpentine. That night I could not sleep at my ease, fearing by lack of cauterization that I should find the wounded on whom I had failed to put the said oil dead or empoisoned, which made me rise very early to visit them, where beyond my

> hope, I found those upon whom I had put the digestive medicament feeling little pain, and their wounds without inflammation or swelling having rested fairly well throughout the night; the others to whom I had applied the said boiling oil, I had found feverish, with great pain and swelling about their wounds. Then I resolved with myself never more to burn thus cruelly poor men wounded with gunshot.

The results of this comparison–one of the first documented clinical trials–were totally convincing. The same could be said of studies investigating the use of penicillin in the treatment of bacterial diseases. Unfortunately, such trials are the exception rather than the rule; progress is usually measured much more slowly.

This chapter introduces a procedure for deciding whether differences we have observed between populations are too large to be attributed to chance alone. A test of hypothesis involving two samples is similar in many respects to a test conducted for a single sample. We begin by specifying a null hypothesis; in most cases, we are interested in testing whether the two population means are equal. We then calculate the probability of obtaining a pair of sample means as discrepant or more discrepant than the observed means, given that the null hypothesis is true. If this probability is sufficiently small, we reject the null hypothesis and conclude that the two means are different. As before, we must specify a level of significance α and state whether we are interested in a one-sided or a two-sided test. The specific form of the analysis depends on the nature of the two sets of observations involved; in particular, we must determine whether the data come from paired or independent samples.

11.1 Paired Samples

The distinguishing characteristic of paired samples is that, for each observation in the first group, there is a corresponding observation in the second group. In the technique known as *self-pairing*, measurements are taken on a single subject at each of two distinct times. One common example of self-pairing is the "before and after" experiment in which each individual is examined both before and after a specified procedure or treatment. A second type of pairing occurs when an investigator matches the subjects in one group with those in a second group so that the members of a pair are as much alike as possible with respect to important characteristics such as age and sex.

Pairing is frequently employed in an attempt to control for extraneous sources of variation. If two measurements are made on the same subject, then a certain amount of the biological variability that exists among individuals is eliminated. The intent of pairing, therefore, is to make a comparison more precise.

Consider the data from a study in which each of 63 adult males with coronary artery disease is subjected to a series of exercise tests on a number of different occasions. On one day, a patient first undergoes an exercise test on a treadmill;

the length of time from the start of the test until the patient experiences angina is recorded. He is then exposed to plain room air for approximately one hour. At the end of this time, he performs a second exercise test; time until the onset of angina is again recorded. The observation of interest is the percent decrease in time to angina between the first and the second tests. If during the first test, for example, a man has an attack of angina after 983 seconds, and during the second test he has an attack after 957 seconds, then his percent decrease in time to angina is

$$\begin{aligned} \frac{983 - 957}{983} &= 0.026 \\ &= 2.6\%. \end{aligned}$$

The unknown population mean of this distribution of percent decreases is μ_1; for the 63 patients in the sample, the mean percent decrease is $\bar{x}_1 = 3.35\%$ [2].

On another day, the same patient undergoes a similar series of tests. This time, however, he is exposed to a mixture of air and carbon monoxide during the interval between the tests. The amount of carbon monoxide added to the air is intended to increase the patient's carboxyhemoglobin level to 4%—a level lower than that typically endured by smokers, but about what might be experienced by an individual sitting in heavy traffic in an area with poor ventilation. Again the observation of interest is the percent decrease in time to angina between the first and second tests. The unknown mean of this distribution is μ_2; the sample mean for the group of 63 subjects is $\bar{x}_2 = 9.63\%$.

For the first ten patients in the study, the percent decreases in time to angina for each of the two occasions are displayed in Figure 11.1. Note that, for eight of the men, the measure increases; for the other two, it decreases. We would like to determine whether there is any evidence of a difference in the percent

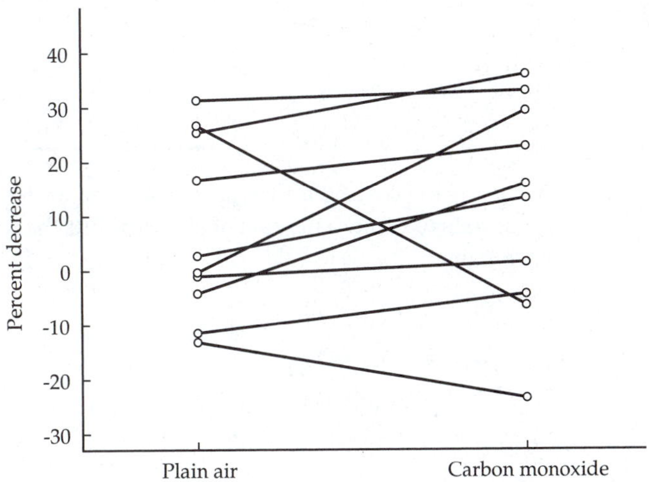

Figure 11.1. Percent decrease in time to angina on two different occasions for each of ten men with coronary artery disease

decrease in time to angina between the trial in which the subjects are exposed to carbon monoxide and the trial in which they are not. Since we believe that excessive exposure to carbon monoxide could not possibly be beneficial to an individual's health, we consider deviations that occur in one direction only; we therefore conduct a one-sided test at the $\alpha = 0.05$ level of significance. The null hypothesis is

$$H_0\colon \mu_1 \geq \mu_2$$

or

$$H_0\colon \mu_1 - \mu_2 \geq 0,$$

and the alternative is

$$H_A\colon \mu_1 - \mu_2 < 0.$$

In this study, each patient is subjected to the same series of tests both with and without exposure to carbon monoxide. This self-pairing eliminates any difficulties that might be introduced by comparing patients who differ with respect to age, weight, or severity of condition. Because the data consist of paired samples, the *paired t-test* is the appropriate method of analysis.

Rather than consider the two sets of observations to be distinct samples, we focus on the **difference** in measurements within each pair. Suppose that our two groups of observations are as follows:

Sample 1	Sample 2
x_{11}	x_{12}
x_{21}	x_{22}
x_{31}	x_{32}
.	.
.	.
x_{n1}	x_{n2}

In these samples, x_{11} and x_{12} are a pair, x_{21} and x_{22} are a pair, and so on. We use these data to create a new set of observations that represent the differences within each pair:

$$\begin{aligned} d_1 &= x_{11} - x_{12} \\ d_2 &= x_{21} - x_{22} \\ d_3 &= x_{31} - x_{32} \\ &\cdot \\ &\cdot \\ d_n &= x_{n1} - x_{n2} \end{aligned}$$

Instead of analyzing the individual observations, we use this difference between pairs as the variable of interest. Since the difference is treated as a single measure, we may apply the procedure for a one-sample test, which was discussed in Chapter 10.

The mean of the set of differences is

$$\bar{d} = \frac{\sum_{i=1}^{n} d_i}{n}$$

and the standard deviation is

$$s_d = \sqrt{\frac{\sum_{i=1}^{n}(d_i - \bar{d})^2}{n-1}}.$$

If we denote the true difference in population means by

$$\delta = \mu_1 - \mu_2,$$

then we can rewrite the null hypothesis as

$$H_0: \delta = 0.$$

H_0 may be tested by computing the statistic

$$t = \frac{\bar{d} - \delta}{s_d/\sqrt{n}};$$

note that $s_d/\sqrt{n}$ is the standard error of $\bar{d}$. If the population of differences is normally distributed and the null hypothesis is true, then this quantity has a t distribution with $n - 1$ degrees of freedom. We compare the outcome of t to the values in Table A.2 to find p, the probability of observing a mean difference as large or larger than $\bar{d}$ given that $\delta = 0$. If $p \leq \alpha$, we reject H_0. If $p > \alpha$, however, we do not reject the null hypothesis.

Returning to the study of males with coronary artery disease, we focus on the difference in measurements for a given individual. Therefore, for each of the 63 men in the study, we calculate the percent decrease in time to angina when he is exposed to carbon monoxide minus the percent decrease when he is exposed to unadulterated air. The mean of these differences is

$$\begin{aligned}\bar{d} &= \frac{\sum_{i=1}^{63} d_i}{63} \\ &= -6.63,\end{aligned}$$

and the standard deviation is

$$s_d = \sqrt{\frac{\sum_{i=1}^{63}(d_i - \bar{d})^2}{63 - 1}}$$
$$= 20.29.$$

If we rewrite the null hypothesis as

$$H_0: \delta \geq 0$$

where $\delta = \mu_1 - \mu_2$, then we can test H_0 using the statistic

$$t = \frac{\bar{d} - \delta}{s_d/\sqrt{n}}$$

or

$$t = \frac{-6.63 - 0}{20.29/\sqrt{63}}$$
$$= -2.59.$$

Referring to Table A.2, we observe that for a t distribution with $63-1 = 62$ degrees of freedom, the area under the curve to the left of $t_{62} = -2.59$ is between 0.005 and 0.01. Therefore, $0.005 < p < 0.01$. Rejecting the null hypothesis at the 0.05 level, we conclude that there is a significant difference between the mean percent decrease in time to angina when the patients are exposed to carbon monoxide and when they are exposed to air alone. Exposure to carbon monoxide increases the percent decrease in time to angina; in other words, exposed patients tend to develop angina more quickly.

We might also be interested in calculating an upper confidence bound for the true difference in population means δ. For a t distribution with 62 degrees of freedom, 95% of the observations lie above -1.671. Therefore,

$$P(t \geq -1.671) = P\left(\frac{\bar{d} - \delta}{s_d/\sqrt{n}} \geq -1.671\right)$$
$$= 0.95.$$

A one-sided 95% confidence interval for δ is

$$\delta \leq \bar{d} + 1.671 \frac{s_d}{\sqrt{n}}$$
$$= -6.63 + 1.671 \frac{20.29}{\sqrt{63}}$$
$$= -2.36.$$

We are 95% confident that the true difference in population means is less than or equal to −2.36%. In other words, we are 95% confident that the **decrease** in time to angina after exposure to carbon monoxide is at least 2.36%.

11.2 Independent Samples

Now suppose that we are given the measurements of serum iron level for two samples of children: one group healthy and the other group suffering from cystic fibrosis. The underlying populations are independent and normally distributed. If the population of diseased children has mean serum iron level μ_1 and the population of healthy children has mean μ_2, we might again be interested in testing the null hypothesis that the two population means are identical. This can be expressed as either

$$H_0: \mu_1 - \mu_2 = 0$$

or

$$H_0: \mu_1 = \mu_2.$$

The alternative hypothesis is

$$H_A: \mu_1 \neq \mu_2.$$

From the normal population with mean μ_1 and standard deviation σ_1 we draw a random sample of size n_1. The mean of this sample is denoted by $\bar{x}_1$ and its standard deviation by s_1. Similarly, we select a random sample of size n_2 from the normal population with mean μ_2 and standard deviation σ_2. The mean of this sample is represented by $\bar{x}_2$ and its standard deviation by s_2. Note that the numbers of observations in each sample—n_1 and n_2—need not be the same.

		Group 1	Group 2
Population	Mean	μ_1	μ_2
	Standard deviation	σ_1	σ_2
Sample	Mean	$\bar{x}_1$	$\bar{x}_2$
	Standard deviation	s_1	s_2
	Sample size	n_1	n_2

Two different scenarios arise in the comparison of independent samples. In the first, the variances of the underlying populations are either known or assumed to be equal. This leads to the *two-sample t-test*, which is omnipresent in the literature. In the second, the variances are not assumed to be the same;

in this case, the standard t-test is no longer valid. Before a test of means is carried out, many individuals believe that a preliminary test of variances should be conducted to distinguish between these two situations. Others object to the test on philosophical grounds; it is highly sensitive to the assumption of normality and has low power in many circumstances where the t-test should be avoided [3]. In addition, a modification of the two-sample test performed without this initial check has been shown to have high power in situations in which it is not known whether the underlying population variances are equal [4]. Since it is generally either unnecessary or ineffective, we do not recommend the use of a preliminary test of variances in this text.

11.2.1 Equal Variances

We first consider the situation in which it is either known or reasonable to assume that the population variances are identical. Recall that, for a single normal population with mean μ and standard deviation σ, the central limit theorem states that the sample mean $\overline{X}$ is approximately normally distributed with mean μ and standard error $\sqrt{\sigma^2/n} = \sigma/\sqrt{n}$. As a result,

$$z = \frac{\bar{x} - \mu}{\sigma/\sqrt{n}}$$

is the outcome of a standard normal random variable. When we are dealing with samples from two independent normal populations, an extension of the central limit theorem says that the difference in sample means $\overline{X}_1 - \overline{X}_2$ is approximately normal with mean $\mu_1 - \mu_2$ and standard error $\sqrt{\sigma_1{}^2/n_1 + \sigma_2{}^2/n_2}$. Since it is assumed that the population variances are equal, we substitute the common value σ^2 for both $\sigma_1{}^2$ and $\sigma_2{}^2$. Consequently, we know that

$$\begin{aligned} z &= \frac{(\bar{x}_1 - \bar{x}_2) - (\mu_1 - \mu_2)}{\sqrt{\sigma^2/n_1 + \sigma^2/n_2}} \\ &= \frac{(\bar{x}_1 - \bar{x}_2) - (\mu_1 - \mu_2)}{\sqrt{\sigma^2[(1/n_1) + (1/n_2)]}} \end{aligned}$$

is the outcome of a standard normal random variable. If the population variance σ^2 is in fact known, then this statistic can be used to test the null hypothesis

$$H_0: \mu_1 = \mu_2.$$

As we have previously noted, it is much more common that the true value of the population variance is not known. In this case, we use the test statistic

$$\mathrm{t} = \frac{(\bar{x}_1 - \bar{x}_2) - (\mu_1 - \mu_2)}{\sqrt{s_p^2\,[(1/n_1) + (1/n_2)]}}.$$

The quantity s_p^2 is a pooled estimate of the common variance σ^2. Under the null hypothesis that the population means are identical, $\mu_1 - \mu_2$ is equal to 0, and the test statistic t has a t distribution with $n_1 + n_2 - 2$ degrees of freedom. We compare the value of this statistic to the critical values in Table A.2 to find p, the probability of observing a discrepancy as large as $\bar{x}_1 - \bar{x}_2$ given that μ_1 is equal to μ_2. If $p \leq \alpha$, we reject the null hypothesis. If $p > \alpha$, then we do not reject H_0.

The pooled estimate of the variance, s_p^2, combines information from both samples to produce a more reliable estimate of σ^2. It can be calculated in two different ways. If we know the values of all the observations in the two samples, we apply the formula

$$s_p^2 = \frac{\sum_{j=1}^{n_1}(x_{j1} - \bar{x}_1)^2 + \sum_{i=1}^{n_2}(x_{i2} - \bar{x}_2)^2}{n_1 + n_2 - 2}.$$

If we are given s_1 and s_2 only, then we must use

$$s_p^2 = \frac{(n_1 - 1)s_1^2 + (n_2 - 1)s_2^2}{n_1 + n_2 - 2}.$$

This second formula demonstrates that s_p^2 is actually a weighted average of the two sample variances s_1^2 and s_2^2, where each variance is weighted by the degrees of freedom associated with it. If n_1 is equal to n_2, then s_p^2 is the simple arithmetic average; otherwise, more weight is given to the variance of the larger sample. Recalling that

$$s_1^2 = \frac{\sum_{j=1}^{n_1}(x_{j1} - \bar{x}_1)^2}{n_1 - 1}$$

and

$$s_2^2 = \frac{\sum_{i=1}^{n_2}(x_{i2} - \bar{x}_2)^2}{n_2 - 1},$$

we can see that the two formulas for computing s_p^2 are equivalent.

As an illustration of the two-sample t-test, consider the distributions of serum iron levels for the population of healthy children and the population of children with cystic fibrosis. Denote the mean serum iron level of the healthy children by μ_1 and the mean iron level of the children with disease by μ_2. The standard deviations of the two populations—σ_1 and σ_2—are unknown but assumed to be equal. We would like to determine whether children with cystic fibrosis have a normal level of iron in their blood; therefore, we test the null hypothesis that the two population means are identical,

$$H_0: \mu_1 = \mu_2.$$

A random sample is selected from each population. The sample of $n_1 =$ 9 healthy children has mean serum iron level $\bar{x}_1 = 18.9$ μmol/l and standard deviation $s_1 = 5.9$ μmol/l; the sample of $n_2 = 13$ children with cystic fibrosis has mean iron level $\bar{x}_2 = 11.9$ μmol/l and standard deviation $s_2 = 6.3$ μmol/l [5]. Is it likely that the observed difference in sample means—18.9 versus 11.9—is the result of chance variation, or should we conclude that the discrepancy is due to a true difference in population means?

Note that the two samples of children were randomly selected from distinct populations; in addition, the population variances are assumed to be equal. The two-sample t-test is thus the appropriate technique to apply. Recall that the null hypothesis states that there is no difference in the underlying population mean iron levels for the two groups of children. Since we wish to detect a difference that could occur in either direction, we conduct a two-sided test and set the significance level at $\alpha = 0.05$. The alternative hypothesis is

$$H_A: \mu_1 \neq \mu_2.$$

We begin by calculating the pooled estimate of the variance;

$$\begin{aligned} s_p^2 &= \frac{(n_1 - 1)\, s_1^2 + (n_2 - 1)\, s_2^2}{n_1 + n_2 - 2} \\ &= \frac{(9 - 1)\,(5.9)^2 + (13 - 1)\,(6.3)^2}{9 + 13 - 2} \\ &= \frac{(8)\,(34.81) + (12)\,(39.69)}{20} \\ &= 37.74. \end{aligned}$$

We next calculate the test statistic

$$\begin{aligned} t &= \frac{(\bar{x}_1 - \bar{x}_2) - (\mu_1 - \mu_2)}{\sqrt{s_p^2\, [(1/n_1) + (1/n_2)]}} \\ &= \frac{(18.9 - 11.9) - 0}{\sqrt{(37.74)\, [(1/9) + (1/13)]}} \\ &= -2.63. \end{aligned}$$

Referring to Table A.2, we observe that for a t distribution with $n_1 + n_2 - 2 = 9 + 13 - 2 = 20$ degrees of freedom, the total area under the curve to the left of $t_{20} = -2.63$ is between 0.005 and 0.01. Therefore, the sum of the areas to the left of $t_{20} = -2.63$ and to the right of $t_{20} = 2.63$ must be between 0.01 and 0.02. Since p is less than 0.05, we reject the null hypothesis

$$H_0: \mu_1 = \mu_2$$

at the 0.05 level of significance. The difference between the mean serum iron level of the healthy children and the mean level of the children with cystic fibrosis is statistically significant; based on these samples, it appears that children with cystic fibrosis suffer from an iron deficiency.

The quantity $\bar{x}_1 - \bar{x}_2$ provides a point estimate for the true difference in population means $\mu_1 - \mu_2$; however, we might also wish to construct a confidence interval. Note that for a t distribution with 20 degrees of freedom, 95% of the observations lie between −2.086 and 2.086. As a result,

$$P\left(-2.086 \leq \frac{(\overline{X}_1 - \overline{X}_2) - (\mu_1 - \mu_2)}{\sqrt{s_p^2\,[(1/n_1) + (1/n_2)]}} \leq 2.086\right) = 0.95.$$

Rearranging terms, we find that the 95% confidence limits for $\mu_1 - \mu_2$ are

$$(\bar{x}_1 - \bar{x}_2) \pm (2.086)\sqrt{s_p^2\left[\frac{1}{n_1} + \frac{1}{n_2}\right]}$$

or

$$(18.9 - 11.9) \pm (2.086)\sqrt{(37.74)\left[\frac{1}{9} + \frac{1}{13}\right]}.$$

Therefore, we are 95% confident that the interval

$$(-12.6\,,\ -1.4)$$

covers $\mu_1 - \mu_2$, the true difference in the mean serum iron levels of the two populations of children. This interval does not contain the value 0; consequently, it is consistent with the test of hypothesis conducted at the 0.05 level.

11.2.2 Unequal Variances

We now turn to the situation in which the variances of the two populations are not assumed to be equal. In this case, a modification of the two-sample t-test must be applied. Instead of using s_p^2 as an estimate of the common variance σ^2, we substitute s_1^2 for σ_1^2 and s_2^2 for σ_2^2. Therefore, the appropriate test statistic is

$$t = \frac{(\bar{x}_1 - \bar{x}_2) - (\mu_1 - \mu_2)}{\sqrt{(s_1^2/n_1) + (s_2^2/n_2)}}.$$

Unlike the case in which we had equal variances, the exact distribution of t is difficult to derive. Therefore, it is necessary to use an approximation [6]. We begin by calculating the quantity

$$v = \frac{[(s_1^2/n_1) + (s_2^2/n_2)]^2}{[(s_1^2/n_1)^2/(n_1 - 1) + (s_2^2/n_2)^2/(n_2 - 1)]};$$

the value of v is rounded down to the nearest integer. Under the null hypothesis, we may approximate the distribution of t by a t distribution with v degrees of freedom. We thus compare the value of the statistic to the critical values in Table A.2 to decide whether or not we should reject H_0.

Suppose that we are interested in investigating the effects of antihypertensive drug treatment on persons over the age of 60 who suffer from isolated systolic hypertension. By definition, individuals with isolated systolic hypertension have a systolic blood pressure greater than 160 mm Hg while their diastolic blood pressure is below 90 mm Hg. Before the start of the study, subjects who had been randomly selected to take the active drug and those chosen to receive a placebo were comparable with respect to blood pressure. After one year on the study, the mean systolic blood pressure of patients receiving the drug is denoted by μ_1 and the mean of those receiving the placebo by μ_2. The standard deviations of the two populations are unknown and not assumed to be equal. We would like to determine whether the mean systolic blood pressures of the patients in these two different groups remain the same; consequently, we test the null hypothesis

$$H_0: \mu_1 = \mu_2.$$

We begin by selecting a random sample from each of the two groups. The sample of $n_1 = 2,308$ individuals receiving the active drug treatment has mean systolic blood pressure $\bar{x}_1 = 142.5$ mm Hg and standard deviation $s_1 = 15.7$ mm Hg; the sample of $n_2 = 2,293$ persons receiving the placebo has mean $\bar{x}_2 = 156.5$ mm Hg and standard deviation $s_2 = 17.3$ mm Hg [7]. We are interested in detecting differences that could occur in either direction and thus conduct a two-sided test at the $\alpha = 0.05$ level of significance. The alternative hypothesis is

$$H_0: \mu_1 \neq \mu_2.$$

Since the two groups of patients were selected from independent populations and the variances were not assumed to be equal, the modified two-sample test should be applied. In this case, the test statistic is

$$\begin{aligned} t &= \frac{(\bar{x}_1 - \bar{x}_2) - (\mu_1 - \mu_2)}{\sqrt{(s_1^2/n_1) + (s_2^2/n_2)}} \\ &= \frac{(142.5 - 156.5) - 0}{\sqrt{[(15.7)^2/2,308] + [(17.3)^2/2,293]}} \\ &= -28.74. \end{aligned}$$

We next calculate the approximate degrees of freedom; since $s_1^2 = (15.7)^2 = 246.49$ and $s_2^2 = (17.3)^2 = 299.29$,

$$
\begin{aligned}
v &= \frac{[(s_1^2/n_1) + (s_2^2/n_2)]^2}{[(s_1^2/n_1)^2/(n_1-1) + (s_2^2/n_2)^2/(n_2-1)]} \\
&= \frac{[(246.49/2,308) + (299.29/2,293)]^2}{[(246.49/2,308)^2/(2,308-1) + (299.29/2,293)^2/(2,293-1)]} \\
&= 4,550.5.
\end{aligned}
$$

Rounding down to the nearest integer, $v = 4,550$. Since a t distribution with 4,550 degrees of freedom is for all practical purposes identical to the standard normal distribution, we may refer to either Table A.1 or Table A.2. In both cases, we find that p is less than 0.001. As a result, we reject the null hypothesis

$$H_0: \mu_1 = \mu_2$$

at the 0.05 level of significance. After one year, individuals who receive the active drug treatment have a lower mean systolic blood pressure than those who receive the placebo.

Once again we might wish to construct a confidence interval for the true difference in population means $\mu_1 - \mu_2$. For a t distribution with 4,550 degrees of freedom—or, effectively, the standard normal distribution—95% of the observations lie between −1.96 and 1.96. Consequently,

$$P\left(-1.96 \le \frac{(\overline{X}_1 - \overline{X}_2) - (\mu_1 - \mu_2)}{\sqrt{(s_1^2/n_1) + (s_2^2/n_2)}} \le 1.96\right) = 0.95.$$

If we rearrange terms, the 95% confidence limits for $\mu_1 - \mu_2$ are

$$(\bar{x}_1 - \bar{x}_2) \pm (1.96)\sqrt{\frac{s_1^2}{n_1} + \frac{s_2^2}{n_2}}$$

or

$$(142.5 - 156.5) \pm (1.96)\sqrt{\frac{(15.7)^2}{2,308} + \frac{(17.3)^2}{2,293}}.$$

Therefore, we are 95% confident that the interval

$$(-15.0\,,\ -13.0)$$

covers $\mu_1 - \mu_2$, the true difference in mean systolic blood pressures for the two populations. Note that the interval does not contain the value 0.

11.3 Further Applications

We now return to the study that investigated the effects of carbon monoxide exposure on patients with coronary artery disease. Previously we compared the mean percent decrease in time to angina when the patients were exposed to a level of carbon monoxide intended to raise their carboxyhemoglobin levels to 4% to the mean percent decrease when they were not exposed. We now wish to compare the percent decreases in time to angina when the same patients are exposed to carbon monoxide targeted to increase their carboxyhemoglobin levels to only 2% and when they are exposed to uncontaminated air. In this part of the study, therefore, each patient is exposed to a lower concentration of carbon monoxide.

Again we would like to know whether the two population means μ_1 and μ_2 are identical. Since we believe that exposure to carbon monoxide cannot be beneficial to an individual's health, we are concerned with deviations that occur in one direction only. We conduct a one-sided test at the $\alpha = 0.05$ level of significance; the null hypothesis is

$$H_0 : \delta \geq 0,$$

where $\delta = \mu_1 - \mu_2$, and the alternative hypothesis is

$$H_A : \delta < 0.$$

Rather than focusing on the two individual sets of observations, we observe the difference in the percent decreases in time to angina for each individual. The mean of these differences is

$$\begin{aligned}\bar{d} &= \frac{\sum_{i=1}^{63} d_i}{63} \\ &= -4.87,\end{aligned}$$

and their standard deviation is

$$\begin{aligned}s_d &= \sqrt{\frac{\sum_{i=1}^{63}(d_i - \bar{d})^2}{63 - 1}} \\ &= 18.90.\end{aligned}$$

Therefore, the test statistic is

$$\mathrm{t} = \frac{\bar{d} - \delta}{s_d/\sqrt{n}}$$

or

$$t = \frac{-4.87 - 0}{18.90/\sqrt{63}}$$
$$= -2.04.$$

For a t distribution with 62 degrees of freedom, $0.01 < p < 0.025$. Therefore, we reject the null hypothesis that there is no difference in means at the 0.05 level. The paired samples suggest that the mean percent decrease in time to angina when the patient is exposed to a low level of carbon monoxide is higher than the mean percent decrease when the patient is not exposed; once again, exposed patients tend to develop angina more quickly.

Rather than work through the paired t-test ourselves, we could have used a computer to perform the calculations. Suppose that the 63 measurements of percent decrease in time to angina when the patient is exposed to carbon monoxide are saved under the variable name `carbon`, and the corresponding measurements when the patient is exposed to uncontaminated air are saved under `air`. To perform the t-test, we must first generate the difference in percent decrease for each of the 63 subjects in the study; call this difference `diff`. We then proceed as if we were conducting a one-sample test. The appropriate output from Stata is provided in Table 11.1. In addition to summary statistics for the differences, the output reports the null hypothesis, the test statistic and its associated degrees of freedom, and the p-value of the test. Note, however, that $p = 0.0451$ is the p-value of the two-sided test. The p-value of the corresponding one-sided test is approximately 0.0451/2, or 0.0226.

Now consider a study that was designed to investigate the effect of lactose consumption on carbohydrate energy absorption among premature infants. In particular, we are interested in determining whether a reduction in lactose intake either increases or decreases absorption. In this study, one group of newborns was fed human milk; another received a formula that contained only half as much lactose. The distributions of carbohydrate energy absorption for the two populations are approximately normal and are assumed to have equal variances; we would like to know whether they also have identical means. Since we are

Table 11.1 Stata output illustrating the paired t-test

Variable	Obs	Mean	Std. Dev.
diff	63	-4.86986	18.90311

```
Ho:        mean  =  0
              t  =  -2.04 with 62 df
    Prob >| t |  =  0.0451
```

concerned with deviations that could occur in either direction, we test the null hypothesis

$$H_0: \mu_1 = \mu_2$$

against the two-sided alternative

$$H_A: \mu_1 \neq \mu_2.$$

A random sample of $n_1 = 8$ infants who were fed human milk has mean energy absorption $\bar{x}_1 = 87.375\%$ and standard deviation $s_1 = 4.56\%$; a sample of $n_2 = 10$ newborns who were given the formula has mean $\bar{x}_2 = 90.140\%$ and standard deviation $s_2 = 4.58\%$ [8]. Since the samples are independent and the underlying population variances are assumed to be equal, we apply the two-sample t-test.

We begin by calculating the pooled estimate of the variance;

$$\begin{aligned} s_p^2 &= \frac{(n_1 - 1)\, s_1^2 + (n_2 - 1)\, s_2^2}{n_1 + n_2 - 2} \\ &= \frac{(8-1)\,(4.56)^2 + (10-1)\,(4.58)^2}{8 + 10 - 2} \\ &= 20.90. \end{aligned}$$

The test statistic is

$$\begin{aligned} t &= \frac{(\bar{x}_1 - \bar{x}_2) - (\mu_1 - \mu_2)}{\sqrt{s_p^2\,[(1/n_1) + (1/n_2)]}} \\ &= \frac{(87.375 - 90.140) - 0}{\sqrt{(20.90)\,[(1/8) + (1/10)]}} \\ &= -1.27. \end{aligned}$$

For a t distribution with $8 + 10 - 2 = 16$ degrees of freedom, the total area under the curve to the left of -1.27 and to the right of 1.27 is greater than 0.10. Therefore, we do not reject the null hypothesis. Based on these samples, lactose intake does not appear to have an effect on carbohydrate energy absorption.

Once again we could have used the computer to conduct the test of hypothesis. Suppose that the eight measurements of carbohydrate energy absorption for the sample of infants who were fed human milk are saved under the variable name `milk` and that the ten measurements for the sample of infants receiving the formula are saved under the name `formula`. Sample output from Stata is contained in Table 11.2. This output includes summary statistics for each of the two variables and the combined sample, the null hypothesis, the test statistic and its associated degrees of freedom, and the exact p-value of the test.

Table 11.2 Stata output displaying the two-sample t-test with equal variances

Variable	Obs	Mean	Std. Dev.
milk	8	87.375	4.563127
formula	10	90.140	4.580442
combined	18	88.911	4.656164

```
Ho:       mean(x)   =   mean(y)      (assuming equal variances)
                t   =   -1.27 with 16 df
     Prob > | t |   =   0.2206
```

Another study was conducted to investigate the risk factors for heart disease among persons suffering from diabetes. One of the characteristics examined was body mass index, a measure of the extent to which an individual is overweight. We wish to determine whether the mean body mass index of the male diabetics is equal to the mean of the female diabetics. In each group, the distribution of indices is approximately normal; the variances are not assumed to be equal. We test the null hypothesis

$$H_0: \mu_1 = \mu_2$$

against the two-sided alternative

$$H_A: \mu_1 \neq \mu_2.$$

A random sample is selected from each population. The $n_1 = 207$ male diabetics have mean body mass index $\bar{x}_1 = 26.4$ kg/m^2 and standard deviation $s_1 = 3.3$ kg/m^2; the $n_2 = 127$ female diabetics have mean body mass index $\bar{x}_2 = 25.4$ kg/m^2 and standard deviation $s_2 = 5.2$ kg/m^2 [9]. Since the samples are independent and the underlying variances are not assumed to be equal, we apply the modified version of the two-sample test.

We begin by calculating the test statistic

$$\begin{aligned} t &= \frac{(\bar{x}_1 - \bar{x}_2) - (\mu_1 - \mu_2)}{\sqrt{(s_1^2/n_1) + (s_2^2/n_2)}} \\ &= \frac{(26.4 - 25.4) - 0}{\sqrt{[(3.3)^2/207] + [(5.2)^2/127]}} \\ &= 1.94. \end{aligned}$$

Table 11.3 Stata output displaying the two-sample test with unequal variances

Variable	Obs	Mean	Std. Dev.
men	207	26.4	3.3
women	127	25.4	5.2
combined	334	26.01976	

```
Ho:      mean(x)  =  mean(y)    (assuming unequal variances)
               t  =  1.94 with 188 df
     Prob >| t |  =  0.0538
```

Next, since $s_1^2 = (3.3)^2 = 10.89$ and $s_2^2 = (5.2)^2 = 27.04$,

$$\begin{aligned} v &= \frac{[(s_1^2/n_1) + (s_2^2/n_2)]^2}{[(s_1^2/n_1)^2/(n_1-1) + (s_2^2/n_2)^2/(n_2-1)]} \\ &= \frac{[(10.89/207) + (27.04/127)]^2}{[(10.89/207)^2/(207-1) + (27.04/127)^2/(127-1)]} \\ &= 188.9. \end{aligned}$$

Rounding down to the nearest integer, $v = 188$. For a t distribution with 188 degrees of freedom, we find that $0.05 < p < 0.10$. The results of this test are borderline; although we would reject the null hypothesis

$$H_0: \mu_1 = \mu_2$$

at the 0.10 level of significance, we would not reject it at the 0.05 level. It appears that males who have diabetes may tend to have a slightly higher body mass index–and therefore be more overweight–than females suffering from this disease.

Again we could have used the computer to perform these calculations. If we saved the 207 measurements of body mass index for male diabetics under the variable name `men` and the 127 measurements for females under the name `women`, then the sample output from Stata would look like Table 11.3. In addition to summary statistics for each of the two variables, the output contains the null hypothesis, the test statistic, and the p-value of the test.

11.4 Review Exercises

1. A study was conducted to investigate whether oat bran cereal helps to lower serum cholesterol in hypercholesterolemic males. Fourteen such individuals were randomly placed on a diet which included either oat bran or corn flakes; after two weeks, their low-density lipoprotein cholesterol levels were recorded.

Each man was then switched to the alternative diet. After a second two week period, the LDL cholesterol level of each individual was again recorded. The data from this study are provided below [10].

	LDL (mmol/l)	
Subject	Corn Flakes	Oat Bran
1	4.61	3.84
2	6.42	5.57
3	5.40	5.85
4	4.54	4.80
5	3.98	3.68
6	3.82	2.96
7	5.01	4.41
8	4.34	3.72
9	3.80	3.49
10	4.56	3.84
11	5.35	5.26
12	3.89	3.73
13	2.25	1.84
14	4.24	4.14

a. What are the appropriate null and alternative hypotheses for a two-sided test?
b. Conduct the test at the 0.05 level of significance.
c. What do you conclude?

2. The following data come from a study that examines the efficacy of saliva cotinine as an indicator for exposure to tobacco smoke. In one part of the study, seven subjects–none of whom were heavy smokers and all of whom had abstained from smoking for at least one week before the study–were required to smoke a single cigarette. Samples of saliva were taken from all individuals 2, 12, 24, and 48 hours after smoking the cigarette. The cotinine levels at 12 hours and at 24 hours are provided below [11].

	Cotinine Levels (nmol/l)	
Subject	After 12 Hours	After 24 Hours
1	73	24
2	58	27
3	67	49
4	93	59
5	33	0
6	18	11
7	147	43

Let μ_{12} represent the population mean cotinine level 12 hours after smoking the cigarette and μ_{24} the mean cotinine level 24 hours after smoking. It is believed that μ_{24} must be lower than μ_{12}.

a. Construct a one-sided 95% confidence interval for the true difference in population means $\mu_{12} - \mu_{24}$.
b. Test the null hypothesis that the population means are identical at the $\alpha = 0.05$ level of significance.

3. A study was conducted to determine whether maternal cigarette smoking has any effect on the bone mineral content of healthy newborns. A sample of 77 infants whose mothers had smoked during pregnancy has mean bone mineral content $\bar{x}_1 = 0.098$ g/cm and standard deviation $s_1 = 0.026$ g/cm; a sample of 161 children whose mothers had not smoked has mean $\bar{x}_2 = 0.095$ g/cm and standard deviation $s_2 = 0.025$ g/cm [12]. Assume that the underlying population variances are equal.

a. State the null and alternative hypotheses of the two-sided test.
b. Conduct the test at the 0.05 level of significance.
c. What do you conclude?

4. In a study of pregnancy-induced hypertension, one group of women was treated with low-dose aspirin, and a second group was given a placebo. A sample consisting of 23 women who received aspirin has mean arterial blood pressure $\bar{x}_1 = 111$ mm Hg and standard deviation $s_1 = 8$ mm Hg; a sample of 24 women who were given the placebo has mean blood pressure $\bar{x}_2 = 109$ mm Hg and standard deviation $s_2 = 8$ mm Hg [13].

a. At the 0.05 level of significance, test the null hypothesis that the two populations of women have the same mean arterial blood pressure.
b. Construct a 90% confidence interval for the true difference in population means.

5. The table below compares the levels of carboxyhemoglobin for a group of nonsmokers and a group of cigarette smokers. Sample means and standard deviations are provided [14]. It is believed that the mean carboxyhemoglobin level of the smokers must be higher than the mean level of the nonsmokers. Do not assume that the underlying population variances are identical.

Group	n	Carboxyhemoglobin (%)
Nonsmokers	121	$\bar{x} = 1.3$, $s = 1.3$
Smokers	75	$\bar{x} = 4.1$, $s = 2.0$

a. What are the null and alternative hypotheses of the one-sided test?
b. Conduct the test at the 0.01 level of significance.
c. What do you conclude?

6. Suppose that you wish to compare the characteristics of tuberculosis meningitis in patients infected with the human immunodeficiency virus and those who are not infected. In particular, you would like to determine whether the two populations have the same mean age. A sample of 37 infected patients has mean age $\bar{x}_1 = 27.9$ years and standard deviation $s_1 = 5.6$ years; a sample of 19 patients who are not infected has mean age $\bar{x}_2 = 38.8$ years and standard deviation $s_2 = 21.7$ years [15].

a. Test the null hypothesis that the two populations of patients have the same mean age at the 0.05 level of significance.
b. Construct a 95% confidence interval for the true difference in population means.

7. Consider the numbers of community hospital beds per 1,000 population that are available in each state in the United States and in the District of Columbia. The data for both 1980 and 1986 are stored on your disk in a data set called `bed` [16]. The values for 1980 are saved under the variable name `bed80`; those for 1986 are saved under `bed86`.

a. Generate descriptive statistics for the number of beds in each year.
b. Since there are two observations for each state–one for 1980 and one for 1986–the data are actually paired. A common error in analyzing this type of data is to ignore the pairing and assume that the samples are independent. Compare the mean number of community hospital beds per 1,000 population in 1980 to the mean number of beds in 1986 using the two-sample t-test.
c. Now carry out the paired t-test, first creating a new variable that is the difference between `bed80` and `bed86`.
d. Comment on the differences between the two tests. Do you reach the same conclusion in each case?

References

[1] Packard, F. R., *The Life and Times of Ambroise Paré, 1510–1590*, New York: Paul B. Hoeber, 1921.

[2] Allred, E. N., Bleecker, E. R., Chaitman, B. R., Dahms, T. E., Gottlieb, S. O., Hackney, J. D., Hayes, D., Pagano, M., Selvester, R. H., Walden, S. M., and Warren, J., "Acute Effects of Carbon Monoxide Exposure on Individuals with Coronary Artery Disease," *Health Effects Institute Research Report Number 25*, November 1989.

[3] Markowski, C. A., and Markowski, E. P., "Conditions for the Effectiveness of a Preliminary Test of Variance," *The American Statistician*, Volume 44, November 1990, 322–326.

[4] Moser, B. K., and Stevens, G. R., "Homogeneity of Variance in the Two-Sample Means Test," *The American Statistician*, Volume 46, February 1992, 19–21.

[5] Zempsky, W. T., Rosenstein, B. J., Carroll, J. A., and Oski, F. A., "Effect of Pancreatic Enzyme Supplements on Iron Absorption," *American Journal of Diseases of Children*, Volume 143, August 1989, 966–972.

[6] Satterthwaite, F. W., "An Approximate Distribution of Estimates of Variance Components," *Biometrics Bulletin*, Volume 2, December 1946, 110–114.

[7] SHEP Cooperative Research Group, "Prevention of Stroke by Antihypertensive Drug Treatment in Older Persons with Isolated Systolic Hypertension: Final Results of the Systolic Hypertension in the Elderly Program (SHEP)," *Journal of the American Medical Association*, Volume 265, June 26, 1991, 3255–3264.

[8] Kien, C. L., Liechty, E. A., and Mullett, M. D., "Effects of Lactose Intake on Nutritional Status in Premature Infants," *Journal of Pediatrics*, Volume 116, March 1990, 446–449.

[9] Barrett-Connor, E. L., Cohn, B. A., Wingard, D. L., and Edelstein, S. L., "Why Is Diabetes Mellitus a Stronger Risk Factor for Fatal Ischemic Heart Disease in Women Than in Men?," *Journal of the American Medical Association*, Volume 265, February 6, 1991, 627–631.

[10] Anderson, J. W., Spencer, D. B., Hamilton, C. C., Smith, S. F., Tietyen, J., Bryant, C. A., and Oeltgen, P., "Oat-Bran Cereal Lowers Serum Total and LDL Cholesterol in Hypercholesterolemic Men," *American Journal of Clinical Nutrition*, Volume 52, September 1990, 495–499.

[11] DiGiusto, E., and Eckhard, I., "Some Properties of Saliva Cotinine Measurements in Indicating Exposure to Tobacco Smoking," *American Journal of Public Health*, Volume 76, October 1986, 1245–1246.

[12] Venkataraman, P. S., and Duke, J. C., "Bone Mineral Content of Healthy, Full-term Neonates: Effect of Race, Gender, and Maternal Cigarette Smoking," *American Journal of Diseases of Children*, Volume 145, November 1991, 1310–1312.

[13] Schiff, E., Barkai, G., Ben-Baruch, G., and Mashiach, S., "Low-Dose Aspirin Does Not Influence the Clinical Course of Women with Mild Pregnancy-Induced Hypertension," *Obstetrics and Gynecology*, Volume 76, November 1990, 742–744.

[14] Jarvis, M. J., Tunstall-Pedoe, H., Feyerabend, C., Vesey, C., and Saloojee, Y., "Comparison of Tests Used to Distinguish Smokers from Nonsmokers," *American Journal of Public Health*, Volume 77, November 1987, 1435–1438.

[15] Berenguer, J., Moreno, S., Laguna, F., Vicente, T., Adrados, M., Ortega, A., González-LaHoz, J., and Bouza, E., "Tuberculosis Meningitis in Patients Infected with the Human Immunodeficiency Virus," *The New England Journal of Medicine*, Volume 326, March 5, 1992, 668–672.

[16] National Center for Health Statistics, *Health United States 1988*, Public Health Service, Hyattsville, Md.: March 1989.

CHAPTER 12

Analysis of Variance

In the previous chapter we covered techniques for determining whether a difference exists between the means of two independent populations. It is not unusual, however, to encounter situations in which we wish to test for a difference among several independent means rather than between only two. The extension of the two-sample t-test to three or more samples is known as the *analysis of variance*.

12.1 One-Way Analysis of Variance

12.1.1 The Problem

When discussing the paired t-test, we examined data from a study that investigates the effect of carbon monoxide exposure on patients with coronary artery disease. The subjects involved in the study were recruited from three different medical centers–the Johns Hopkins University School of Medicine, the Rancho Los Amigos Medical Center, and the St. Louis University School of Medicine. Before combining the subjects into one large group to conduct the analysis, we can first examine some baseline characteristics to ensure that the patients from the various centers are in fact comparable.

One characteristic that we might wish to consider is pulmonary function before the start of the study; if the patients from one medical center begin with measures of forced expiratory volume in 1 second that are much larger–or much smaller–than those from the other centers, then the results of the analysis may be affected. Therefore, given that the populations of patients in the three centers have mean baseline FEV_1 measurements μ_1, μ_2, and μ_3 respectively, we would like to test the null hypothesis that the population means are identical. This may be expressed as

$$H_0: \mu_1 = \mu_2 = \mu_3.$$

The alternative hypothesis is that at least one of the population means differs from the others.

In general, we are interested in comparing the means of k different populations. Suppose that the k populations are independent and normally distributed. We begin by drawing a random sample of size n_1 from the normal population with mean μ_1 and standard deviation σ_1. The mean of this sample is denoted by $\bar{x}_1$ and its standard deviation by s_1. Similarly, we select a random sample of size n_2 from the normal population with mean μ_2 and standard deviation σ_2, and so on for the remaining populations. The numbers of observations in each sample need not be the same.

		Group 1	Group 2	...	Group k
Population	Mean	μ_1	μ_2	...	μ_k
	Standard deviation	σ_1	σ_2	...	σ_k
Sample	Mean	$\bar{x}_1$	$\bar{x}_2$	...	$\bar{x}_k$
	Standard deviation	s_1	s_2	...	s_k
	Sample size	n_1	n_2	...	n_k

For the study investigating the effects of carbon monoxide exposure on individuals with coronary artery disease, the FEV_1 distributions of patients associated with each of the three medical centers comprise distinct populations. From the population of FEV_1 measurements for the patients at Johns Hopkins University, we select a sample of size $n_1 = 21$. From the population at Rancho Los Amigos we draw a sample of size $n_2 = 16$, and from the one at St. Louis University we select a sample of size $n_3 = 23$. The data, along with their sample means and standard deviations, are provided in Table 12.1 [1].

Presented with these data, we might attempt to compare the three population means by evaluating all possible pairs of sample means using the two-sample t-test. For a total of three groups, the number of tests required is $\binom{3}{2} = 3$. We would compare group 1 to group 2, group 1 to group 3, and group 2 to group 3. We assume that the variances of the underlying populations are all equal, or

$$\sigma_1^2 = \sigma_2^2 = \sigma_3^2.$$

The pooled estimate of the common variance, which we denote s_W^2, combines information from all three samples; in particular,

$$s_W^2 = \frac{(n_1 - 1)s_1^2 + (n_2 - 1)s_2^2 + (n_3 - 1)s_3^2}{n_1 + n_2 + n_3 - 3}.$$

This quantity is simply an extension of s_p^2, the pooled estimate of the variance used in the two-sample t-test.

Table 12.1 Forced expiratory volume in 1 second for patients with coronary artery disease sampled at three different medical centers

Johns Hopkins	Rancho Los Amigos	St. Louis
3.23	3.22	2.79
3.47	2.88	3.22
1.86	1.71	2.25
2.47	2.89	2.98
3.01	3.77	2.47
1.69	3.29	2.77
2.10	3.39	2.95
2.81	3.86	3.56
3.28	2.64	2.88
3.36	2.71	2.63
2.61	2.71	3.38
2.91	3.41	3.07
1.98	2.87	2.81
2.57	2.61	3.17
2.08	3.39	2.23
2.47	3.17	2.19
2.47		4.06
2.74		1.98
2.88		2.81
2.63		2.85
2.53		2.43
		3.20
		3.53
$\bar{x}_1 = 2.63$ liters	$\bar{x}_2 = 3.03$ liters	$\bar{x}_3 = 2.88$ liters
$s_1 = 0.496$ liters	$s_2 = 0.523$ liters	$s_3 = 0.498$ liters

Performing all possible pairs of tests is not a problem if the number of populations is relatively small. In the instance where $k = 3$, there are only three such tests. If $k = 10$, however, the process becomes much more complicated. In this case, we would have to perform $\binom{10}{2} = 45$ different pairwise tests.

More important, another problem that arises when all possible two-sample tests are conducted is that this procedure is likely to lead to an incorrect conclusion. Suppose that the three population means are in fact equal and that we conduct all three pairwise tests. Assume that the tests are independent and set the significance level at 0.05 for each one. By the multiplicative rule, the probability of failing to reject a null hypothesis of no difference in all three instances would be

$$\begin{aligned} \text{P(fail to reject in all three tests)} &= (1 - 0.05)^3 \\ &= (0.95)^3 \\ &= 0.857. \end{aligned}$$

Consequently, the probability of rejecting the null hypothesis in at least one of the tests would be

$$\begin{aligned} \text{P(reject in at least one test)} &= 1 - 0.857 \\ &= 0.143. \end{aligned}$$

Since we know that the null hypothesis is true in each case, 0.143 is the probability of committing a type I error. As we can see, the combined probability of a type I error for the set of three tests is much larger than 0.05. In reality the problem is even more complex; since each of the t-tests is conducted using the same set of data, we cannot assume that they are all independent. We would like to be able to use a testing procedure in which the overall probability of committing a type I error is equal to some predetermined level α. The one-way analysis of variance is such a technique.

12.1.2 Sources of Variation

As its name implies, the one-way analysis of variance procedure is dependent on estimates of dispersion. When working with several different populations, we can compute two measures of variance: the variation of the individual values around their population means and the variation of the population means around the overall mean. If the variability within the k different populations is small relative to the variability among their respective means, this suggests that the population means are in fact different.

To test the null hypothesis

$$H_0: \mu_1 = \mu_2 = \cdots = \mu_k$$

for a set of k populations, we first need to find a measure of the variability of the individual observations around their population means. The pooled estimate of the common variance σ^2 is one such measure; if we let $n = n_1 + n_2 + \ldots + n_k$, then

$$\begin{aligned} s_W^2 &= \frac{(n_1 - 1)s_1^2 + (n_2 - 1)s_2^2 + \cdots + (n_k - 1)s_k^2}{n_1 + n_2 + \cdots + n_k - k} \\ &= \frac{(n_1 - 1)s_1^2 + (n_2 - 1)s_2^2 + \cdots + (n_k - 1)s_k^2}{n - k}. \end{aligned}$$

This quantity is simply a weighted average of the k individual sample variances. Note that the subscript W refers to the "within-groups" variability.

We next need an expression that estimates the extent to which the population means vary around the overall mean. If the null hypothesis is true and the means are identical, then this quantity also estimates the common variance σ^2. In particular,

$$s_B^2 = \frac{n_1(\bar{x}_1 - \bar{x})^2 + n_2(\bar{x}_2 - \bar{x})^2 + \cdots + n_k(\bar{x}_k - \bar{x})^2}{k - 1}.$$

The terms $(\bar{x}_i - \bar{x})^2$ are the squared deviations of the sample means $\bar{x}_i$ from the grand mean $\bar{x}$. The *grand mean* is defined as the overall average of the n observations that make up the k different samples; consequently,

$$\begin{aligned}\bar{x} &= \frac{n_1\bar{x}_1 + n_2\bar{x}_2 + \cdots + n_k\bar{x}_k}{n_1 + n_2 + \cdots + n_k} \\ &= \frac{n_1\bar{x}_1 + n_2\bar{x}_2 + \cdots + n_k\bar{x}_k}{n}.\end{aligned}$$

The subscript B denotes the "between-groups" variability.

Now that we have the two estimates of the variance, we would like to be able to answer the following question: Do the sample means vary around the grand mean more than the individual observations vary around the sample means? If they do, then this implies that the corresponding population means are in fact different. To test the null hypothesis that the population means are identical, we use the test statistic

$$F = \frac{s_B^2}{s_W^2}.$$

Under the null hypothesis, both s_W^2 and s_B^2 estimate the common variance σ^2, and F is close to 1. If there is a difference among populations, then the between-groups variance exceeds the within-groups variance, and F is greater than 1. Under H_0, the ratio F has an F distribution with $k-1$ and $n-k$ degrees of freedom; the degrees of freedom correspond to the numerator and the denominator, respectively. We represent this distribution using the notation $F_{n_1-1,\, n_2-1}$. If we have only two independent samples, the *F-test* reduces to the two-sample t-test.

The F distribution is similar to the t in that it is not unique; there is a different F distribution for each possible pair of values $n_1 - 1$ and $n_2 - 1$. Unlike the t distribution, however, the F distribution cannot assume negative values. In addition, it is skewed to the right. The extent to which it is skewed is determined by the degrees of freedom. The F distribution with 4 and 2 degrees of freedom is pictured in Figure 12.1.

Table A.3 in Appendix A is a table of critical values computed for the family of F distributions. Only selected percentiles are included–in this case, the upper 10.0, 5.0, 2.5, 1.0, 0.05, and 0.01% of the distributions. The degrees of freedom for the numerator are displayed across the top of the table, and the degrees of freedom for the denominator are listed down the left-hand side. For any given combination, the corresponding entry in the table represents the value of $F_{n_1-1,\, n_2-1}$ that cuts off the specified area in the upper tail of the distribution. Given an F distribution with 4 and 2 degrees of freedom, for example, the table shows that $F_{4,\,2} = 19.25$ cuts off the upper 5% of the curve.

Referring back to the FEV_1 data collected for patients from three different medical centers, we are interested in testing

$$H_0\colon \mu_1 = \mu_2 = \mu_3,$$

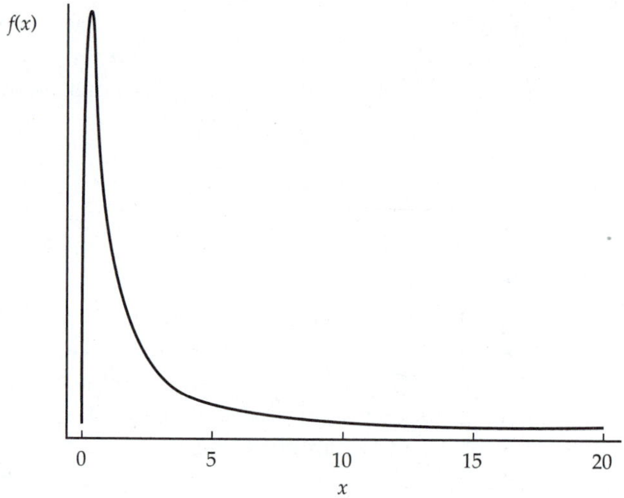

Figure 12.1. The F distribution with 4 and 2 degrees of freedom

the null hypothesis that the mean forced expiratory volumes in 1 second for each of the three centers are identical. To begin, we compute the estimate of the within-groups variance;

$$\begin{aligned} s_W^2 &= \frac{(n_1 - 1)s_1^2 + (n_2 - 1)s_2^2 + (n_3 - 1)s_3^2}{n_1 + n_2 + n_3 - 3} \\ &= \frac{(21 - 1)(0.496)^2 + (16 - 1)(0.523)^2 + (23 - 1)(0.498)^2}{21 + 16 + 23 - 3} \\ &= 0.254 \text{ liters}^2. \end{aligned}$$

Since

$$\begin{aligned} \bar{x} &= \frac{n_1\bar{x}_1 + n_2\bar{x}_2 + n_3\bar{x}_3}{n_1 + n_2 + n_3} \\ &= \frac{21(2.63) + 16(3.03) + 23(2.88)}{21 + 16 + 23} \\ &= 2.83 \text{ liters}, \end{aligned}$$

the estimate of the between-groups variance is

$$\begin{aligned} s_B^2 &= \frac{n_1(\bar{x}_1 - \bar{x})^2 + n_2(\bar{x}_2 - \bar{x})^2 + n_3(\bar{x}_3 - \bar{x})^2}{3 - 1} \\ &= \frac{21(2.63 - 2.83)^2 + 16(3.03 - 2.83)^2 + 23(2.88 - 2.83)^2}{3 - 1} \\ &= 0.769 \text{ liters}^2. \end{aligned}$$

Therefore, the test statistic is

$$
\begin{aligned}
F &= \frac{s_B^2}{s_W^2} \\
&= \frac{0.769}{0.254} \\
&= 3.028.
\end{aligned}
$$

For an F distribution with $k - 1 = 3 - 1 = 2$ and $n - k = 60 - 3 = 57$ degrees of freedom, $0.05 < p < 0.10$. Although we would reject the null hypothesis at the 0.10 level, we do not reject it at the 0.05 level. There may possibly be some difference among the mean FEV_1 measurements for these three populations.

12.2 Multiple Comparisons Procedures

As we have seen, one-way analysis of variance may be used to test the null hypothesis that k population means are identical,

$$H_0: \mu_1 = \mu_2 = \cdots = \mu_k.$$

What happens, however, if we reject H_0? Although we are able to conclude that the population means are not all equal, we cannot be more specific than this. We do not know whether all of the means are different from one another or if only some of them are different. Once we reject the null hypothesis, therefore, we often want to conduct additional tests to find out where the differences lie.

Many different techniques for conducting multiple comparisons exist; they typically involve testing each pair of means individually. In the previous section we mentioned one possible approach: we could perform a series of $\binom{k}{2}$ two-sample t-tests. As we noted, however, performing multiple tests increases the probability that we will commit a type I error. We can avoid this problem by being more conservative in our individual comparisons; by reducing the individual α levels, we can ensure that the overall level of significance is kept at a predetermined level.

The significance levels for each of the individual comparisons depend on the number of tests being conducted; the greater the number of tests, the smaller they must be. To set the overall probability of committing a type I error at 0.05, for example, we should use

$$\alpha^* = \frac{0.05}{\binom{k}{2}}$$

as the significance level for an individual comparison. This modification is known as the *Bonferroni correction*.

For the case in which we have $k = 3$ populations, a total of $\binom{3}{2} = 3$ tests are required. If we wish to set the overall level of significance at 0.10, therefore, we must use

$$\alpha^* = \frac{0.10}{3} = 0.033$$

for each individual test.

To conduct a test of the null hypothesis

$$H_0: \mu_i = \mu_j,$$

we calculate

$$t_{ij} = \frac{\bar{x}_i - \bar{x}_j}{\sqrt{s_W^2\,[(1/n_i) + (1/n_j)]}},$$

the test statistic for a two-sample t-test. Note that, instead of using the data from only two samples to estimate the common variance, we take advantage of the additional information and use all k samples. Under the null hypothesis, t_{ij} has a t distribution with $n - k$ degrees of freedom.

For the comparison of baseline FEV_1 values among the three medical centers, we begin by considering populations 1 and 2, the patients at Johns Hopkins and those at Rancho Los Amigos. In this case,

$$\begin{aligned} t_{12} &= \frac{\bar{x}_1 - \bar{x}_2}{\sqrt{s_W^2\,[(1/n_1) + (1/n_2)]}} \\ &= \frac{2.63 - 3.03}{\sqrt{0.254\,[(1/21) + (1/16)]}} \\ &= -2.39. \end{aligned}$$

For a t distribution with $n - k = 60 - 3 = 57$ degrees of freedom, $p = 0.02$. Therefore, we reject the null hypothesis at the 0.033 level and conclude that μ_1 is not equal to μ_2.

We next compare populations 1 and 3, the group at Johns Hopkins and the group at St. Louis. The test statistic is

$$\begin{aligned} t_{13} &= \frac{\bar{x}_1 - \bar{x}_3}{\sqrt{s_W^2\,[(1/n_1) + (1/n_3)]}} \\ &= \frac{2.63 - 2.88}{\sqrt{0.254\,[(1/21) + (1/23)]}} \\ &= -1.64. \end{aligned}$$

Since $p > 0.10$, we do not have sufficient evidence to conclude that μ_1 and μ_3 differ.

Finally, we compare the patients at Rancho Los Amigos and those at St. Louis;

$$
\begin{aligned}
t_{23} &= \frac{\bar{x}_2 - \bar{x}_3}{\sqrt{s_W^2\,[(1/n_2) + (1/n_3)]}} \\
&= \frac{3.03 - 2.88}{\sqrt{0.254\,[(1/16) + (1/23)]}} \\
&= 0.91.
\end{aligned}
$$

Once again, $p > 0.10$. We are unable to reject the null hypothesis that μ_2 is equal to μ_3. In summary, we find that the mean baseline FEV_1 measurement for the patients at Johns Hopkins is somewhat lower than the mean for the patients at Rancho Los Amigos; we cannot make any further distinctions.

One disadvantage of the Bonferroni multiple comparisons procedure is that it can suffer from a lack of power. It is highly conservative and may fail to detect a difference in means that actually exists. However, there are many other competing multiple comparisons procedures that could be used instead [2]. The appropriate technique to apply in a given situation depends on both the characteristics of the data and the questions that we wish to address.

12.3 Further Applications

A study was conducted that followed three groups of overweight males for a period of one year. The first group decreased its energy intake by dieting but did not participate in an exercise program. The second group exercised regularly but did not alter its eating habits. The third group changed neither diet nor level of physical activity. At the end of one year, total change in body weight was measured for each individual. Among these three populations, is there any difference in mean change in body weight?

We begin by selecting a random sample from each population. The sample of 42 men who dieted has mean change in body weight $\bar{x}_1 = -7.2$ kg and standard deviation $s_1 = 3.7$ kg. The sample of 47 men who participated in an exercise program has mean $\bar{x}_2 = -4.0$ kg and standard deviation $s_2 = 3.9$ kg. Finally, the sample of 42 men who neither dieted nor exercised has mean $\bar{x}_3 = 0.6$ kg and standard deviation $s_3 = 3.7$ kg [3].

We are interested in testing the null hypothesis that the mean changes in total body weight are identical for the three populations, or

$$H_0: \mu_1 = \mu_2 = \mu_3.$$

We assume that the underlying population variances are equal. It should be noted, however, that the one-way analysis of variance procedure is relatively insensitive to departures from this assumption.

To conduct the test, we begin by computing an estimate of the within-groups variance;

$$\begin{aligned} s_W^2 &= \frac{(n_1 - 1)s_1^2 + (n_2 - 1)s_2^2 + (n_3 - 1)s_3^2}{n_1 + n_2 + n_3 - 3} \\ &= \frac{(42-1)(3.7)^2 + (47-1)(3.9)^2 + (42-1)(3.7)^2}{42 + 47 + 42 - 3} \\ &= 14.24 \text{ kg}^2. \end{aligned}$$

Since the grand mean of the data is

$$\begin{aligned} \bar{x} &= \frac{n_1\bar{x}_1 + n_2\bar{x}_2 + n_3\bar{x}_3}{n_1 + n_2 + n_3} \\ &= \frac{42(-7.2) + 47(-4.0) + 42(0.6)}{42 + 47 + 42} \\ &= -3.55 \text{ kg}, \end{aligned}$$

the estimate of the between-groups variance is

$$\begin{aligned} s_B^2 &= \frac{n_1(\bar{x}_1 - \bar{x})^2 + n_2(\bar{x}_2 - \bar{x})^2 + n_3(\bar{x}_3 - \bar{x})^2}{3 - 1} \\ &= \frac{42(-7.2 + 3.55)^2 + 47(-4.0 + 3.55)^2 + 42(0.6 + 3.55)^2}{3 - 1} \\ &= 646.20 \text{ kg}^2. \end{aligned}$$

Consequently, the test statistic is

$$\begin{aligned} F &= \frac{s_B^2}{s_W^2} \\ &= \frac{646.20}{14.24} \\ &= 45.38. \end{aligned}$$

For an F distribution with $k - 1 = 3 - 1 = 2$ and $n - k = 131 - 3 = 128$ degrees of freedom, $p < 0.001$. Therefore, we reject the null hypothesis and conclude that the mean changes in total body weight are not identical for all three populations.

Now that we have determined that the population means are not all equal, we would like to find out where the specific differences lie. One way that we can do this is by applying the Bonferroni multiple comparisons procedure. If we wish to set the overall probability of committing a type I error at 0.05, then the significance level for an individual comparison is

$$\begin{aligned} \alpha^* &= \frac{0.05}{\binom{3}{2}} \\ &= \frac{0.05}{3} \\ &= 0.0167. \end{aligned}$$

To compare the mean changes in total body weight for the three populations of overweight males, we begin by considering the group of men who dieted and those who were involved in an exercise program. To test the null hypothesis

$$H_0: \mu_1 = \mu_2,$$

we calculate the test statistic

$$\begin{aligned} t_{12} &= \frac{\bar{x}_1 - \bar{x}_2}{\sqrt{s_W^2\,[(1/n_1) + (1/n_2)]}} \\ &= \frac{-7.2 - (-4.0)}{\sqrt{14.24\,[(1/42) + (1/47)]}} \\ &= -3.99. \end{aligned}$$

For a t distribution with $n - k = 131 - 3 = 128$ degrees of freedom, $p < 0.001$. Therefore, we reject the null hypothesis at the 0.0167 level of significance and conclude that μ_1 differs from μ_2.

We next compare the men who dieted to those who neither dieted nor exercised. In this case, the test statistic is

$$\begin{aligned} t_{13} &= \frac{\bar{x}_1 - \bar{x}_3}{\sqrt{s_W^2\,[(1/n_1) + (1/n_3)]}} \\ &= \frac{-7.2 - 0.6}{\sqrt{14.24\,[(1/42) + (1/42)]}} \\ &= -9.47. \end{aligned}$$

Since $p < 0.001$, we again reject the null hypothesis of equal means and conclude that μ_1 differs from μ_3.

Finally we compare the men who exercised to those who did not change their lifestyles in any way;

$$\begin{aligned} t_{23} &= \frac{\bar{x}_2 - \bar{x}_3}{\sqrt{s_W^2\,[(1/n_2) + (1/n_3)]}} \\ &= \frac{-4.0 - 0.6}{\sqrt{14.24\,[(1/47) + (1/42)]}} \\ &= -5.74. \end{aligned}$$

Again we find that $p < 0.001$; as a result, we conclude that μ_2 differs from μ_3. In summary, all three of the population means are different. The mean change in total body weight is greater for the men who exercised than for those who made no changes in their lifestyle. Furthermore, the mean change is greater for those who dieted than for those who exercised.

Rather than work through all these computations ourselves, we could have used a computer to carry out the analysis. If we saved the 131 changes in body weight under the variable name `wtchange` and the group indicators—where 1 represents dieting, 2 represents exercise, and 3 represents no change in lifestyle—under the name `group`, the Stata output would look like Table 12.2. The top portion of the output is called an *analysis of variance*, or *ANOVA*, *table*. On the

Table 12.2 Stata output illustrating the one-way analysis of variance and the Bonferroni multiple comparisons procedure

Analysis of Variance

Source	SS	df	MS	F	Prob > F
Between groups	1292.40	2	646.20	45.38	0.0000
Within groups	1822.72	128	14.24		
Total	3115.12	130	23.96		

Comparison of wtchange by Group (Bonferroni)

Row Mean - Col Mean	1	2
2	-3.20 0.000	
3	-7.80 0.000	-4.60 0.000

far right, the ANOVA table lists the test statistic F and its associated p-value. The column labeled MS, or mean squares, contains the between- and within-groups estimates of the variance. The columns labeled SS and df contain the numerators and denominators of these estimates, respectively. Note that each of the numerators is actually a sum of squared deviations from the mean; with s_B^2 we are concerned with the deviations of the sample means from the grand mean, and with s_W^2 we focus on the deviations of the individual observations from the sample means. Consequently, the between- and within-groups estimates of the variance may be thought of as averages of squared deviations.

The bottom portion of Table 12.2 contains the results of the Bonferroni multiple comparisons procedure. For each of the three possible pairwise tests, the table lists the difference in sample means $\bar{x}_i - \bar{x}_j$ and the appropriate p-value.

12.4 Review Exercises

1. Consider the F distribution with 8 and 16 degrees of freedom.

a. What proportion of the area under the curve lies to the right of $F = 2.09$?
b. What value of F cuts off the upper 1% of the distribution?
c. What proportion of the area under the curve lies to the left of $F = 4.52$?

2. A study of patients with insulin-dependent diabetes was conducted to investigate the effects of cigarette smoking on renal and retinal complications. Before examining the results of the study, you wish to compare the baseline measures of systolic blood pressure across four different subgroups: nonsmokers,

current smokers, exsmokers, and tobacco chewers. A sample is selected from each subgroup; the relevant data are provided below [4]. Means and standard deviations are measured in millimeters of mercury.

	n	$\bar{x}$	s
Nonsmokers	269	115	13.4
Current smokers	53	114	10.1
Exsmokers	28	118	11.6
Tobacco chewers	9	126	12.2

a. Calculate the estimate of the within-groups variance.
b. Calculate the estimate of the between-groups variance.
c. At the 0.05 level of significance, test the null hypothesis that the mean systolic blood pressures of the four groups are identical.
d. What do you conclude?
e. If you find that the population means are not all equal, use the Bonferroni multiple comparisons procedure to determine where the differences lie.
f. What is the significance level of each individual test?

3. One of the goals of the Edinburgh Artery Study is to investigate the risk factors for peripheral arterial disease among persons 55 to 74 years of age. You wish to compare mean LDL cholesterol levels, measured in mmol/liter, among four different populations of subjects: patients with intermittent claudication or interruptions in movement, those with major asymptomatic disease, those with minor asymptomatic disease, and individuals with no evidence of disease at all. Samples are selected from each population; summary statistics are provided below [5].

	n	$\bar{x}$	s
Intermittent claudication	73	6.22	1.62
Major asymptomatic disease	105	5.81	1.43
Minor asymptomatic disease	240	5.77	1.24
No disease	1,080	5.47	1.31

a. Test the null hypothesis that the mean LDL cholesterol levels are the same for each of the four populations.
b. What are the degrees of freedom associated with this test?
c. What do you conclude?

4. For the previously mentioned study that investigates the effect of carbon monoxide exposure on patients with coronary artery disease, baseline measures of pulmonary function were examined across medical centers. Another characteristic that you might wish to investigate is age. The relevant data are saved on your disk in a data set called `cad`. Values of age are saved under the variable name `age`, and indicators of center are saved under `center`.

a. What is the between-groups estimate of the variance?
b. What is the within-groups estimate of the variance?
c. Test the null hypothesis that the mean ages for each of the three centers are identical.
d. What do you conclude?

References

[1] Allred, E. N., Bleecker, E. R., Chaitman, B. R., Dahms, T. E., Gottlieb, S. O., Hackney, J. D., Hayes, D., Pagano, M., Selvester, R. H., Walden, S. M., and Warren, J., "Acute Effects of Carbon Monoxide Exposure on Individuals with Coronary Artery Disease," *Health Effects Institute Research Report Number 25*, November 1989.

[2] Ott, L., *An Introduction to Statistical Methods and Data Analysis*, North Scituate, MA: Duxbury, 1977.

[3] Wood, P. D., Stefanick, M. L., Dreon, D. M., Frey-Hewitt, B., Garay, S. C., Williams, P. T., Superko, H. R., Fortmann, S. P., Albers, J. J., Vranizan, K. M., Ellsworth, N. M., Terry, R. B., and Haskell, W. L., "Changes in Plasma Lipids and Lipoproteins in Overweight Men During Weight Loss Through Dieting as Compared with Exercise," *The New England Journal of Medicine*, Volume 319, November 3, 1988, 1173–1179.

[4] Chase, H. P., Garg, S. K., Marshall, G., Berg, C. L., Harris, S., Jackson, W. E., and Hamman, R. E., "Cigarette Smoking Increases the Risk of Albuminuria Among Subjects with Type I Diabetes," *Journal of the American Medical Association*, Volume 265, February 6, 1991, 614–617.

[5] Fowkes, F. G. R., Housley, E., Riemersma, R. A., Macintyre, C. C. A., Cawood, E. H. H., Prescott, R. J., and Ruckley, C. V., "Smoking, Lipids, Glucose Intolerance, and Blood Pressure as Risk Factors for Peripheral Atherosclerosis Compared with Ischemic Heart Disease in the Edinburgh Artery Study," *American Journal of Epidemiology*, Volume 135, February 15, 1992, 331–340.

CHAPTER 13

Nonparametric Methods

For all the statistical tests that we have studied up to this point, the populations from which the data were sampled were assumed to be either normally distributed or approximately so. In fact, this property is necessary for the tests to be valid. Since the forms of the underlying distributions are assumed to be known and only the values of certain parameters—such as the means and the standard deviations—are not, these tests are said to be *parametric.* If the data do not conform to the assumptions made by such traditional techniques, then nonparametric methods of statistical inference should be used instead. Nonparametric techniques make fewer assumptions about the nature of the underlying distributions. As a result, they are sometimes called *distribution-free methods*.

Nonparametric tests of hypotheses follow the same general procedure as the parametric tests that we have already studied. We begin by making some claim about the underlying populations in the form of a null hypothesis; we then calculate the value of a test statistic using the data contained in a random sample of observations. Depending on the magnitude of this statistic, we either reject or do not reject the null hypothesis.

13.1 The Sign Test

The *sign test* may be used to compare samples of observations when the populations from which they are drawn are not independent. In this respect, it is similar to the paired t-test. Like the t-test, it does not examine the two groups individually; instead, it focuses on the difference in values for each pair. However, it does not require that the population of differences be normally distributed. The sign test is used to evaluate the null hypothesis that in the underlying population of differences among pairs, the median difference is equal to 0.

Consider a study designed to investigate the amount of energy expended by patients with cystic fibrosis. We would like to compare resting energy expenditure for persons suffering from the disease to resting energy expenditure for individuals who are healthy. To do this, we conduct a two-sided test of the null hypothesis that the median difference is 0.

To carry out the sign test, we begin by selecting a random sample of pairs of observations. Table 13.1 contains the measurements of resting energy expenditure for a sample of 13 patients with cystic fibrosis and 13 healthy individuals matched on age, sex, height, and weight [1]. We then calculate the difference of each pair of observations. If the difference is greater than 0, the pair is assigned a plus sign; if it is less than 0, it receives a minus sign. Differences of 0 are excluded from the analysis, and the number of pairs remaining is denoted by n.

We next count the number of plus signs in the sample; this total is denoted by D. Under the null hypothesis, we would expect to have approximately equal numbers of plus and minus signs. Equivalently, the probability that a particular difference is positive is equal to 1/2. If a plus sign is considered to be a "success," then the n plus and minus signs can be thought of as the outcomes of a Bernoulli random variable with probability of success $p = 1/2$. The total number of plus signs D is a binomial random variable with parameters n and p. Therefore, the mean number of plus signs is $np = n/2$, and the standard deviation is $\sqrt{np(1-p)} = \sqrt{n/4}$.

If D is either much larger or much smaller than $n/2$, we would want to reject H_0. We evaluate the null hypothesis by considering the test statistic

$$z_+ = \frac{D - (n/2)}{\sqrt{(n/4)}}.$$

When the sample size n is large, z_+ follows an approximate normal distribution with mean 0 and standard deviation 1.

For the data in Table 13.1, there are $D = 11$ plus signs. In addition,

$$\begin{aligned} \frac{n}{2} &= \frac{13}{2} \\ &= 6.5, \end{aligned}$$

and

$$\begin{aligned} \sqrt{\frac{n}{4}} &= \sqrt{\frac{13}{4}} \\ &= \sqrt{3.25} \\ &= 1.80. \end{aligned}$$

Table 13.1 Resting energy expenditure (REE) for patients with cystic fibrosis and healthy individuals matched on age, sex, height, and weight

	REE (kcal/day)			
Pair	**CF**	**Healthy**	**Difference**	**Sign**
1	1,153	996	157	+
2	1,132	1,080	52	+
3	1,165	1,182	−17	−
4	1,460	1,452	8	+
5	1,634	1,162	472	+
6	1,493	1,619	−126	−
7	1,358	1,140	218	+
8	1,453	1,123	330	+
9	1,185	1,113	72	+
10	1,824	1,463	361	+
11	1,793	1,632	161	+
12	1,930	1,614	316	+
13	2,075	1,836	239	+

Therefore,

$$\begin{aligned} z_+ &= \frac{D - (n/2)}{\sqrt{(n/4)}} \\ &= \frac{11 - 6.5}{1.80} \\ &= 2.50. \end{aligned}$$

The area under the standard normal curve to the right of $z = 2.50$ and to the left of $z = -2.50$ is 0.012; since p is less than 0.05, we reject the null hypothesis and conclude that the median difference among pairs is not equal to 0. Because most of the differences are positive, we can infer that resting energy expenditure is higher among persons with cystic fibrosis than it is among healthy individuals. This could be due to a number of factors, including differences in metabolism and the increased effort required to breathe.

If the sample size n is small, then the test statistic z_+ does not have a standard normal distribution. In this case, we use a different procedure to evaluate H_0. Recall that, under the null hypothesis, D is a binomial random variable with parameters n and $p = 1/2$. Therefore, we can use the binomial distribution to calculate the probability of observing D positive differences–or some number more extreme–given that H_0 is true.

For the resting energy expenditure data in Table 13.1, we found $D = 11$ plus signs. Under the null hypothesis that the median difference is 0, we would expect only 13/2 or 6.5. The probability of observing 11 or more plus signs is

$$\begin{aligned}
P(D \geq 11) &= P(D = 11) + P(D = 12) + P(D = 13) \\
&= \binom{13}{11}(0.5)^{11}(0.5)^{13-11} + \binom{13}{12}(0.5)^{12}(0.5)^{13-12} \\
&\quad + \binom{13}{13}(0.5)^{13}(0.5)^{13-13} \\
&= 0.0095 + 0.0016 + 0.0001 \\
&= 0.0112.
\end{aligned}$$

This is the p-value of the one-sided test; we are only considering the case in which D is larger than 6.5. The p-value of the corresponding two-sided test is approximately $2(0.0112) = 0.0224$. Once again, we would reject the null hypothesis at the 0.05 level of significance.

13.2 The Wilcoxon Signed-Rank Test

Although the sign test frees us from having to make any assumptions about the underlying distribution of differences, it also ignores some potentially useful information: the magnitude of these differences. As a result, the sign test is not often used in practice. Instead, the *Wilcoxon signed-rank test* can be used to compare two samples from populations that are not independent. Like the sign test, it does not examine the two groups individually; instead, it focuses on the difference in values for each pair of observations. However, it does take into account the magnitudes of these differences as well as their signs. The Wilcoxon signed-rank test is used to evaluate the null hypothesis that, in the underlying population of differences among pairs, the median difference is equal to 0.

Suppose that we would like to investigate the use of the drug amiloride as a therapy for patients with cystic fibrosis. It is believed that the drug may help to improve air flow in the lungs and thereby delay the loss of pulmonary function often associated with the disease. Forced vital capacity (FVC) is the volume of air that a person can expel in 6 seconds; we would like to compare the reduction in FVC which occurs over a 25-week period of treatment with the drug to that which occurs over a similar period of time during treatment with a placebo.

To conduct the Wilcoxon signed-rank test, we proceed as follows. We begin by selecting a random sample of n pairs of observations. Table 13.2 contains the measurements of FVC reduction for a sample of 14 patients with cystic fibrosis [2]. We next calculate the difference of each pair of observations, and, ignoring the signs of these differences, rank their absolute values from smallest to largest. A difference of 0 is not ranked; it is eliminated from the analysis, and the sample

Table 13.2 Reduction in forced vital capacity (FVC) for a sample of patients with cystic fibrosis

Subject	Reduction in FVC (ml) Placebo	Drug	Difference	Rank	Signed Rank	
1	224	213	11	1	1	
2	80	95	−15	2		−2
3	75	33	42	3	3	
4	541	440	101	4	4	
5	74	−32	106	5	5	
6	85	−28	113	6	6	
7	293	445	−152	7		−7
8	−23	−178	155	8	8	
9	525	367	158	9	9	
10	−38	140	−178	10		−10
11	508	323	185	11	11	
12	255	10	245	12	12	
13	525	65	460	13	13	
14	1,023	343	680	14	14	
					86	−19

size is reduced by one. Tied observations are assigned an average rank; if the two smallest differences both take the value 11, for instance, then each would receive a rank of $(1 + 2)/2 = 1.5$. Finally, we assign each rank either a plus or a minus sign depending on the sign of the difference. For example, the difference in Table 13.2 that has the second smallest absolute value is −15; therefore, this observation receives rank 2. Because the difference itself is negative, however, the observation's signed rank is −2.

The next step is to compute the sum of the positive ranks and the sum of the negative ranks. Ignoring the signs, we denote the smaller sum by T. Under the null hypothesis that the median of the underlying population of differences is equal to 0, we would expect the sample to have approximately equal numbers of positive and negative ranks. Furthermore, the sum of the positive ranks should be comparable to the sum of the negative ranks in magnitude. We test this hypothesis by considering the statistic

$$z_T = \frac{T - \mu_T}{\sigma_T}$$

where

$$\mu_T = \frac{n(n+1)}{4}$$

is the mean sum of the ranks and

$$\sigma_T = \sqrt{\frac{n(n+1)(2n+1)}{24}}$$

is the standard deviation of T [3]. When the sample size n is large,

$$z_T = \frac{T - \mu_T}{\sigma_T}$$

follows an approximate normal distribution with mean 0 and standard deviation 1.

The differences in Table 13.2 are displayed in Figure 13.1 in the form of a box plot; recall that the line running through the center of the box represents the median of the observations. The sum of the positive ranks is 86, and the sum of the negative ranks is −19. Ignoring the signs, therefore, $T = 19$. Also,

$$\begin{aligned} \mu_T &= \frac{n(n+1)}{4} \\ &= \frac{14(14+1)}{4} \\ &= 52.5, \end{aligned}$$

and

$$\begin{aligned} \sigma_T &= \sqrt{\frac{n(n+1)(2n+1)}{24}} \\ &= \sqrt{\frac{14(14+1)[2(14)+1]}{24}} \\ &= 15.93. \end{aligned}$$

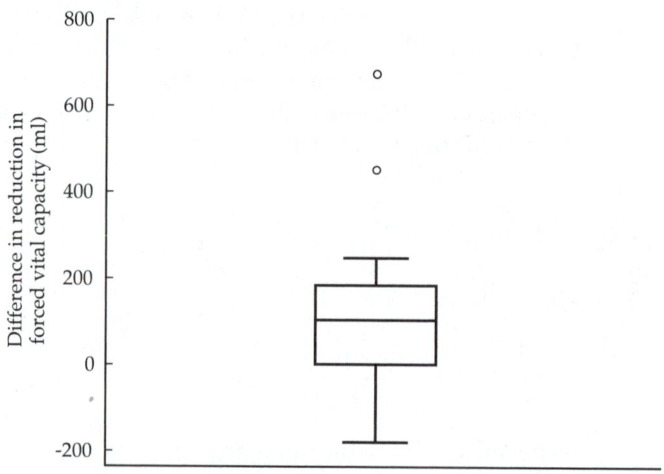

Figure 13.1. Differences in reduction in forced vital capacity for a sample of patients with cystic fibrosis

Solving for z_T, we find that

$$
\begin{aligned}
z_T &= \frac{T - \mu_T}{\sigma_T} \\
&= \frac{19 - 52.5}{15.93} \\
&= -2.10.
\end{aligned}
$$

The area under the standard normal curve to the left of $z = -2.10$ and to the right of $z = 2.10$ is $2(0.018) = 0.036$. Since p is less than $\alpha = 0.05$, we reject the null hypothesis and conclude that the median difference is not equal to 0. Most of the differences are positive; this suggests that the reduction in forced vital capacity is greater during treatment with the placebo than it is during treatment with the drug. In other words, use of the drug does diminish the loss of pulmonary function.

If n is small, then the test statistic z_T cannot be assumed to follow a standard normal distribution. In this case, tables are available that help us to determine whether we should reject the null hypothesis. Table A.4 in Appendix A displays the distribution function of the smaller sum of ranks T for samples of size n less than or equal to 12. The possible values of T, represented by T_0, are listed down the left-hand side of the table; the sample sizes are displayed across the top. For each combination of T_0 and n, the entry in the table is the probability that T is less than or equal to T_0. If $n = 8$, for instance, then the probability that T is less than or equal to 5 is 0.0391. This is the p-value of the one-sided test. The p-value of the appropriate two-sided test is approximately $2(0.0391) = 0.0782$.

13.3 The Wilcoxon Rank Sum Test

The *Wilcoxon rank sum test* is used to compare two samples that have been drawn from independent populations. Consequently, it is a nonparametric counterpart to the two-sample t-test. Unlike the t-test, it does not require that the underlying populations be normally distributed or that their variances be equal. It does, however, assume that the distributions have the same general shape. The Wilcoxon rank sum test evaluates the null hypothesis that the medians of the two populations are identical.

Consider the distributions of normalized mental age scores for two populations of children suffering from phenylalaninemia (PKU). The members of the first group have average daily serum phenylalanine levels below 10.0 mg/dl; those in the second group have average levels above 10.0 mg/dl. It has been suggested that an elevated level of serum phenylalanine increases a child's chance of mental deficiency. Therefore, we would like to determine whether the median normalized mental age scores for the two populations of children are identical. We conduct a two-sided test at the $\alpha = 0.05$ level of significance.

To carry out the Wilcoxon rank sum test, we begin by selecting an independent random sample from each of the populations. Table 13.3 displays the samples taken from the two populations of children with PKU; there are 21 children from the low exposure group and 18 children from the high exposure group [4]. The values of normalized mental age score for each group are pictured in Figure 13.2. We next combine the two samples into one large group, order the observations from smallest to largest, and assign a rank to each one. If there are tied observations, then we assign an average rank to all measurements with the same value. Note, for instance, that two of the children in the sample have a normalized mental age score of 37.0 months. Since these observations are fourth and fifth in the ordered list of 39 measurements, we assign an average rank of $(4+5)/2 = 4.5$ to each one. Similarly, three subjects have a normalized mental age score of 51.0 months; these observations each receive a rank of $(22+23+24)/3 = 23$.

The next step is to find the sum of the ranks corresponding to each of the original samples. The smaller of the two sums is denoted by W. Under the null hypothesis that the two underlying populations have identical medians, we would

Table 13.3 Normalized mental age scores for two samples of children suffering from phenylalaninemia

Low Exposure (< 10.0 mg/dl)		High Exposure (≥ 10.0 mg/dl)	
nMA (mos)	**Rank**	**nMA (mos)**	**Rank**
34.5	2.0	28.0	1.0
37.5	6.0	35.0	3.0
39.5	7.0	37.0	4.5
40.0	8.0	37.0	4.5
45.5	11.5	43.5	9.0
47.0	14.5	44.0	10.0
47.0	14.5	45.5	11.5
47.5	16.0	46.0	13.0
48.7	19.5	48.0	17.0
49.0	21.0	48.3	18.0
51.0	23.0	48.7	19.5
51.0	23.0	51.0	23.0
52.0	25.5	52.0	25.5
53.0	28.0	53.0	28.0
54.0	31.5	53.0	28.0
54.0	31.5	54.0	31.5
55.0	34.5	54.0	31.5
56.5	36.0	55.0	34.5
57.0	37.0		313.0
58.5	38.5		
58.5	38.5		
	467.0		

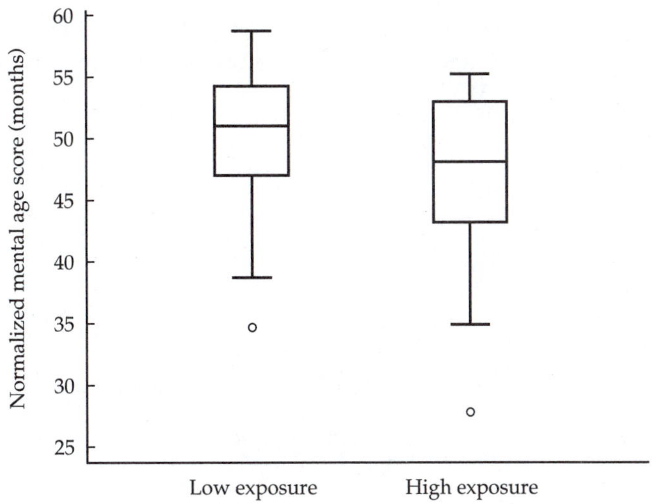

Figure 13.2. Normalized mental age scores for two samples of children suffering from phenylalaninemia

expect the average ranks for each of the samples to be approximately equal. We test this hypothesis by calculating the statistic

$$z_W = \frac{W - \mu_W}{\sigma_W}$$

where

$$\mu_W = \frac{n_S(n_S + n_L + 1)}{2}$$

is the mean sum of the ranks and

$$\sigma_W = \sqrt{\frac{n_S\, n_L(n_S + n_L + 1)}{12}}$$

is the standard deviation of W [3]. In these equations, n_S represents the number of observations in the sample that has the smaller sum of ranks and n_L the number of observations in the sample with the larger sum. For large values of n_S and n_L,

$$z_W = \frac{W - \mu_W}{\sigma_W}$$

follows an approximate normal distribution with mean 0 and standard deviation 1.

For the data in Table 13.3, the sum of the ranks in the low exposure group is 467, and the sum of the ranks in the high exposure group is 313; therefore, $W = 313$. In addition,

$$\begin{aligned}\mu_W &= \frac{n_S(n_S + n_L + 1)}{2} \\ &= \frac{18(18 + 21 + 1)}{2} \\ &= 360,\end{aligned}$$

and

$$\begin{aligned}\sigma_W &= \sqrt{\frac{n_S\, n_L(n_S + n_L + 1)}{12}} \\ &= \sqrt{\frac{18(21)(18 + 21 + 1)}{12}} \\ &= 35.5.\end{aligned}$$

Substituting these values into the equation for the test statistic, we have

$$\begin{aligned}z_W &= \frac{W - \mu_W}{\sigma_W} \\ &= \frac{313 - 360}{35.5} \\ &= -1.32.\end{aligned}$$

Since $p = 2(0.093) = 0.186$ is greater than 0.05, we do not reject the null hypothesis. The samples do not provide evidence of a difference in median normalized mental age scores for the two populations; the children with high serum phenylalanine exposure have achieved a level of mental functioning that is comparable to the level of the children with low exposure.

If n_S and n_L are very small, then z_W cannot be assumed to follow a standard normal distribution. In this case, Table A.5 in Appendix A contains the distribution functions of the smaller sum of ranks W for sample sizes n_S and n_L that are each less than or equal to 10. The possible values of W, denoted by W_0, are displayed on the left-hand sides of the tables; values of n_1 and n_2 appear across the top. In this case, n_2 is the larger of the two sample sizes n_S and n_L, whereas n_1 is the smaller. For each combination of W_0, n_1, and n_2, the entry in the table corresponds to the probability that W is less than or equal to W_0. Suppose that $n_S = 5$ and $n_L = 4$. Therefore, $n_1 = 4$ and $n_2 = 5$. The probability that W is less than or equal to 13 is 0.0556. This is the p-value of the one-sided test; the p-value of the corresponding two-sided test is $2(0.0556) = 0.1112$.

13.4 Advantages and Disadvantages of Nonparametric Methods

Nonparametric techniques have several advantages over traditional methods of statistical inference. One advantage is that they do not incorporate all the restrictive assumptions characteristic of parametric tests. They do not require that the underlying populations be normally distributed, for instance. At most, the populations should have the same basic shape. In addition, because nonparametric tests deal with ranks rather than the actual values of the observations, they can be performed relatively quickly for small samples. Their use of ranks makes nonparametric techniques less sensitive to measurement error than traditional tests and also permits the use of ordinal rather than continuous measurements. Since it does not make sense to calculate either a mean or a standard deviation for ordinal data, parametric tests are usually not appropriate.

Nonparametric methods also have a number of disadvantages. If the assumptions underlying a parametric test are satisfied, then the nonparametric test is less powerful than the comparable parametric technique. This means that if the null hypothesis is false, the nonparametric test would typically require a larger sample to provide sufficient evidence to reject it. If the data do come from an underlying normal population, then the power of the Wilcoxon tests is approximately 95% of that for the t-tests; if a t-test requires 19 observations to achieve a particular power, a Wilcoxon test would require 20 observations to achieve the same power. In addition, the hypotheses tested by nonparametric techniques tend to be less specific than those tested by traditional methods. Because they rely on ranks rather than on the actual values of the observations, nonparametric tests do not use all the information known about a distribution. This, of course, presumes that our information about the underlying population is correct. Finally, if a large proportion of the observations are tied, then σ_T and σ_W overestimate the standard deviations of T and W respectively. To compensate for this, a correction term must be added to the calculations [3].

13.5 Further Applications

A study was conducted to investigate the use of extracorporeal membrane oxygenation (ECMO) in the treatment of newborns with neonatal respiratory failure. It is thought that the use of this procedure may decrease the output of an infant's left ventricle, thereby decreasing the amount of blood pumped to the body. We would like to compare left ventricular dimension before and during the treatment. We may not be willing to make the assumption that the population of differences is normally distributed; therefore, we use the Wilcoxon signed-rank test to evaluate the null hypothesis that the median difference is equal to 0.

Table 13.4 provides the relevant data for a sample of 15 infants suffering from respiratory distress [5]. We calculate the difference for each pair of observations and, ignoring the signs of these differences, rank their absolute values from smallest to largest. Since a difference of 0 is not ranked, the sample size is reduced to $n = 11$. Each rank is then assigned either a plus or a minus sign depending on the sign of the difference itself.

If we ignore the signs, the smaller sum of ranks is $T = 18.5$. In addition, the mean sum of the ranks is

$$\begin{aligned}\mu_T &= \frac{n(n+1)}{4} \\ &= \frac{11(11+1)}{4} \\ &= 33,\end{aligned}$$

Table 13.4 Left ventricular dimension for a sample of infants suffering from neonatal respiratory failure

	LVD (centimeters)					
Subject	Before ECMO	During ECMO	Difference	Rank	Signed Rank	
1	1.6	1.6	0.0	–		
2	2.0	2.0	0.0	–		
3	1.2	1.2	0.0	–		
4	1.6	1.6	0.0	–		
5	1.6	1.5	0.1	2.5	2.5	
6	1.7	1.6	0.1	2.5	2.5	
7	1.6	1.5	0.1	2.5	2.5	
8	1.6	1.7	−0.1	2.5		−2.5
9	1.6	1.4	0.2	5.5	5.5	
10	1.7	1.5	0.2	5.5	5.5	
11	1.0	1.3	−0.3	8.0		−8.0
12	1.5	1.8	−0.3	8.0		−8.0
13	1.5	1.8	−0.3	8.0		−8.0
14	1.4	1.8	−0.4	10.0		−10.0
15	1.5	2.0	−0.5	11.0		−11.0
					18.5	−47.5

and the standard deviation is

$$\begin{aligned}\sigma_T &= \sqrt{\frac{n(n+1)(2n+1)}{24}} \\ &= \sqrt{\frac{11(11+1)(22+1)}{24}} \\ &= 11.25.\end{aligned}$$

Therefore, we find the test statistic to be

$$\begin{aligned}z_T &= \frac{T - \mu_T}{\sigma_T} \\ &= \frac{18.5 - 33}{11.25} \\ &= -1.29.\end{aligned}$$

The p-value of this test is 2(0.099), or 0.198. We do not reject the null hypothesis; the sample fails to provide evidence that the median difference is not equal to 0. Therefore, treatment with ECMO does not appear to affect the output of the infants' left ventricles.

For the data in Table 13.4, we end up with a sample of size 11. Since n is small, we cannot be sure that z_T has a standard normal distribution. Recall that $T = 18.5$; to be conservative, we round T up to 19. Referring to Table A.4, we see that the p-value of the two-sided test is $2(0.1202) = 0.2404$. Again we would not reject the null hypothesis.

Rather than work through the calculations ourselves, we could use a computer to conduct the signed-rank test. If we saved the 15 measures of left ventricular dimension before treatment with ECMO under the variable name `before` and those during treatment under `during`, the sample output would look like Table 13.5. The output includes the null hypothesis, the sums of the positive and negative ranks, the test statistic z_T, and the exact p-value of the two-sided test.

Table 13.5 Stata output displaying the Wilcoxon signed-rank test

```
Test:  Equality of distributions (Wilcoxon signed-ranks)

  Result of before - (during)
  Sum of positive ranks = 18.5
  Sum of negative ranks = 47.5

  z-statistic    -1.29
  Prob > |z|     0.198
```

Carbon monoxide diffusing capacity, denoted Dl_{CO}, is a measure of lung function that has been tested as a possible diagnostic tool for detecting emphysema. Consider the distributions of CO-diffusing capacity for the population of healthy individuals and the population of patients with emphysema. Using a two-sided test conducted at the $\alpha = 0.05$ level of significance, we evaluate the null hypothesis that the two populations have the same median Dl_{CO} by means of the Wilcoxon rank sum test.

Table 13.6 provides the data for random samples of 13 individuals who have emphysema and 23 persons who do not [6]. The observations are ordered from smallest to largest, and a rank is assigned to each one. The smaller sum of the ranks corresponds to the sample of persons who have emphysema. Therefore, $W = 168$. In addition,

$$\begin{aligned}\mu_W &= \frac{n_S(n_S + n_L + 1)}{2} \\ &= \frac{13(13 + 23 + 1)}{2} \\ &= 240.5\end{aligned}$$

is the mean sum of the ranks, and

$$\begin{aligned}\sigma_W &= \sqrt{\frac{n_S\, n_L(n_S + n_L + 1)}{12}} \\ &= \sqrt{\frac{13(23)(13 + 23 + 1)}{12}} \\ &= 30.36\end{aligned}$$

is the standard deviation of W. Under the null hypothesis that the two underlying populations have identical medians, the test statistic is

$$\begin{aligned}z_W &= \frac{W - \mu_W}{\sigma_W} \\ &= \frac{168 - 240.5}{30.36} \\ &= -2.39.\end{aligned}$$

The p-value of the test is $2(0.008) = 0.016$. Therefore, we reject the null hypothesis at the 0.05 level. The samples provide evidence that the median CO-diffusing capacity of the population of individuals who have emphysema is different from the median of the population of persons who do not. In general, individuals suffering from emphysema have lower carbon monoxide diffusing capacities.

Again, we can perform the calculations using a computer. If we saved the 13 measures of CO-diffusing capacity for individuals with emphysema under the

Table 13.6 Carbon monoxide diffusing capacity for samples of individuals with and without emphysema

Emphysema		No Emphysema	
Dl_{CO}	Rank	Dl_{CO}	Rank
7.51	2	6.19	1
10.81	3	12.11	5
11.75	4	14.12	8
12.59	6	15.50	11
13.47	7	15.52	12
14.18	9	16.56	13
15.25	10	17.06	14
17.40	15	19.59	18
17.75	16	20.21	19
19.13	17	20.35	20
20.93	21	21.05	22
25.73	28	21.41	23
26.16	30	23.39	24
	168	23.60	25
		24.05	26
		25.59	27
		25.79	29
		26.29	31
		29.60	32
		30.88	33
		31.42	34
		32.66	35
		36.16	36
			498

variable name `emph` and the 23 values for persons without emphysema under `noemph`, the sample output for this test would look like Table 13.7. The output includes the null hypothesis, the value of W, the mean sum of the ranks, the test statistic z_W, and the p-value of the two-sided test.

Table 13.7 Stata output illustrating the Wilcoxon rank sum test

```
Test:  Equality of medians (Two-Sample Wilcoxon Rank-Sum)

   Sum of ranks:   168     (emph == 1)
   Expected sum:   240.5

   z-statistic    -2.39
   Prob > |z|     0.0160
```

13.6 Review Exercises

1. Refer to the resting energy expenditure data for patients with cystic fibrosis and matched healthy individuals that was provided in Table 13.1.

a. Using the Wilcoxon signed-rank test, evaluate the null hypothesis that the median of the population of differences is equal to 0.
b. Compare the results of the signed-rank test to those obtained when the sign test was used.

2. Nineteen individuals with asthma were enrolled in a study investigating the respiratory effects of sulphur dioxide. During the study, two measurements were obtained for each subject. The first is the increase in specific airway resistance (SAR)—a measure of bronchoconstriction—from the time when the individual is at rest until after he or she has been exercising for 5 minutes; the second is the increase in SAR for the same subject after he or she has undergone a similar exercise test conducted in an atmosphere of 0.25 ppm sulphur dioxide [7].

	Increase in SAR ($cm\ H_20 \times sec$)	
Subject	**Air**	**SO_2**
1	0.82	0.72
2	0.86	1.05
3	1.86	1.40
4	1.64	2.30
5	12.57	13.49
6	1.56	0.62
7	1.28	2.41
8	1.08	2.32
9	4.29	8.19
10	1.37	6.33
11	14.68	19.88
12	3.64	8.87
13	3.89	9.25
14	0.58	6.59
15	9.50	2.17
16	0.93	9.93
17	0.49	13.44
18	31.04	16.25
19	1.66	19.89

a. At the $\alpha = 0.05$ level of significance, test the null hypothesis that the median difference in increase in specific airway resistance for the two occasions is equal to 0.
b. What do you conclude?

3. Suppose that you have a set of 12 measurements, 6 drawn from one population and 6 from another. The two populations are not independent. You wish

to use the signed-rank test to evaluate the null hypothesis that, in the underlying population of differences, the median difference is equal to 0.

a. If $T = 2$, what is the p-value of the two-sided test?
b. Suppose that the sum of the positive ranks is 15 and the sum of the negative ranks is -6. What is the p-value of the two-sided test?

4. The following data are taken from a study that compares adolescents who have bulimia to healthy adolescents with similar body compositions and levels of physical activity. The data consist of measures of daily caloric intake for random samples of 23 bulimic adolescents and 15 healthy ones [8].

Daily Caloric Intake (kcal/kg)				
Bulimic			Healthy	
15.9	18.9	25.1	20.7	30.6
16.0	19.6	25.2	22.4	33.2
16.5	21.5	25.6	23.1	33.7
17.0	21.6	28.0	23.8	36.6
17.6	22.9	28.7	24.5	37.1
18.1	23.6	29.2	25.3	37.4
18.4	24.1	30.9	25.7	40.8
18.9	24.5		30.6	

a. Test the null hypothesis that the median daily caloric intake of the population of individuals suffering from bulimia is equal to the median caloric intake of the healthy population. Conduct a two-sided test at the 0.05 level of significance.
b. Do you believe that adolescents with bulimia require a lower daily caloric intake than do healthy adolescents?

5. The characteristics of low birth weight children dying of sudden infant death syndrome were examined for both females and males. The ages at time of death for samples of 12 girls and 17 boys are provided [9].

Age (days)		
Females	Males	
53	46	115
56	52	133
60	58	134
60	59	167
78	77	175
87	78	
102	80	
117	81	
134	84	
160	103	
277	114	

a. Test the null hypothesis that the median ages at death are identical for the two populations.
b. What do you conclude?

6. Suppose that you have two independent populations and wish to use the rank sum test to evaluate the null hypothesis that their medians are identical. You select a sample of size 4 from the first population and a sample of size 5 from the second.

a. If $W = 16$, then what is the p-value of the two-sided test?
b. If the sum of the ranks in the first sample is 12 and the sum in the second sample is 33, what is the p-value of the two-sided test?

7. The community hospital bed data that was analyzed by means of the t-test in Chapter 11 will now be used to conduct a nonparametric analysis. Recall that the observations are in the data set `bed`.

a. Construct a pair of box plots for the numbers of community hospital beds per 1,000 population in 1980 and in 1986.
b. Use the Wilcoxon signed-rank test to determine whether the median difference in the number of beds is equal to 0.
c. Compare the median number of hospital beds in 1980 to the median number in 1986 using the Wilcoxon rank sum test.
d. What conclusions do you reach for each test?
e. Comment on the differences between the two Wilcoxon tests. Which was the appropriate one to use in this situation?
f. Compare the results obtained using nonparametric techniques to those obtained with the parametric t-tests in Chapter 11. Consider the underlying assumptions of the tests.

References

[1] Spicher, V., Roulet, M., and Schutz, Y., "Assessment of Total Energy Expenditure in Free-Living Patients with Cystic Fibrosis," *Journal of Pediatrics*, Volume 118, June 1991, 865–872.

[2] Knowles, M. R., Church, N. L., Waltner, W. E., Yankaskas, J. R., Gilligan, P., King, M., Edwards, L. J., Helms, R. W., and Boucher, R. C., "A Pilot Study of Aerosolized Amiloride for the Treatment of Lung Disease in Cystic Fibrosis," *The New England Journal of Medicine*, Volume 322, April 26, 1990, 1189–1194.

[3] Hollander, M., and Wolfe, D. A., *Nonparametric Statistical Methods*, New York: Wiley, 1973.

[4] Wrona, R. M., "A Clinical Epidemiologic Study of Hyperphenylalaninemia," *American Journal of Public Health*, Volume 69, July 1979, 673–679.

[5] Burch, K. D., Covitz, W., Lovett, E. J., Howell, C., and Kanto, W. P., "The Significance of Ductal Shunting During Extracorporeal Membrane Oxygenation," *Journal of Pediatric Surgery*, Volume 24, September 1989, 855–859.

[6] Morrison, N. J., Abboud, R. T., Ramadan, F., Miller, R. R., Gibson, N. N., Evans, K. G., Nelems, B., and Müller, N. L., "Comparison of Single Breath Carbon Monoxide Diffusing Capacity and Pressure-Volume Curves in Detecting Emphysema," *American Review of Respiratory Disease*, Volume 139, May 1989, 1179–1187.

[7] Bethel, R. A., Sheppard, D., Geffroy, B., Tam, E., Nadel, J. A., and Boushey, H. A., "Effect of 0.25 ppm Sulphur Dioxide on Airway Resistance in Freely Breathing, Heavily Exercising, Asthmatic Subjects," *American Review of Respiratory Disease*, Volume 31, April 1985, 659–661.

[8] Gwirtsman, H. E., Kaye, W. H., Obarzanek, E., George, D. T., Jimerson, D. C., and Ebert, M. H., "Decreased Caloric Intake in Normal-Weight Patients with Bulimia: Comparison with Female Volunteers," *American Journal of Clinical Nutrition*, Volume 49, January 1989, 86–92.

[9] Walker, A. M., Jick, H., Perera, D. R., Thompson, R. S., and Knauss, T. A., "Diphtheria–Tetanus–Pertussis Immunization and Sudden Infant Death Syndrome," *American Journal of Public Health*, Volume 77, August 1987, 945–951.

CHAPTER 14

Inference on Proportions

In previous chapters we applied the techniques of statistical inference to continuous or measured data. In particular, we investigated the properties of population means and–in the case of nonparametric methods–medians. We now extend the methodology of statistical inference to include enumerated data, or counts. The basic underlying principles remain the same, and the normal distribution again plays a key role.

When studying counts, we are usually concerned with the proportion of times that an event occurs rather than the number of times. In the mid-19th century, for example, the Vienna General Hospital–the Allgemeines Krankenhaus of the University of Vienna–had two obstetrical divisions [1]. Every year, approximately 3,500 babies were delivered in each division. There were two major differences between the groups, however. In the first division, all deliveries were supervised by obstetricians and medical students; in the second, they were overseen by midwives. Furthermore, the proportion of women who died of puerperal fever in the first division was between 0.17 and 0.23, whereas the proportion who died in the second division was about 0.017.

Ignac Semmelweiss, the assistant to the professor of obstetrics, was convinced that this tenfold difference in proportions was not due to chance alone. His research led him to conclude that the discrepancy existed because the obstetricians and students dissected several cadavers a day. Since the germ theory of diseases had not yet been proposed, proper hygiene was not practiced; individuals went freely from dissections to deliveries without taking sanitary precautions. Believing that this practice was the root of the problem, Semmelweiss changed the procedure. He insisted that obstetricians wash their hands in a chlorine solution before being allowed to attend a delivery. In the subsequent year, the proportions of women who died were 0.012 in the first division and 0.013 in the second. Unfortunately, Semmelweiss was ahead of his time. His conclusions were not generally accepted; in fact, his discovery caused him to lose his position. Would such a discrepancy in proportions be ignored today, or would we accept that the two divisions are in fact different? To address this issue, we investigate some aspects of the variability of proportions.

14.1 Normal Approximation to the Binomial Distribution

The binomial distribution provides the foundation for the analysis of proportions. Recall that if we have a sequence of n independent Bernoulli trials—each of which results in one of two mutually exclusive outcomes, often designated "success" and "failure"—and each trial has a constant probability of success p, then the total number of successes X is a binomial random variable. The probability distribution of X, represented by the formula

$$P(X = x) = \binom{n}{x} p^x (1-p)^{n-x},$$

can be used to make statements about the possible outcomes of the random variable. In particular, we may use this expression to compute the probabilities that are associated with specified outcomes x.

Suppose that we select a random sample of 30 individuals from the population of adults in the United States. The probability that a member of this population currently smokes either cigarettes, cigars, or a pipe is equal to 0.29 [2]; therefore, the total number of smokers in the sample is a binomial random variable with parameters $n = 30$ and $p = 0.29$. For a given sample of size 30, what is the probability that six or fewer of its members smoke? Applying the additive rule of probability, we find that

$$\begin{aligned} P(X \leq 6) &= P(X = 0) + P(X = 1) + P(X = 2) + P(X = 3) \\ &\quad + P(X = 4) + P(X = 5) + P(X = 6) \\ &= (1)(0.29)^0(0.71)^{30} + (30)(0.29)^1(0.71)^{29} \\ &\quad + (435)(0.29)^2(0.71)^{28} + (4060)(0.29)^3(0.71)^{27} \\ &\quad + (27,405)(0.29)^4(0.71)^{26} + (142,506)(0.29)^5(0.71)^{25} \\ &\quad + (593,775)(0.29)^6(0.71)^{24} \\ &= 0.190. \end{aligned}$$

We are also able to calculate the probabilities associated with the outcomes of a binomial random variable X using an approximate procedure that is based on the normal distribution. When the sample size is large, the binomial distribution is cumbersome to work with, and so this alternative procedure can be more convenient. In Chapter 7 we saw that, as the sample size n grows larger, the shape of a binomial distribution increasingly resembles that of a normal distribution. Furthermore, the mean number of successes per sample is np, and the variance is

$np(1-p)$. If n is sufficiently large, therefore, we can approximate the distribution of X by a normal distribution with the same mean and variance as the binomial. A widely used criterion states that n is "sufficiently large" if both np and $n(1-p)$ are greater than or equal to 5. In this case,

$$Z = \frac{X - np}{\sqrt{np(1-p)}}$$

is approximately normal with mean 0 and standard deviation 1.

Using the normal approximation, we would like to find the proportion of samples of size 30 in which at most six individuals are current smokers. Since $np = 30(0.29) = 8.7$ and $n(1-p) = 30(0.71) = 21.3$ are both greater than 5, we calculate the standard normal deviate

$$\begin{aligned} z &= \frac{6 - (30)(0.29)}{\sqrt{30(0.29)(0.71)}} \\ &= -1.09. \end{aligned}$$

The area under the standard normal curve that lies to the left of $z = -1.09$ is 0.138. This is the probability that at most six of the individuals smoke. Note that the normal approximation provides only an adequate estimate of the exact binomial probability 0.190.

It has been shown that a better approximation to the binomial distribution can be obtained using the ratio

$$Z = \frac{X - np + 0.5}{\sqrt{np(1-p)}}$$

when $X < np$ and

$$Z = \frac{X - np - 0.5}{\sqrt{np(1-p)}}$$

when $X > np$. The factor 0.5 that appears in the numerators of these expressions is known as a *continuity correction*. It is used to compensate for the fact that a discrete binomial distribution is being approximated by a continuous normal distribution.

Suppose that we again wish to find the proportion of samples of size 30 in which at most six individuals are current smokers. Applying the continuity correction, we find that

$$\begin{aligned} z &= \frac{6 - (30)(0.29) + 0.5}{\sqrt{30(0.29)(0.71)}} \\ &= -0.89. \end{aligned}$$

The area under the standard normal curve to the left of $z = -0.89$ is 0.187. In this example, the use of a continuity correction results in a much better approximation of the exact binomial probability 0.190.

14.2 Sampling Distribution of a Proportion

Suppose that we are interested in estimating the proportion of times that a particular event occurs in a given population rather than the number of times. For example, we might wish to make a statement about the proportion of patients who survive five years after being diagnosed with lung cancer based on a sample drawn from this group. If the sample is of size n, and x of its members are alive five years after diagnosis, we could estimate the population proportion p by

$$\hat{p} = \frac{x}{n}.$$

The sample proportion $\hat{p}$, or p-hat, is the maximum likelihood estimator of p. Recall that this is the value of the parameter that is most likely to have produced the observed sample. Before we use $\hat{p}$, however, we should first investigate some of its properties.

In the population of patients who have been diagnosed with lung cancer, we may represent five-year survival by a 1 and death by a 0. The mean is equal to the proportion of 1s in the population, or p. The standard deviation is $\sqrt{p(1-p)}$. Suppose that we randomly select a sample of size n from the population and denote the proportion of 1s in the sample by $\hat{p}_1$. Similarly, we could select a second sample of size n and denote the proportion of 1s in this new sample by $\hat{p}_2$. If we were to continue to select samples of size n indefinitely, we would end up with a sampling distribution of proportions. According to the central limit theorem, the distribution of sample proportions has the following three properties:

1. The mean of the sampling distribution is the population mean p.
2. The standard deviation of the distribution of sample proportions is equal to $\sqrt{p(1-p)/n}$. As in the case of the mean, this quantity is known as the *standard error*.
3. The shape of the sampling distribution is approximately normal provided that n is sufficiently large.

Because the distribution of sample proportions is approximately normal with mean p and standard deviation $\sqrt{p(1-p)/n}$, we know that

$$Z = \frac{\hat{p} - p}{\sqrt{p(1-p)/n}}$$

is normally distributed with mean 0 and standard deviation 1. As a result, we are able to use tables of the standard normal distribution to make inference about the value of a population proportion.

For example, consider five-year survival among patients who have been diagnosed with lung cancer. The mean proportion of individuals surviving is $p = 0.10$; the standard deviation is $\sqrt{p(1-p)} = \sqrt{0.10(1-0.10)} = 0.30$ [3]. If we select repeated samples of size 50 from this population, what fraction will have a sample proportion of 0.20 or above? It is again recommended that np and $n(1-p)$ be greater than or equal to 5. Since $np = 50(0.10) = 5$ and $n(1-p) = 50(0.90) = 45$, the central limit theorem states that the distribution of sample proportions is approximately normal with mean $p = 0.10$ and standard error $\sqrt{p(1-p)/n} = \sqrt{0.10(1-0.10)/50} = 0.0424$. The quantity that we are interested in is the area under this curve that lies to the right of $p = 0.20$. Note that

$$Z = \frac{\hat{p} - 0.10}{\sqrt{0.10(1-0.10)/50}}$$

is a standard normal random variable. If $\hat{p} = 0.20$, then

$$\begin{aligned} z &= \frac{0.20 - 0.10}{\sqrt{0.10(1-0.10)/50}} \\ &= 2.36. \end{aligned}$$

Referring to Table A.1, we observe that the area under the standard normal curve that lies to the right of $z = 2.36$ is 0.009. Only about 0.9% of the samples will have a sample proportion of 0.20 or larger.

The continuity correction could be applied in this example as well. As long as n is reasonably large, however, its effect is negligible. Therefore, we omit it here and in the remainder of the chapter.

14.3 Confidence Intervals

To construct a confidence interval for a population proportion, we follow the same procedure as we did for a population mean. We begin by drawing a sample of size n, computing the sample proportion $\hat{p}$, and using $\hat{p}$ as a point estimate of p. Applying the results of the previous section, we then note that

$$Z = \frac{\hat{p} - p}{\sqrt{p(1-p)/n}}$$

is a standard normal random variable if n is sufficiently large. We know that 95% of its possible outcomes lie between -1.96 and 1.96. Therefore,

$$P\left(-1.96 \leq \frac{\hat{p} - p}{\sqrt{p(1-p)/n}} \leq 1.96\right) = 0.95$$

and

$$P\left(\hat{p} - 1.96\sqrt{\frac{p(1-p)}{n}} \leq p \leq \hat{p} + 1.96\sqrt{\frac{p(1-p)}{n}}\right) = 0.95.$$

The terms $\hat{p} - 1.96\sqrt{p(1-p)/n}$ and $\hat{p} + 1.96\sqrt{p(1-p)/n}$ are 95% confidence limits for the population proportion p. Note, however, that these quantities depend on the value of p. Since p is unknown, we must estimate it using the sample proportion $\hat{p}$. Consequently,

$$\left(\hat{p} - 1.96\sqrt{\frac{\hat{p}(1-\hat{p})}{n}}\,,\ \hat{p} + 1.96\sqrt{\frac{\hat{p}(1-\hat{p})}{n}}\right)$$

is an approximate 95% confidence interval for p. This interval has a 95% chance of covering the population proportion before a sample is selected. Using similar reasoning, we are also able to construct a one-sided confidence interval for p.

Consider the distribution of five-year survival for individuals under the age of 40 who have been diagnosed with lung cancer. This distribution has an unknown population mean p. In a randomly selected sample of 52 persons, only six survive five years [4]. Therefore,

$$\begin{aligned} \hat{p} &= \frac{x}{n} \\ &= \frac{6}{52} \\ &= 0.115 \end{aligned}$$

is a point estimate for p. Since $n\hat{p} = 52(0.115) = 5.98$ and $n(1-\hat{p}) = 52(0.885) = 46.02$, the sample size is large enough to justify the use of the normal approximation; an approximate 95% confidence interval for p is

$$\left(0.115 - 1.96\sqrt{\frac{0.115(1-0.115)}{52}}\,,\ 0.115 + 1.96\sqrt{\frac{0.115(1-0.115)}{52}}\right)$$

or

$$(0.028\,,\ 0.202).$$

We are 95% confident that these limits cover the true proportion of individuals under the age of 40 who survive five years.

14.4 Hypothesis Testing

As mentioned above, the distribution of five-year survival for individuals under the age of 40 who have been diagnosed with lung cancer has an unknown population proportion p. We do know, however, that the proportion of patients surviving five years among those who are over 40 years of age at the time of diagnosis is 8.2% [4]. Is it possible that the proportion surviving in the under-40 population is 0.082 as well? To determine whether this is the case, we conduct a statistical test of hypothesis.

We begin by making a claim about the value of the population proportion p. If we wish to test whether the fraction of lung cancer patients surviving at least five years after diagnosis is the same among persons under 40 years of age as it is among those over the age of 40, then the null hypothesis is

$$H_0: p = 0.082.$$

For a two-sided test conducted at the 0.01 level of significance, the alternative hypothesis would be

$$H_A: p \neq 0.082.$$

We next draw a sample of dichotomous observations from the underlying population and find the probability of observing a sample proportion as extreme as $\hat{p}$, given that the true population mean is p. We do this by calculating the test statistic

$$z = \frac{\hat{p} - p}{\sqrt{p(1-p)/n}}.$$

If n is sufficiently large and the null hypothesis is true, this ratio is normally distributed with mean 0 and standard deviation 1. Depending on its magnitude, we either reject or do not reject H_0.

For the sample of 52 persons younger than 40 who have been diagnosed with lung cancer, $\hat{p} = 0.115$. Therefore, the test statistic is

$$\begin{aligned} z &= \frac{\hat{p} - p}{\sqrt{p(1-p)/n}} \\ &= \frac{0.115 - 0.082}{\sqrt{0.082(1-0.082)/52}} \\ &= 0.87. \end{aligned}$$

According to Table A.1, the p-value of the test is 0.384. Since this is greater than the level of significance 0.01, we do not reject the null hypothesis. These samples do not provide evidence that the proportions of lung cancer patients who survive five years differ in the two age groups.

14.5 Sample Size Estimation

When designing a study, we often wish to determine the sample size n that is necessary to draw a particular inference. When we are dealing with proportions, the calculations are a little more complex than those based on means; however, the reasoning is quite similar. For instance, suppose that we wish to test the null hypothesis

$$H_0: p \leq 0.082$$

at the $\alpha = 0.01$ level of significance. Once again, p is the proportion of lung cancer patients under 40 years of age at diagnosis who survive at least five years. Although we previously conducted a two-sided test, we are now concerned only with values of p that are greater than 0.082. If the true population proportion is as large as 0.200, we want to risk only a 5% chance of failing to reject the null hypothesis; therefore, β is equal to 0.05, and the power of the test is 0.95. How large a sample would we require?

Since $\alpha = 0.01$ and we are conducting a one-sided test, we begin by noting that H_0 would be rejected for $z \geq 2.32$. Therefore, we let

$$2.32 = \frac{\hat{p} - 0.082}{\sqrt{0.082(1 - 0.082)/n}}.$$

Solving for $\hat{p}$,

$$\hat{p} = 0.082 + 2.32\sqrt{\frac{0.082(1 - 0.082)}{n}}.$$

We would reject the null hypothesis if $\hat{p}$ is greater than this value.

We now focus on the desired power of the test. If the true proportion of patients surviving for five years is 0.200, then we want to reject the null hypothesis with probability 0.05. The value of z that corresponds to $\beta = 0.05$ is $z = -1.645$; therefore,

$$-1.645 = \frac{\hat{p} - 0.200}{\sqrt{0.200(1 - 0.200)/n}},$$

and

$$\hat{p} = 0.200 - 1.645\sqrt{\frac{0.200(1 - 0.200)}{n}}.$$

Using the preceding expressions for $\hat{p}$ to solve for n,

$$\begin{aligned} n &= \left[\frac{2.32\sqrt{0.082(1 - 0.082)} + 1.645\sqrt{0.200(1 - 0.200)}}{0.200 - 0.082}\right]^2 \\ &= 120.4. \end{aligned}$$

A sample of size 121 would be required to reject the null hypothesis.

In general, if the probability of committing a type I error is $\alpha = 0.01$, and the probability of making a type II error is $\beta = 0.05$, then the sample size n is

$$n = \left[\frac{2.32\sqrt{p_0(1-p_0)} + 1.645\sqrt{p_1(1-p_1)}}{p_1 - p_0} \right]^2 .$$

Note that p_0 is the hypothesized population proportion, and p_1 is the alternative. The magnitudes of these proportions—along with the values of α and β—determine the value of n. If we are interested in conducting a two-sided test, then we must make one adjustment to the preceding formula. For $\alpha = 0.01$, the null hypothesis would be rejected for $z \geq 2.58$. As a result, the sample size n would be approximately

$$n = \left[\frac{2.58\sqrt{p_0(1-p_0)} + 1.645\sqrt{p_1(1-p_1)}}{p_1 - p_0} \right]^2 .$$

14.6 Comparison of Two Proportions

Just as we did for the mean, we can again generalize the procedure of hypothesis testing to accommodate the comparison of two proportions. Most often we are interested in testing the null hypothesis that the proportions from two independent populations are identical. To conduct the test, we draw a random sample of size n_1 from the population with mean p_1. If there are x_1 successes in the sample, then

$$\hat{p}_1 = \frac{x_1}{n_1}.$$

Similarly, we select a sample of size n_2 from the population with mean p_2; therefore,

$$\hat{p}_2 = \frac{x_2}{n_2}.$$

We then calculate the probability of obtaining a pair of sample proportions as discrepant or more discrepant than those observed given that the null hypothesis is true.

If the population proportions are in fact equal, then the two samples can be combined to estimate this common parameter; in particular,

$$\begin{aligned} \hat{p} &= \frac{n_1\hat{p}_1 + n_2\hat{p}_2}{n_1 + n_2} \\ &= \frac{x_1 + x_2}{n_1 + n_2}. \end{aligned}$$

The quantity $\hat{p}$ is a weighted average of the sample proportions $\hat{p}_1$ and $\hat{p}_2$.

Under the null hypothesis, the estimate of the standard error of $\hat{p}_1 - \hat{p}_2$ takes the form $\sqrt{\hat{p}(1-\hat{p})[(1/n_1)+(1/n_2)]}$. To perform a test for independent proportions, we calculate the statistic

$$z = \frac{(\hat{p}_1 - \hat{p}_2) - (p_1 - p_2)}{\sqrt{\hat{p}(1-\hat{p})[(1/n_1)+(1/n_2)]}}.$$

If n_1 and n_2 are sufficiently large, the test statistic has a normal distribution with mean 0 and standard deviation 1. A commonly used although conservative criterion is that each of the quantities $n_1\hat{p}$, $n_1(1-\hat{p})$, $n_2\hat{p}$ and $n_2(1-\hat{p})$ be greater than or equal to 5. We compare the value of the statistic to the critical values in Table A.1 to find the p-value of the test; we then either reject or do not reject the null hypothesis.

In a study investigating morbidity and mortality among pediatric victims of motor vehicle accidents, information regarding the effectiveness of seat belts was collected over an 18-month period. Two random samples were selected, one from the population of children who were wearing a seat belt at the time of the accident and the other from the population who were not. We would like to test

$$H_0: p_1 = p_2,$$

the null hypothesis that the proportions of children who die as a result of the accident are identical in the two populations. We conduct a two-sided test at the 0.05 level of significance; the alternative hypothesis is

$$H_A: p_1 \neq p_2.$$

In the sample of 123 children who were wearing a seat belt at the time of the accident, three died [5]. Therefore,

$$\begin{aligned} \hat{p}_1 &= \frac{x_1}{n_1} \\ &= \frac{3}{123} \\ &= 0.024. \end{aligned}$$

In the sample of 290 children who were not wearing a seat belt, 13 died. Consequently,

$$\begin{aligned} \hat{p}_2 &= \frac{x_2}{n_2} \\ &= \frac{13}{290} \\ &= 0.045. \end{aligned}$$

Is this discrepancy in sample proportions too large to be attributed to chance?

If the population proportions are in fact equal, then their common value p is estimated by

$$\begin{aligned}\hat{p} &= \frac{x_1 + x_2}{n_1 + n_2} \\ &= \frac{3 + 13}{123 + 290} \\ &= 0.039.\end{aligned}$$

Substituting these estimates into the expression for the test statistic, we find that

$$\begin{aligned}z &= \frac{(\hat{p}_1 - \hat{p}_2) - (p_1 - p_2)}{\sqrt{\hat{p}(1-\hat{p})[(1/n_1) + (1/n_2)]}} \\ &= \frac{(0.024 - 0.045) - 0}{\sqrt{(0.039)(1 - 0.039)[(1/123) + (1/290)]}} \\ &= -1.01.\end{aligned}$$

According to Table A.1, the p-value of the test is 0.312. Therefore, we are unable to reject the null hypothesis. These samples do not provide evidence that the proportions of children dying differ in the two populations.

The quantity $\hat{p}_1 - \hat{p}_2$ provides a point estimate for the true difference in population proportions; we might also wish to construct a confidence interval. Using the normal approximation, the 95% confidence limits for the difference $p_1 - p_2$ are

$$\left(\hat{p}_1 - \hat{p}_2 \pm 1.96\sqrt{\frac{\hat{p}_1(1-\hat{p}_1)}{n_1} + \frac{\hat{p}_2(1-\hat{p}_2)}{n_2}} \right).$$

Note that the standard error of the difference in proportions is not the same as that used in the test of significance. In the hypothesis test, the standard error

$$\sqrt{\hat{p}(1-\hat{p})[(1/n_1) + (1/n_2)]}$$

was employed based on the assumption that the null hypothesis was true. This assumption is not necessary in the calculation of confidence limits.

Since $\hat{p}_1 = 0.024$ and $\hat{p}_2 = 0.045$, a 95% confidence interval for $p_1 - p_2$ is

$$\left(0.024 - 0.045 \pm 1.96\sqrt{\frac{0.024(1-0.024)}{123} + \frac{0.045(1-0.045)}{290}} \right)$$

or

$$(-0.057, 0.015).$$

We are 95% confident that this interval covers the true difference in the proportions of children dying in each population. Note that the interval contains the value 0; in this case, it is consistent with the test of hypothesis conducted at the 0.05 level.

14.7 Further Applications

Suppose that we are interested in investigating the cognitive abilities of children weighing less than 1,500 grams at birth. Although their birth weights are extremely low, many of these children exhibit normal growth patterns during the first year of life. A small group does not. These children suffer from perinatal growth failure, a condition that prevents them from developing properly. One indicator of perinatal growth failure is that, during the first several months of life, the infant has a head circumference measurement that is far below normal.

We are interested in examining the relationship between perinatal growth failure and subsequent cognitive ability. In particular, we would like to estimate the proportion of children suffering from this condition who, when they reach 8 years of age, have an IQ score that is below 70. We begin by constructing a 99% confidence interval for p. Using the normal approximation, we know that

$$\left(\hat{p} - 2.58\sqrt{\frac{\hat{p}(1-\hat{p})}{n}}\,,\ \hat{p} + 2.58\sqrt{\frac{\hat{p}(1-\hat{p})}{n}}\right)$$

has roughly a 99% chance of covering p before a sample is selected.

A random sample of 33 infants with perinatal growth failure is chosen; at the age of 8 years, eight children have an IQ below 70 [6]. Therefore,

$$\begin{aligned}\hat{p} &= \frac{x}{n} \\ &= \frac{8}{33} \\ &= 0.242.\end{aligned}$$

An approximate 99% confidence interval for p is

$$\left(0.242 - 2.58\sqrt{\frac{0.242(1-0.242)}{33}}\,,\ 0.242 + 2.58\sqrt{\frac{0.242(1-0.242)}{33}}\right)$$

or

$$(0.050\,,\ 0.434).$$

We are 99% confident that these limits cover the true proportion of children who suffered from perinatal growth failure as infants and subsequently have an IQ score below 70.

Although we do not know the true value of p for this population, we do know that 3.2% of the children who exhibited normal growth have an IQ score below 70 when they reach school age. We would like to know whether this is also true of the children who suffered from perinatal growth failure. Since we are concerned with deviations that could occur in either direction, we conduct a two-sided test of the null hypothesis

$$H_0: p = 0.032$$

at the 0.01 level of significance. The alternative hypothesis is

$$H_A: p \neq 0.032.$$

Recall that $\hat{p} = 0.242$. If the true population proportion is 0.032, then what is the probability of selecting a sample with a proportion as high as 0.242? To answer this question, we calculate the test statistic

$$\begin{aligned} z &= \frac{\hat{p} - p}{\sqrt{p(1-p)/n}} \\ &= \frac{0.242 - 0.032}{\sqrt{0.032(1-0.032)/33}} \\ &= 6.85. \end{aligned}$$

According to Table A.1, the p-value of the test is less than 0.001. We reject the null hypothesis and conclude that, among children who experienced perinatal growth failure, the proportion having an IQ score below 70 is larger than 0.032.

Another study was conducted to investigate the nonclinical factors associated with the method of surgical treatment received for early-stage breast cancer. While some patients undergo a modified radical mastectomy, others have a partial mastectomy accompanied by radiation therapy. We are interested in determining whether the age of the patient affects the type of treatment that she receives. In particular, we would like to know whether the proportions of women under the age of 55 are identical in the two treatment groups. Consequently, we test the null hypothesis

$$H_0: p_1 = p_2$$

against the two-sided alternative

$$H_A: p_1 \neq p_2.$$

A random sample of 658 women who underwent a partial mastectomy and subsequent radiation therapy contains 292 women under the age of 55; a sample of 1,580 women who received a modified radical mastectomy contains 397 women under 55 [7]. Therefore,

$$\begin{aligned}\hat{p}_1 &= \frac{x_1}{n_1} \\ &= \frac{292}{658} \\ &= 0.444,\end{aligned}$$

and

$$\begin{aligned}\hat{p}_2 &= \frac{x_2}{n_2} \\ &= \frac{397}{1,580} \\ &= 0.251.\end{aligned}$$

If the two proportions p_1 and p_2 are identical, then an estimate of their common value p is

$$\begin{aligned}\hat{p} &= \frac{x_1 + x_2}{n_1 + n_2} \\ &= \frac{292 + 397}{658 + 1,580} \\ &= 0.308.\end{aligned}$$

Therefore, the test statistic is

$$\begin{aligned}z &= \frac{(\hat{p}_1 - \hat{p}_2) - (p_1 - p_2)}{\sqrt{\hat{p}(1 - \hat{p})[(1/n_1) + (1/n_2)]}} \\ &= \frac{(0.444 - 0.251) - 0}{\sqrt{(0.308)(1 - 0.308)[(1/658) + (1/1,580)]}} \\ &= 9.01.\end{aligned}$$

Table A.1 indicates that the p-value of the test is less than 0.001. We reject the null hypothesis and conclude that women who undergo a modified radical mastectomy tend to be older than those who receive a partial mastectomy and radiation therapy.

Since $\hat{p}_1 = 0.444$ and $\hat{p}_2 = 0.251$, $\hat{p}_1 - \hat{p}_2 = 0.193$ is a point estimate for the true difference in population proportions. A 95% confidence interval for $p_1 - p_2$ is

$$\left(0.444 - 0.251 \pm 1.96\sqrt{\frac{0.444(1-0.444)}{658} + \frac{0.251(1-0.251)}{1,580}}\right)$$

or

$$(0.149\,,\,0.237).$$

We are 95% confident that the true difference in the proportions of women under the age of 55 is between 0.149 and 0.237.

14.8 Review Exercises

1. Suppose that you select a random sample of 40 children from the population of newborn infants in Mexico. The probability that a child in this population weighs at most 2,500 grams is 0.15 [8].

   a. For the sample of size 40, what is the probability that four or fewer of the infants weigh at most 2,500 grams? Compute the exact binomial probability.
   b. Using the normal approximation to the binomial distribution, estimate the probability that four or fewer of the children weigh at most 2,500 grams.
   c. Do these two methods provide consistent results?

2. A study was conducted to investigate the relationship between maternal smoking during pregnancy and the presence of congenital malformations in the child. Among children who suffer from an abnormality other than Downs syndrome or an oral cleft, 32.8% have mothers who smoked during pregnancy [9]. This proportion is homogeneous for children with various types of defects.

   a. If you were to select repeated samples of size 25 from this population, what could you say about the distribution of sample proportions? List three properties.
   b. Among the samples of size 25, what fraction has a sample proportion of 0.45 or more?
   c. What fraction has a sample proportion of 0.20 or less?
   d. What value of p cuts off the lower 10% of the distribution?

3. Now consider the children born with an oral cleft. In a random sample of 27 such infants, 15 have mothers who smoked during pregnancy [9].

a. Construct a 95% confidence interval for the population proportion p.
b. You would like to know whether the proportion of mothers who smoked during pregnancy for children with an oral cleft is identical to the proportion of mothers who smoked for children with other types of malformations. What is the null hypothesis of the appropriate test?
c. What is the alternative hypothesis?
d. Conduct the test at the 0.01 level of significance.
e. What do you conclude?
f. If the true population proportion is as low as 0.250, then you want to risk only a 10% chance of failing to reject the null hypothesis. If you are conducting a one-sided test at the 0.01 level of significance, how large a sample would be required?

4. Suppose that you are interested in determining whether the advice given by a physician during a routine physical examination is effective in encouraging patients to stop smoking. In a study of current smokers, one group of patients was given a brief talk about the hazards of smoking and was encouraged to quit. A second group received no advice pertaining to smoking. All patients were given a follow-up exam. In a sample of 114 patients who had received the advice, 11 reported that they had quit smoking; in a sample of 96 patients who had not, 7 quit smoking [10].

a. Estimate the true difference in population proportions $p_1 - p_2$.
b. At the 0.05 level of significance, test the null hypothesis that the proportions of patients who quit smoking are identical in the two populations.
c. Do you believe that the advice given by physicians is effective? Why or why not?
d. Construct a 95% confidence interval for the difference $p_1 - p_2$.

5. A study was conducted to investigate the use of community-based treatment programs among Medicaid beneficiaries suffering from severe mental illness. The study involved assigning a sample of 311 patients to a prepaid medical plan and a sample of 310 patients to the traditional Medicaid program. After a specified period of time, the number of persons in each group who had visited a community crisis center in the previous three months was determined. Among the individuals assigned to the prepaid plan, 13 had visited a crisis center; among those receiving traditional Medicaid, 22 had visited a center [11].

a. For each group, estimate the proportion of patients who had visited a community crisis center in the previous three months.
b. At the 0.10 level of significance, test the null hypothesis that the proportions are identical in the two populations.
c. What do you conclude?

References

[1] Nuland, S. B., *Doctors: The Biography of Medicine*, New York: Vintage Books, 1989.

[2] Centers for Disease Control, "The Surgeon General's 1989 Report on Reducing the Health Consequences of Smoking: 25 Years of Progress," *Morbidity and Mortality Weekly Report Supplement*, March 24, 1989.

[3] Myers, M. H., and Gloecker-Ries, L. A., "Cancer Patient Survival Rates: SEER Program Results for 10 Years of Follow-Up," *Ca–A Cancer Journal for Clinicians*, Volume 39, January-February 1989, 21–32.

[4] Jubelirer, S. J., and Wilson, R. A., "Lung Cancer in Patients Younger Than 40 Years of Age," *Cancer*, Volume 67, March 1, 1991, 1436–1438.

[5] Osberg, J. S., and DiScala, C., "Morbidity Among Pediatric Motor Vehicle Crash Victims: The Effectiveness of Seat Belts," *American Journal of Public Health*, Volume 82, March 1992, 422–425.

[6] Hack, M., Breslau, N., Weissman, B., Aram, D., Klein, N., and Borawski, E., "Effect of Very Low Birth Weight and Subnormal Head Size on Cognitive Abilities at School Age," *The New England Journal of Medicine*, Volume 325, July 25, 1991, 231–237.

[7] Satariano, E. R., Swanson, G. M., and Moll, P. P., "Nonclinical Factors Associated with Surgery Received for Treatment of Early-Stage Breast Cancer," *American Journal of Public Health*, Volume 82, February 1992, 195–198.

[8] United Nations Children's Fund, *The State of the World's Children 1991*, New York: Oxford University Press.

[9] Khoury, M. J., Weinstein, A., Panny, S., Holtzman, N. A., Lindsay, P. K., Farrel, K., and Eisenberg, M., "Maternal Cigarette Smoking and Oral Clefts: A Population-Based Study," *American Journal of Public Health*, Volume 77, May 1987, 623–625.

[10] Folsom, A. R., and Grim, R. H., "Stop Smoking Advice by Physicians: A Feasible Approach?," *American Journal of Public Health*, Volume 77, July 1987, 849–850.

[11] Christianson, J. B., Lurie, N., Finch, M., Moscovice, I. S., and Hartley, D., "Use of Community-Based Mental Health Programs by HMOs: Evidence from a Medicaid Demonstration," *American Journal of Public Health*, Volume 82, June 1992, 790–796.

CHAPTER 15

Contingency Tables

In the previous chapter we used the normal approximation to the binomial distribution to conduct tests of hypotheses for a pair of independent proportions. However, we could have achieved the same results using an alternative technique. When working with nominal data that have been grouped into categories, we often arrange the counts in a tabular format known as a *contingency table*. In the simplest case, two dichotomous random variables are involved; the rows of the table represent the outcomes of one variable, and the columns represent the outcomes of the other. The entries in the table are the counts that correspond to a particular combination of categories.

15.1 The Chi-Square Test

15.1.1 2×2 Tables

Consider the 2×2 table below, which displays the results of a study investigating the effectiveness of bicycle safety helmets in preventing head injury [1]. The data consist of a random sample of 793 individuals who were involved in bicycle accidents during a specified one-year period.

	Wearing Helmet		
Head Injury	**Yes**	**No**	**Total**
Yes	17	218	235
No	130	428	558
Total	147	646	793

Of the 793 individuals who were involved in bicycle accidents, 147 were wearing safety helmets at the time of the incident, and 646 were not. Among those wearing helmets, 17 suffered head injuries requiring the attention of a doctor, whereas the remaining 130 did not; among the individuals not wearing safety helmets, 218 sustained serious head injuries, and 428 did not. The entries in the table—17, 130, 218, and 428—are thus the observed counts within each combination of categories.

To examine the effectiveness of bicycle safety helmets, we wish to know whether there is an association between the incidence of head injury and the use of helmets among individuals who have been involved in an accident. To determine this, we test the null hypothesis

H_0: the proportion of persons suffering head injuries among the population of individuals wearing a safety helmet at the time of the accident is equal to the proportion of persons sustaining head injuries among those not wearing a helmet

against the alternative

H_A: the proportions of persons suffering head injuries are not identical in the two populations.

We conduct the test at the $\alpha = 0.05$ level of significance.

The first step in carrying out the test is to calculate the expected count for each cell of the contingency table. Under the null hypothesis, the proportions of individuals experiencing head injuries among those wearing helmets and those not wearing helmets are identical; therefore, we are able to ignore the two separate categories and treat all 793 individuals as a single homogeneous sample. In this sample, the overall proportion of persons sustaining head injuries is

$$\frac{235}{793} = 29.6\%,$$

and the proportion of persons not experiencing head injuries is

$$\frac{558}{793} = 70.4\%.$$

As a result, of the 147 individuals wearing a safety helmet at the time of the accident, we would expect that 29.6%, or

$$\begin{aligned} 147\,(0.296) &= 147\left(\frac{235}{793}\right) \\ &= 43.6, \end{aligned}$$

suffer head injuries, and 70.4%, or

$$\begin{aligned} 147\,(0.704) &= 147\left(\frac{558}{793}\right) \\ &= 103.4, \end{aligned}$$

do not. Similarly, among the 646 bicyclists not wearing a safety helmet, we would expect that 29.6%, or

$$\begin{aligned} 646\,(0.296) &= 646\left(\frac{235}{793}\right) \\ &= 191.4, \end{aligned}$$

sustain head injuries, whereas 70.4%, or

$$\begin{aligned} 646\,(0.704) &= 646\left(\frac{558}{793}\right) \\ &= 454.6, \end{aligned}$$

do not. In general, the expected count for a given cell in the table is equal to the row total multiplied by the column total divided by the table total. To avoid roundoff error, we usually compute expected counts to a fraction of a person. Thus, for the four categories in the original table, the expected counts are

	Wearing Helmet		
Head Injury	**Yes**	**No**	**Total**
Yes	43.6	191.4	235.0
No	103.4	454.6	558.0
Total	147.0	646.0	793.0

In general, if a 2 × 2 table of observed frequencies for a sample of size n can be represented as follows,

	Variable 2		
Variable 1	**Yes**	**No**	**Total**
Yes	a	b	$a+b$
No	c	d	$c+d$
Total	$a+c$	$b+d$	n

then the corresponding table of expected counts is

	Variable 2		
Variable 1	Yes	No	Total
Yes	$(a+b)(a+c)/n$	$(a+b)(b+d)/n$	$a+b$
No	$(c+d)(a+c)/n$	$(c+d)(b+d)/n$	$c+d$
Total	$a+c$	$b+d$	n

The row and column totals in the table of expected counts are identical to those in the observed table. These marginal totals have been held fixed by design; we calculate the cell entries that would have been expected given that there is no association between the row and column classifications **and** that the number of individuals within each group remains constant.

The *chi-square test* compares the observed frequencies in each category, represented by O, with the expected frequencies given that the null hypothesis is true, denoted by E. It is used to determine whether the deviations between the observed and the expected counts, $O - E$, are too large to be attributed to chance. Since there is more than one cell in the table, these deviations must be combined in some way. To perform the test for the data in a contingency table with r rows and c columns, we calculate the sum

$$X^2 = \sum_{i=1}^{rc} \frac{(O_i - E_i)^2}{E_i}$$

where rc is the number of cells in the table. The probability distribution of this sum is approximated by a *chi-square* (χ^2) *distribution* with $(r-1)(c-1)$ degrees of freedom. For instance, a 2×2 table has $(2-1)(2-1) = 1$ degree of freedom; a 3×4 table has $(3-1)(4-1) = 6$ degrees of freedom. To ensure that the sample size is large enough to make this approximation valid, no cell should have an expected count less than 1, and no more than 20% of the cells should have an expected count less than 5 [2].

Like the F distribution, the chi-square distribution is not symmetric. A chi-square random variable cannot be negative; it assumes values from zero to infinity and is skewed to the right. Once again, however, the total area beneath the curve is equal to one. As with the t and F distributions, there is a different chi-square distribution for each possible value of the degrees of freedom. The distributions with small degrees of freedom are highly skewed; as the number of degrees of freedom increases, the distributions become less skewed and more symmetric. This is illustrated in Figure 15.1.

Table A.6 in the appendix is a condensed table of areas for the chi-square distribution with various degrees of freedom. For a particular value of df, the entry in the table is the outcome of χ^2_{df} that cuts off the specified area in the upper tail of the distribution. Given a chi-square distribution with 1 degree of freedom, for instance, $\chi^2_1 = 3.84$ cuts off the upper 5% of the area under the curve.

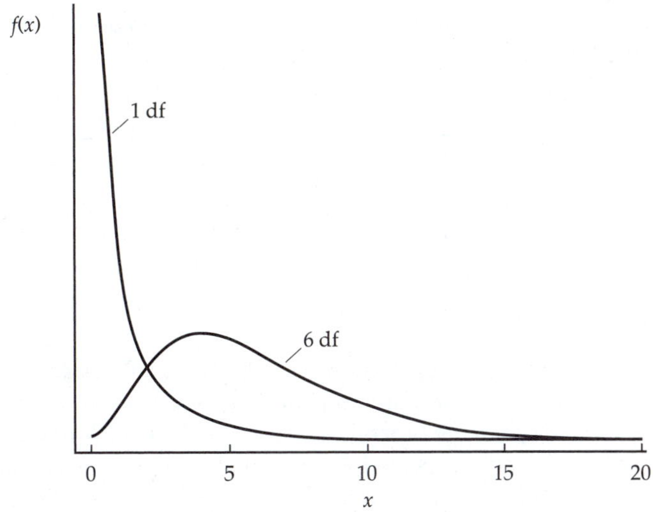

Figure 15.1. Chi-square distributions with 1 and 6 degrees of freedom

Since all the expected counts for the data pertaining to bicycle safety helmets are greater than 5, we may proceed with the chi-square test. The next step is to calculate the sum

$$X^2 = \sum_{i=1}^{rc} \frac{(O_i - E_i)^2}{E_i}.$$

For a 2 × 2 contingency table, this test statistic has an approximate chi-square distribution with (2 − 1)(2 − 1) = 1 degree of freedom. Note that we are using discrete observations to estimate χ^2, a continuous distribution. The approximation is quite good for tables with many degrees of freedom but may not be valid for 2 × 2 tables that have df = 1. We therefore often apply a correction factor in this situation. When we are working with a 2 × 2 table, the appropriate test statistic to use is

$$X^2 = \sum_{i=1}^{4} \frac{(|O_i - E_i| - 0.5)^2}{E_i},$$

where $|O_i - E_i|$ is the absolute value of the difference between O_i and E_i. The term −0.5 in the numerator is often referred to as the *Yates' correction*. Although it has been used extensively in the past, many investigators have begun to question its validity. Some believe that it results in an overly conservative test that may fail to reject a null hypothesis that is false [3].

Substituting the observed and expected counts for the bicycle injury data into the corrected equation, we find that

$$\begin{aligned} X^2 &= \frac{(|17 - 43.6| - 0.5)^2}{43.6} + \frac{(|130 - 103.4| - 0.5)^2}{103.4} \\ &\quad + \frac{(|218 - 191.4| - 0.5)^2}{191.4} + \frac{(|428 - 454.6| - 0.5)^2}{454.6} \\ &= 15.62 + 6.59 + 3.56 + 1.50 \\ &= 27.27. \end{aligned}$$

For a χ^2 distribution with 1 degree of freedom, the p-value is less than 0.001. Although we are considering only one tail of the distribution, the test is two-sided; large outcomes of $(O_i - E_i)^2$ can result when the observed value is bigger than the expected value and also when it is smaller. Since $p < \alpha$, we reject the null hypothesis and conclude that the proportions of individuals suffering head injuries are not identical in the two populations . Among persons involved in a bicycle accident, the use of a safety helmet reduces the incidence of head injury.

If we represent a 2 × 2 table in the general format below,

	Variable 2		
Variable 1	**Yes**	**No**	**Total**
Yes	a	b	$a + b$
No	c	d	$c + d$
Total	$a + c$	$b + d$	n

then the test statistic X^2 may also be expressed as

$$X^2 = \frac{n\,[|ad - bc| - (n/2)]^2}{(a + c)(b + d)(a + b)(c + d)}.$$

Because there is no need to compute a table of expected counts, this form of the statistic is more convenient computationally. However, it does not have the same straightforward interpretation as a measure of the deviations between observed and expected counts.

15.1.2 $r \times c$ Tables

In the case of a 2 × 2 table, the chi-square test for independent proportions is equivalent to the normal approximation to the binomial distribution. The chi-square test is used more often in practice, however. In addition to being relatively easy to calculate, it can be generalized to accommodate the comparison of three or more proportions. In this situation, we arrange the data as an $r \times c$ contingency table where r is the number of rows and c is the number of columns.

Consider the following data taken from a study that investigates the accuracy of death certificates. In two different hospitals, the results of 575 autopsies were compared to the causes of death listed on the certificates. One of the hospitals that participated in the study was a community hospital, labeled A; the other was a university hospital, labeled B. The data are displayed in the form of a 2×3 contingency table [4].

	Certificate Status			
Hospital	**Confirmed Accurate**	**Inaccurate No change**	**Incorrect Recoding**	**Total**
A	157	18	54	229
B	268	44	34	346
Total	425	62	88	575

Of the 575 death certificates considered, 425 were confirmed to be accurate, 62 either lacked information or contained inaccuracies but did not require recoding of the underlying cause of death, and 88 were incorrect and required recoding. We would like to determine whether the results of the study suggest different practices in completing death certificates at the two hospitals. To do this, we test the null hypothesis

H_0: in each category of certificate status, the proportions of death certificates in Hospital A are identical

against the alternative that the proportions are not identical. We use the chi-square test and set the significance level at $\alpha = 0.05$.

The first step in carrying out the test is to calculate the expected count for each cell in the table. Recall that the expected count is equal to the row total multiplied by the column total divided by the table total. For instance, among death certificates confirmed to be accurate, we expect that

$$\frac{229 \times 425}{575} = 169.3$$

would be in Hospital A, whereas

$$\frac{346 \times 425}{575} = 255.7$$

would be in Hospital B. For all six categories in the table, the expected counts are as follows:

Hospital	Certificate Status			Total
	Confirmed Accurate	Inaccurate No change	Incorrect Recoding	
A	169.3	24.7	35.0	229.0
B	255.7	37.3	53.0	346.0
Total	425.0	62.0	88.0	575.0

We next calculate the sum

$$X^2 = \sum_{i=1}^{6} \frac{(O_i - E_i)^2}{E_i}.$$

Since we are working with a 2×3 table, we need not include the continuity correction factor. Therefore,

$$\begin{aligned} X^2 &= \frac{(157-169.3)^2}{169.3} + \frac{(18-24.7)^2}{24.7} + \frac{(54-35.0)^2}{35.0} \\ &\quad + \frac{(268-255.7)^2}{255.7} + \frac{(44-37.3)^2}{37.3} + \frac{(34-53.0)^2}{53.0} \\ &= 0.89 + 1.82 + 10.31 + 0.59 + 6.81 + 1.20 \\ &= 21.62. \end{aligned}$$

For a χ^2 distribution with $(2-1)(3-1) = 2$ degrees of freedom, $p < 0.001$. Since p is less than α, we reject the null hypothesis and conclude that the proportions of death certificates in Hospital A are not identical for the three different categories. Referring back to the contingency table, it appears that Hospital A contains a larger proportion of death certificates that are incorrect and require recoding.

15.2 McNemar's Test

What do we do if we are dealing with paired rather than independent samples? As with continuous data, the distinguishing characteristic of paired samples for counts is that each observation in the first group has a corresponding observation in the second group. In this situation, we use *McNemar's test* to evaluate hypotheses about the data.

Consider the following data taken from a study which investigates acute myocardial infarction among Navajo Indians. In the study, 144 victims of acute myocardial infarction were age- and sex-matched with 144 individuals free of heart disease. The members of each pair were then asked whether they had ever been diagnosed as having diabetes. The results are presented below [5].

Diabetes	MI		Total
	Yes	No	
Yes	46	25	71
No	98	119	217
Total	144	144	288

Among the individuals who had experienced an acute myocardial infarction, 46 had been diagnosed as having diabetes; among those who had not, only 25 suffered from diabetes. We would like to know whether this sample provides evidence that the proportion of diabetics is identical in the two groups.

The preceding table has the same format as the 2×2 contingency table used in the comparison of proportions from independent samples; therefore, we might be tempted to apply the chi-square test. Note, however, that we have a total of 288 observations but only 144 pairs. Each matched pair in the study provides two responses: one for the individual who suffered an MI and one for the individual who did not. Since the chi-square test disregards the paired nature of the data, it is inappropriate in this situation. We must somehow take the pairing into account in our analysis.

Suppose that we were to look again at the raw data from the study but this time classify it in the following manner:

MI	No MI		Total
	Diabetes	No diabetes	
Diabetes	9	37	46
No diabetes	16	82	98
Total	25	119	144

Of the 46 individuals who had experienced an acute myocardial infarction and who were diabetics, 9 were matched with persons free of heart disease who had diabetes and 37 were not. Of the 98 MI victims who did not suffer from diabetes, 16 were paired with diabetics and 82 were not. Each entry in the table corresponds to a combination of responses for a matched pair rather than an individual person.

We are now able to conduct a proper paired analysis. However, we must first change the statement of the null hypothesis to fit the new format of the data. Instead of testing whether the proportion of diabetics among individuals who have experienced an acute myocardial infarction is equal to the proportion of diabetics among those who have not, we test

H_0: there are equal numbers of pairs in which the victim of acute myocardial infarction is a diabetic and the matched individual free of heart disease is not and in which the person who has not experienced an MI is a diabetic but the individual with heart disease is not, or, more concisely,

H_0: there is no association between diabetes and the occurrence of an acute myocardial infarction.

The alternative hypothesis is that an association exists. We conduct this test at the $\alpha = 0.05$ level of significance.

The *concordant pairs*–or the pairs of responses in which either two diabetics or two nondiabetics are matched–provide no information for testing H_0. Therefore, we discard these data and focus only on the *discordant pairs*, or the pairs of responses in which a person who has diabetes is paired with an individual who does not.

Let r represent the number of pairs in which the victim of acute myocardial infarction suffers from diabetes and the individual free of heart disease does not, and s the number of pairs in which the person who did not experience an MI is a diabetic but the individual who did have an MI is not. If the difference between r and s is large, then we would want to reject the null hypothesis of no association. To conduct McNemar's test, we calculate the statistic

$$X^2 = \frac{[|r - s| - 1]^2}{(r + s)}.$$

This ratio has an approximate chi-square distribution with 1 degree of freedom. For the given data, $r = 37$ and $s = 16$. Therefore,

$$\begin{aligned} X^2 &= \frac{[|37 - 16| - 1]^2}{(37 + 16)} \\ &= \frac{[21 - 1]^2}{53} \\ &= 7.55. \end{aligned}$$

For a chi-square distribution with 1 degree of freedom, $0.001 < p < 0.01$. Since p is less than α, we reject the null hypothesis. For the given population of Navajo Indians, therefore, victims of acute myocardial infarction are more likely to suffer from diabetes than the healthy individuals matched on age and sex.

15.3 The Odds Ratio

Although the chi-square test allows us to determine whether an association exists between two nominal random variables, it does not provide a measure of the strength of this association. Methods are available, however, for quantifying the magnitude of the effect using the observations in a single sample. For a 2×2 table, one such measure is the *odds ratio*, or *relative odds*.

In Chapter 6 we stated that, if an event occurs with probability p, the odds in favor of the event are $p/(1-p)$ to 1. If an event occurs with probability 1/2, for instance, then the odds in favor of the event are $(1/2)/(1-1/2) = (1/2)/(1/2) = 1$ to 1. Conversely, if the odds in favor of an event are a to b, then the probability that the event occurs is $a/(a+b)$.

If we have two dichotomous random variables that represent a "disease" and an "exposure," then the odds ratio is defined as the odds in favor of disease among exposed individuals divided by the odds in favor of disease among the unexposed, or

$$\text{OR} = \frac{\text{P(disease} \mid \text{exposed)}/[1 - \text{P(disease} \mid \text{exposed)}]}{\text{P(disease} \mid \text{unexposed)}/[1 - \text{P(disease} \mid \text{unexposed)}]}.$$

Suppose that our data–which consist of a sample of n individuals–are arranged in the form of a 2×2 contingency table:

	Exposed	Unexposed	Total
Disease	a	b	$a+b$
No disease	c	d	$c+d$
Total	$a+c$	$b+d$	n

In this case, we would estimate that

$$\text{P(disease} \mid \text{exposed)} = \frac{a}{a+c},$$

and

$$\text{P(disease} \mid \text{unexposed)} = \frac{b}{b+d};$$

therefore,

$$\begin{aligned} 1 - \text{P(disease} \mid \text{exposed)} &= 1 - \frac{a}{a+c} \\ &= \frac{c}{a+c}, \end{aligned}$$

and

$$\begin{aligned} 1 - \text{P(disease} \mid \text{unexposed)} &= 1 - \frac{b}{b+d} \\ &= \frac{d}{b+d}. \end{aligned}$$

Using these results, we can express the estimator of the odds ratio as

$$\begin{aligned}\widehat{\text{OR}} &= \frac{[a/(a+c)]/[c/(a+c)]}{[b/(b+d)]/[d/(b+d)]} \\ &= \frac{a/c}{b/d} \\ &= \frac{ad}{bc}.\end{aligned}$$

This estimator is simply the cross-product ratio of the entries in the 2 × 2 table.

The relative odds is not the same as the relative risk; *relative risk* is the probability of disease among exposed individuals divided by the probability of disease among the unexposed. The estimator of the relative risk is

$$\begin{aligned}\widehat{\text{RR}} &= \frac{\text{P(disease} \mid \text{exposed)}}{\text{P(disease} \mid \text{unexposed)}} \\ &= \frac{a/(a+c)}{b/(b+d)} \\ &= \frac{a(b+d)}{b(a+c)}.\end{aligned}$$

When we are dealing with rare events—or values of a and b such that the product ab is nearly equal to 0—the relative risk can be approximated by the odds ratio:

$$\begin{aligned}\widehat{\text{RR}} &= \frac{ab+ad}{ba+bc} \\ &\approx \frac{ad}{bc} \\ &= \widehat{\text{OR}}.\end{aligned}$$

Because the sampling distribution of $\widehat{\text{OR}}$ is more straightforward than that of $\widehat{\text{RR}}$, it is generally preferable to work with the relative odds.

Consider the data taken from a study that attempts to determine whether the use of electronic fetal monitoring during labor affects the frequency of cæsarean section deliveries. Of the 5,824 infants included in the study, 2,850 were electronically monitored, and 2,974 were not [6]. The outcomes are as follows:

Cæsarean Delivery	EFM Exposure		Total
	Yes	No	
Yes	358	229	587
No	2,492	2,745	5,237
Total	2,850	2,974	5,824

The odds of being delivered by cæsarean section in the group that was monitored relative to the group that was not monitored are estimated by

$$\begin{aligned}\widehat{\text{OR}} &= \frac{(358)(2,745)}{(229)(2,492)} \\ &= 1.72.\end{aligned}$$

The data suggest that the odds of being delivered by cæsarean section for fetuses that are electronically monitored during labor are 1.72 times greater than the odds for fetuses that are not monitored; therefore, there does appear to be a moderate association between the use of electronic fetal monitoring and the eventual method of delivery. This does not imply, however, that electronic monitoring somehow **causes** a cæsarean delivery; it is possible that the fetuses at higher risk are the ones monitored.

The cross-product ratio is simply a point estimate of the strength of association between two dichotomous random variables. To gauge the uncertainty in this estimate, we must calculate a confidence interval as well; the width of the interval reflects the amount of variability in our estimator $\widehat{\text{OR}}$. Recall that the derivation of the expression for a 95% confidence interval for a mean μ,

$$\left(\bar{x} - 1.96\,\frac{\sigma}{\sqrt{n}},\ \bar{x} + 1.96\,\frac{\sigma}{\sqrt{n}}\right),$$

rests on the assumption that the underlying population values are normally distributed. When computing a confidence interval for the odds ratio, we must make the same assumption of underlying normality. A problem arises, however, in that the probability distribution of the odds ratio is skewed to the right; although the relative odds cannot take on a negative value, it can assume any positive value between 0 and infinity. In contrast, the probability distribution of the natural logarithm of the odds ratio is more symmetric and approximately normal. Therefore, when calculating a confidence interval for the odds ratio, we typically work in the log scale. To ensure that the sample size is large enough, the expected value of each entry in the table should be at least 5.

The expression for a 95% confidence interval for the natural logarithm of the odds ratio is

$$(\ln(\widehat{\text{OR}}) - 1.96\,\text{se}[\ln(\widehat{\text{OR}})],\ \ln(\widehat{\text{OR}}) + 1.96\,\text{se}[\ln(\widehat{\text{OR}})]).$$

Therefore, to compute a confidence interval for ln(OR), we must first know the standard error of this quantity. For a 2×2 table arranged in the following manner,

	Exposed	Unexposed	Total
Disease	a	b	$a+b$
No disease	c	d	$c+d$
Total	$a+c$	$b+d$	n

the standard error of $\ln(\widehat{\text{OR}})$ is estimated by

$$\widehat{\text{se}}[\ln(\widehat{\text{OR}})] = \sqrt{\frac{1}{a}+\frac{1}{b}+\frac{1}{c}+\frac{1}{d}}.$$

If any one of the entries in the table is equal to 0, the standard error is undefined. In this case, adding 0.5 to each of the values will correct the situation. Thus, the modified estimatees of the standard error is

$$\widehat{\text{se}}[\ln(\widehat{\text{OR}})] = \sqrt{\frac{1}{(a+0.5)}+\frac{1}{(b+0.5)}+\frac{1}{(c+0.5)}+\frac{1}{(d+0.5)}}.$$

The appropriate estimate can be substituted into the expression above. To find a 95% confidence interval for the odds ratio itself, we take the antilogarithm of the upper and lower limits to get

$$\left(e^{\ln(\widehat{\text{OR}})-1.96\,\widehat{\text{se}}[\ln(\widehat{\text{OR}})]},\ e^{\ln(\widehat{\text{OR}})+1.96\,\widehat{\text{se}}[\ln(\widehat{\text{OR}})]}\right).$$

For the data examining the relationship between electronic fetal monitoring status and the method of delivery, the log of the estimated odds ratio is

$$\ln(\widehat{\text{OR}}) = \ln(1.72) = 0.542.$$

The estimated standard error of $\ln(\widehat{\text{OR}})$ is

$$\begin{aligned}\widehat{\text{se}}[\ln(\widehat{\text{OR}})] &= \sqrt{\frac{1}{a}+\frac{1}{b}+\frac{1}{c}+\frac{1}{d}} \\ &= \sqrt{\frac{1}{358}+\frac{1}{229}+\frac{1}{2492}+\frac{1}{2745}} \\ &= 0.089.\end{aligned}$$

Therefore, a 95% confidence interval for the log of the odds ratio is

$$(0.542 - 1.96(0.089),\ 0.542 + 1.96(0.089))$$

or

$$(0.368\,,\, 0.716).$$

This means that a 95% confidence interval for the odds ratio itself is

$$(e^{0.368}\,,\, e^{0.716})$$

or

$$(1.44\,,\, 2.05).$$

We are 95% confident that the odds of delivery by cæsarean section among fetuses that are monitored during labor are between 1.44 and 2.05 times the odds of fetuses that are not monitored. Note that this interval does not include the value 1. An odds ratio of 1 would mean that fetuses that are monitored and those that are not monitored have identical odds of cæsarean delivery.

15.4 Berkson's Fallacy

Although the odds ratio is a useful measure of the strength of association between two dichotomous random variables, it is valid only if the sample of observations from which it is estimated is random. This point is sometimes forgotten; a restricted sample, such as a sample of patients from a single hospital, is usually much easier to obtain. This restricted sample is then used to make inference about the population as a whole.

In one study, the investigators surveyed 2,784 individuals–257 of whom were hospitalized–and determined whether each subject suffered from either a disease of the circulatory system or a respiratory illness, or both [7]. If they had limited their questioning to the 257 hospitalized patients only, the results would have been as follows:

Circulatory Disease	Respiratory Disease Yes	Respiratory Disease No	Total
Yes	7	29	36
No	13	208	221
Total	20	237	257

An estimate of the relative odds of having respiratory illness among individuals who suffer from a disease of the circulatory system versus those who do not is

$$\widehat{OR} = \frac{(7)(208)}{(29)(13)} = 3.86;$$

the chi-square test of the null hypothesis that there is no association between the two diseases yields

$$\begin{aligned} X^2 &= \frac{(|7 - 2.8| - 0.5)^2}{2.8} + \frac{(|29 - 33.2| - 0.5)^2}{33.2} \\ &\quad + \frac{(|13 - 17.2| - 0.5)^2}{17.2} + \frac{(|208 - 203.8| - 0.5)^2}{203.8} \\ &= 4.89 + 0.41 + 0.80 + 0.07 \\ &= 6.17. \end{aligned}$$

For a chi-square distribution with 1 degree of freedom, $0.01 < p < 0.025$. We reject the null hypothesis at the 0.05 level and conclude that individuals with a disease of the circulatory system are more likely to suffer from respiratory illness than individuals without.

Now consider the entire sample of 2,784 individuals, which consists of both hospitalized and nonhospitalized subjects:

Circulatory Disease	Respiratory Disease Yes	Respiratory Disease No	Total
Yes	22	171	193
No	202	2,389	2,591
Total	224	2,560	2,784

For these data, an estimate of the odds ratio is

$$\widehat{OR} = \frac{(22)(2,389)}{(171)(202)} = 1.52.$$

This value is much lower than the estimate of the relative odds calculated for the hospitalized patients only. In addition, the value of the chi-square test statistic is

$$
\begin{aligned}
X^2 &= \frac{(|22 - 15.53| - 0.5)^2}{15.53} + \frac{(|171 - 177.47| - 0.5)^2}{177.47} \\
&\quad + \frac{(|202 - 208.47| - 0.5)^2}{208.47} + \frac{(|2,389 - 2,382.53| - 0.5)^2}{2,382.53} \\
&= 2.29 + 0.20 + 0.17 + 0.01 \\
&= 2.67.
\end{aligned}
$$

For a chi-square distribution with 1 degree of freedom, $p > 0.10$. We can no longer reject the null hypothesis that there is no association between circulatory and respiratory diseases at the 0.05 level.

Why do the conclusions drawn from these two samples differ so drastically? To answer this question, we must consider the rates of hospitalization that occur within each of the four disease subgroups. Among the 22 individuals suffering from both circulatory and respiratory disease, 7 are hospitalized. Therefore, the rate of hospitalization for this subgroup is

$$\frac{7}{22} = 31.8\%.$$

The rate of hospitalization among subjects with respiratory illness alone is

$$\frac{13}{202} = 6.4\%.$$

Among individuals with circulatory disease only, the rate is

$$\frac{29}{171} = 17.0\%,$$

and among persons suffering from neither disease, the rate of hospitalization is

$$\frac{208}{2,389} = 8.7\%.$$

Thus, individuals with both circulatory and respiratory diseases are much more likely to be hospitalized than individuals in any of the three other subgroups. Also, subjects with circulatory disease are more likely to be hospitalized than those with respiratory illness. If we sample only patients who are hospitalized, therefore, our conclusions will be biased. We are more likely to select an individual who is suffering from both diseases than a person in any of the other subgroups and to select a person with circulatory disease as opposed to one with respiratory problems. As a result, we observe an association that does not actually exist. This kind of spurious relationship among variables—which is evident only because of the way in which the individuals were chosen—is known as *Berkson's fallacy*.

15.5 Further Applications

When we are presented with independent samples of nominal data that have been grouped into categories, the chi-square test can be used to determine whether the proportions of some event of interest are identical in the various groups. For example, consider the following data taken from a study investigating an outbreak of gastroenteritis following a lunch served in the cafeteria of a U.S. high school. Among a sample of 263 students who bought lunch in the school cafeteria on the day in question, 225 ate a prepared sandwich and 38 did not [8]. The number of cases of gastroenteritis in each group are displayed below.

	Ate Sandwich		
	Yes	**No**	**Total**
Ill	109	4	113
Not ill	116	34	150
Total	225	38	263

We would like to test the null hypothesis that there is no association between the consumption of a sandwich and the onset of gastroenteritis, or

H_0: the proportion of students becoming ill among those who ate the sandwiches is equal to the proportion becoming ill among those who did not eat the sandwiches,

at the $\alpha = 0.05$ level of significance. The alternative hypothesis is that an association does exist. To begin, we must calculate the expected count in each cell of the 2×2 table. Under the null hypothesis, the proportions of students developing gastroenteritis are identical in the two populations. Treating all 263 students as a single sample, the overall proportion of students becoming ill is

$$\frac{113}{263} = 43.0\%$$

and of those not becoming ill is

$$\frac{150}{263} = 57.0\%.$$

Therefore, among the 225 students who ate the sandwiches, we would expect that 43%, or

$$225\left(\frac{113}{263}\right) = 96.7,$$

become ill and that 57%, or

$$225\left(\frac{150}{263}\right) = 128.3,$$

do not. Similarly, among the 38 students who did not eat the sandwiches, we expect that

$$38\left(\frac{113}{263}\right) = 16.3$$

become ill and

$$38\left(\frac{150}{263}\right) = 21.7$$

do not. Thus the table of expected counts is as follows:

	Ate Sandwich		
	Yes	**No**	**Total**
Ill	96.7	16.3	113.0
Not ill	128.3	21.7	150.0
Total	225.0	38.0	263.0

Since the expected count in each cell is greater than 5, we proceed with the chi-square test by calculating the statistic

$$\begin{aligned} X^2 &= \sum_{i=1}^{4} \frac{(|O_i - E_i| - 0.5)^2}{E_i} \\ &= \frac{(|109 - 96.7| - 0.5)^2}{96.7} + \frac{(|4 - 16.3| - 0.5)^2}{16.3} \\ &\quad + \frac{(|116 - 128.3| - 0.5)^2}{128.3} + \frac{(|34 - 21.7| - 0.5)^2}{21.7} \\ &= 1.44 + 8.54 + 1.09 + 6.42 \\ &= 17.49. \end{aligned}$$

For a chi-square distribution with 1 degree of freedom, $p < 0.001$. Since p is less than α, we reject H_0 and conclude that the proportions of students developing gastroenteritis are not identical in the two groups. Among students eating lunch at the high school on the day in question, the consumption of a prepared sandwich is associated with an increased risk of illness.

Table 15.1 Stata output displaying the chi-square test

Ill	Ate Yes	No	Total
Yes	109	4	113
No	116	34	150
Total	225	38	263

```
Pearson chi2(1) = 19.0742     Pr = 0.000
```

Most computer packages are able to perform the chi-square test. The relevant output from Stata is contained in Table 15.1. The output consists of the observed 2×2 contingency table, the test statistic X^2, and the p-value of the test. Note that the test statistic X^2 is somewhat different from the value that we calculated by hand; this is because Stata does not apply the continuity correction factor.

Now consider the situation in which we have paired samples of discrete data. The following data come from a study that examines changes in smoking status over a two-year period [9]. In 1980, a sample of 2,110 adults over the age of 18 were asked to identify themselves as smokers or nonsmokers. In 1982, the same 2,110 individuals were again asked whether they were currently smokers or nonsmokers. Of the 717 individuals who smoked in 1980, 620 were still smoking in 1982, and 97 had stopped. Of the 1,393 nonsmokers in 1980, 1,317 remained nonsmokers in 1982, and 76 had begun to smoke. Each entry in the table corresponds to the paired response of a single individual.

	1982		
1980	**Smoker**	**Nonsmoker**	**Total**
Smoker	620	97	717
Nonsmoker	76	1,317	1,393
Total	696	1,414	2,110

We would like to test the null hypothesis that there is no association between smoking status and year, or, more formally,

H_0: among the individuals who changed their smoking status between 1980 and 1982, equal numbers switched from smokers to nonsmokers and from nonsmokers to smokers.

The alternative hypothesis is that there was a tendency to change in a specified direction. We conduct McNemar's test at the $\alpha = 0.05$ level of significance.

Note that we have $r = 97$ pairs in which a smoker becomes a nonsmoker and $s = 76$ pairs in which a nonsmoker becomes a smoker. To evaluate H_0, we calculate the test statistic

$$\begin{aligned} X^2 &= \frac{[|r - s| - 1]^2}{(r + s)} \\ &= \frac{[|97 - 76| - 1]^2}{(97 + 76)} \\ &= 2.31. \end{aligned}$$

For a chi-square distribution with 1 degree of freedom, $p > 0.10$. Since p is greater than α, we cannot reject the null hypothesis. The sample does not provide evidence that there is an association between smoking status and year.

In both of the previous examples, we determined whether an association exists between two nominal random variables; however, we did not measure the strength of the association. For a 2×2 table, one way in which the effect can be quantified is by means of an odds ratio. For the data from the study that examined the outbreak of gastroenteritis in a U.S. high school, the odds of becoming ill among those who ate a prepared sandwich relative to those who did not are estimated by

$$\begin{aligned} \widehat{OR} &= \frac{(109)(34)}{(4)(116)} \\ &= 7.99. \end{aligned}$$

The odds of students who ate a prepared sandwich becoming ill are 7.99 times greater than the odds of students who did not eat a sandwich; therefore, there does appear to be a strong association between eating the sandwiches and a student's health.

To measure the uncertainty in this estimate, we can calculate a confidence interval for the true population odds ratio. Note that the log of $\widehat{OR}$ is

$$\begin{aligned} \ln(\widehat{OR}) &= \ln(7.99) \\ &= 2.078; \end{aligned}$$

the estimated standard error of $\ln(\widehat{OR})$ is

$$\begin{aligned} \widehat{se}[\ln(\widehat{OR})] &= \sqrt{\frac{1}{109} + \frac{1}{4} + \frac{1}{116} + \frac{1}{34}} \\ &= 0.545. \end{aligned}$$

Therefore, a 95% confidence interval for the log of the odds ratio is

$$(2.078 - 1.96(0.545)\,,\ 2.078 + 1.96(0.545))$$

or

$$(1.010, 3.146),$$

and a 95% confidence interval for the odds ratio itself is

$$(e^{1.010}, e^{3.146})$$

or

$$(2.75, 23.24).$$

We are 95% confident that the odds of students who ate a sandwich becoming ill are between 2.75 and 23.24 times larger than the odds of students who did not eat a sandwich.

15.6 Review Exercises

1. Consider the chi-square distribution with 2 degrees of freedom.

   a. What proportion of the area under the curve lies to the right of $\chi^2 = 9.21$?
   b. What proportion of the area lies to the right of $\chi^2 = 7.38$?
   c. What value of χ^2 cuts off the upper 10% of the distribution?

2. The following data come from a study designed to investigate drinking problems among college students. In 1983, a group of students were asked whether they had ever driven an automobile while drinking. In 1987, after the legal drinking age was raised, a different group of college students were asked the same question [10].

Drove While Drinking	Year		
	1983	1987	Total
Yes	1,250	991	2,241
No	1,387	1,666	3,053
Total	2,637	2,657	5,294

   a. Test the null hypothesis that there is no association between driving while drinking and calendar year.
   b. What do you conclude about the behavior of college students?

3. In a study of HIV infection among women entering the New York State prison system, 475 inmates were cross-classified with respect to HIV seropositivity and their history of intravenous drug use. These data are stored on your disk in a data set called `prison` [11]. The indicators of seropositivity are saved under the variable name `hiv` and those of intravenous drug use under `ivdu`.

a. Test the null hypothesis that there is no association between history of intravenous drug use and HIV seropositivity.
b. Estimate the relative odds of being HIV positive for women who have used intravenous drugs versus those who have not.

4. In a survey conducted in Italy, physicians with different specialties were questioned regarding the surgical treatment of early breast cancer. In particular, they were asked whether they would recommend radical surgery regardless of a patient's age (R), conservative surgery only for younger patients (CR), or conservative surgery regardless of age (C). The results of this survey are presented below [12].

Physician Specialty	Surgery			
	R	CR	C	Total
Internal	6	22	42	70
Surgery	23	61	127	211
Radiotherapy	2	3	54	59
Oncology	1	12	43	56
Gynecology	1	12	31	44
Total	33	110	297	440

a. Test the null hypothesis that there is no association between physician specialty and recommended treatment.
b. What do you conclude?

5. The following table compiles data from six studies designed to investigate the accuracy of death certificates. The results of 5,373 autopsies were compared to the causes of death listed on the certificates. Of those considered, 3,726 certificates were confirmed to be accurate, 783 either lacked information or contained inaccuracies but did not require recoding of the underlying cause of death, and 864 were incorrect and required recoding [13].

Date of Study	Certificate Status			
	Confirmed Accurate	Inaccurate No change	Incorrect Recoding	Total
1955–1965	2,040	367	327	2,734
1970	149	60	48	257
1970–1971	288	25	70	383
1975–1977	703	197	252	1,152
1977–1978	425	62	88	575
1980	121	72	79	272
Total	3,726	783	864	5,373

a. Do you believe that the results are homogeneous across studies?

b. It should be noted that autopsies are not performed at random; in fact, many are done because the cause of death listed on the certificate is uncertain. What problems might arise if we attempt to use the results of these studies to make inference about the population as a whole?

6. In a study of intraobserver variability in assessing cervical smears, 3,325 slides were screened for the presence or absence of abnormal squamous cells. Each slide was screened by a particular observer and then rescreened six months later by the same observer. The results of this study are shown below [14].

First Screening	Second Screening: Present	Second Screening: Absent	Total
Present	1,763	489	2,252
Absent	403	670	1,073
Total	2,166	1,159	3,325

a. Do these data support the null hypothesis that there is no association between the time of screening and diagnosis?

The data could also be displayed in the following manner:

	Screening: First	Screening: Second	Total
Present	2,252	2,166	4,418
Absent	1,073	1,159	2,232
Total	3,325	3,325	6,650

b. Is there anything wrong with this presentation?

7. Suppose that you are interested in investigating the association between retirement and heart disease. In one study, 127 victims of cardiac arrest were matched with 127 healthy individuals; retirement status was ascertained for each subject [15].

Healthy	Cardiac Arrest: Retired	Cardiac Arrest: Not retired	Total
Retired	27	12	39
Not retired	20	68	88
Total	47	80	127

a. Test the null hypothesis that there is no association between retirement status and cardiac arrest.
b. What do you conclude?

8. In response to a study suggesting a link between lack of circumcision and cervical cancer, an investigation was conducted to assess the accuracy of reported circumcision status. Before registering at a cancer institute, all patients were asked to fill out a questionnaire; for males, the data requested included circumcision status. This information was confirmed by interview. Subsequently, all patients received a complete physical examination during which the physician noted whether the man was circumcised or not. The data collected over a two-month period are presented below [16].

	Patient Statement		
Examination	Circumcised	Not circumcised	Total
Circumcised	37	47	84
Not circumcised	19	89	108
Total	56	136	192

Does there appear to be an association between the results of the examination and the patient's response?

9. In a study of the risk factors for invasive cervical cancer that was conducted in Germany, the following data were collected relating smoking status to the presence or absence of cervical cancer [17].

	Smoker	Nonsmoker	Total
Cancer	108	117	225
No cancer	163	268	431
Total	271	385	656

a. Estimate the relative odds of invasive cervical cancer for smokers versus nonsmokers.
b. Calculate a 95% confidence interval for the population odds ratio.
c. Test the null hypothesis that there is no association between smoking status and the presence of cervical cancer at the 0.05 level of significance.

10. In France, a study was conducted to investigate potential risk factors for ectopic pregnancy. Of the 279 women who had experienced an ectopic pregnancy, 28 had suffered from pelvic inflammatory disease. Of the 279 women who had not, 6 had suffered from pelvic inflammatory disease [18].

a. Construct a 2×2 contingency table for these data.

b. Estimate the relative odds of suffering an ectopic pregnancy for women who have had pelvic inflammatory disease versus women who have not.
c. Find a 99% confidence interval for the population relative odds.

References

[1] Thompson, R. S., Rivara, F. P., and Thompson, D. C., "A Case-Control Study of the Effectiveness of Bicycle Safety Helmets," *The New England Journal of Medicine*, Volume 320, May 25, 1989, 1361–1367.
[2] Cochran, W. G., "Some Methods for Strengthening the Common χ^2 Test," *Biometrics*, Volume 10, December 1954, 417–451.
[3] Grizzle, J. E., "Continuity Correction in the χ^2 Test for 2×2 Tables," *The American Statistician*, Volume 21, October 1967, 28–32.
[4] Schottenfeld, D., Eaton, M., Sommers, S. C., Alonso, D. R., and Wilkinson, C., "The Autopsy as a Measure of Accuracy of the Death Certificate," *Bulletin of the New York Academy of Medicine*, Volume 58, December 1982, 778–794.
[5] Coulehan, J. L., Lerner, G., Helzlsouer, K., Welty, T. K., and McLaughlin, J., "Acute Myocardial Infarction Among Navajo Indians, 1976–1983," *American Journal of Public Health*, Volume 76, April 1986, 412–414.
[6] McCusker, J., Harris, D. R., and Hosmer, D. W., "Association of Electronic Fetal Monitoring During Labor with Cæsarean Section Rate and with Neonatal Morbidity and Mortality," *American Journal of Public Health*, Volume 78, September 1988, 1170–1174.
[7] Roberts, R. S., Spitzer, W. O., Delmore, T., and Sackett, D. L., "An Empirical Demonstration of Berkson's Bias," *Journal of Chronic Diseases*, Volume 31, February 1978, 119–128.
[8] Gross, T. P., Conde, J. G., Gary, G. W., Harting, D., Goeller, D., and Israel, E., "An Outbreak of Acute Infectious Nonbacterial Gastroenteritis in a High School in Maryland," *Public Health Reports*, Volume 104, March–April 1989, 164–169.
[9] Kirscht, J. P., Brock, B. M., and Hawthorne, V. M., "Cigarette Smoking and Changes in Smoking Among a Cohort of Michigan Adults, 1980-82," *American Journal of Public Health*, Volume 77, April 1987, 501–502.
[10] Engs, R. C., and Hanson, D. J., "University Students' Drinking Patterns and Problems: Examining the Effects of Raising the Purchase Age," *Public Health Reports*, Volume 103, November–December 1988, 667–673.
[11] Smith, P. F., Mikl, J., Truman, B. I., Lessner, L., Lehman, J. S., Stevens, R. W., Lord, E. A., Broaddus, R. K., and Morse, D. L., "HIV Infection Among Women Entering the New York State Correctional System," *American Journal of Public Health Supplement*, Volume 81, May 1991, 35–40.
[12] Liberati, A., Apolone, G., Nicolucci, A., Confalonieri, C., Fossati, R., Grilli, R., Torri, V., Mosconi, P., and Alexanian, A., "The Role of Attitudes, Beliefs, and Personal Characteristics of Italian Physicians in the Surgical Treatment

of Early Breast Cancer," *American Journal of Public Health*, Volume 81, January 1991, 38–42.

[13] Kircher, T., Nelson, J., and Burdo, H., "The Autopsy as a Measure of Accuracy of the Death Certificate," *The New England Journal of Medicine*, Volume 313, November 14, 1985, 1263–1269.

[14] Klinkhamer, P. J. J. M., Vooijs, G. P., and de Haan, A. F. J., "Intraobserver and Interobserver Variability in the Quality Assessment of Cervical Smears," *Acta Cytologica*, Volume 33, March–April 1989, 215–218.

[15] Siscovick, D. S., Strogatz, D. S., Weiss, N. S., and Rennert, G., "Retirement and Primary Cardiac Arrest in Males," *American Journal of Public Health*, Volume 80, February 1990, 207–208.

[16] Lilienfeld, A. M., and Graham, S., "Validity of Determining Circumcision Status by Questionnaire as Related to Epidemiological Studies of Cancer of the Cervix," *Journal of the National Cancer Institute*, Volume 21, October 1958, 713–720.

[17] Nischan, P., Ebeling, K., and Schindler, C., "Smoking and Invasive Cervical Cancer Risk: Results from a Case-Control Study," *American Journal of Epidemiology*, Volume 128, July 1988, 74–77.

[18] Coste, J., Job-Spira, N., Fernandez, H., Papiernik, E., and Spira, A., "Risk Factors for Ectopic Pregnancy: A Case-Control Study in France, with Special Focus on Infectious Factors," *American Journal of Epidemiology*, Volume 133, May 1, 1991, 839–849.

CHAPTER 16

Multiple 2 × 2 Tables

When the relationship between a pair of dichotomous random variables is being investigated, it is sometimes examined in two or more populations. As a result, the data to be analyzed could consist of a number of 2×2 contingency tables. In some cases these tables originate from different studies; more often, they are the results of a single investigation that have been subclassified, or *stratified*, by some factor that is believed to influence the outcome. In either event, it is possible to make inference about the relationship between the two variables by examining the association in each table separately. In many cases, however, it is more useful to be able to combine the information across tables to make a single, unifying statement.

16.1 Simpson's Paradox

Consider the following data taken from a study that investigates the relationship between smoking and aortic stenosis. Since gender is associated with both of these variables, we suspect that it might influence the observed relationship between them. Therefore, we begin by examining the effect among males and that among females separately. The stratified data are presented below [1].

Males

Aortic Stenosis	Smoker		
	Yes	**No**	**Total**
Yes	37	25	62
No	24	20	44
Total	61	45	106

Females

Aortic Stenosis	Smoker Yes	Smoker No	Total
Yes	14	29	43
No	19	47	66
Total	33	76	109

For males, the odds of developing aortic stenosis among smokers relative to nonsmokers are estimated by

$$\widehat{OR}_M = \frac{(37)(20)}{(25)(24)} = 1.23.$$

For females, the odds of aortic stenosis among smokers relative to nonsmokers are estimated by

$$\widehat{OR}_F = \frac{(14)(47)}{(29)(19)} = 1.19.$$

We observe the same trend in each population; for both males and females, the odds of developing aortic stenosis are larger among smokers than they are among nonsmokers. It is possible that these two quantities are actually estimating the same population value. Consequently, we might attempt to combine the information in the tables to make a single statement summarizing the relationship between smoking and aortic stenosis. If the two individual tables are summed, the results are as follows.

Aortic Stenosis	Smoker Yes	Smoker No	Total
Yes	51	54	105
No	43	67	110
Total	94	121	215

For all individuals in the study, regardless of gender, the odds of developing aortic stenosis among the group of smokers relative to the nonsmokers are

$$\widehat{OR} = \frac{(51)(67)}{(54)(43)} = 1.47.$$

If the influence of gender is ignored, then the strength of the association between smoking and aortic stenosis appears greater than it is for either males or females alone. This phenomenon is an example of *Simpson's paradox*. Simpson's paradox occurs when either the magnitude or the direction of the relationship between two variables is influenced by the presence of a third factor. In this case, gender is a confounder in the relationship between exposure and disease; by failing to control for its effect, the true magnitude of the association has been made to appear greater than it actually is.

16.2 The Mantel-Haenszel Method

Now consider the following data from a study that investigates the relationship between the consumption of caffeinated coffee and nonfatal myocardial infarction among adult males under the age of 55. The study provides exposure and disease information for two samples of men: a group of 1,559 smokers and a group of 937 nonsmokers [2].

Smokers

Myocardial Infarction	Coffee		
	Yes	No	Total
Yes	1,011	81	1,092
No	390	77	467
Total	1,401	158	1,559

Nonsmokers

Myocardial Infarction	Coffee		
	Yes	No	Total
Yes	383	66	449
No	365	123	488
Total	748	189	937

Among the smokers, the relative odds of suffering a nonfatal myocardial infarction for males drinking caffeinated coffee versus males drinking no coffee at all are estimated by

$$\begin{aligned}\widehat{OR}_S &= \frac{(1,011)(77)}{(390)(81)} \\ &= 2.46.\end{aligned}$$

Among nonsmokers, the relative odds of experiencing a nonfatal myocardial infarction among coffee drinkers versus nondrinkers are estimated by

$$\begin{aligned}\widehat{OR}_{NS} &= \frac{(383)(123)}{(365)(66)} \\ &= 1.96.\end{aligned}$$

In both populations we observe that the odds of suffering a myocardial infarction are greater among coffee drinkers. It is possible that the two odds ratios are actually estimating the same population value and differ only because of sampling variability. If this is the case, we would like to be able to combine the information in the two tables to make a single overall statement about the relationship between myocardial infarction and caffeinated coffee.

We have already seen that if smoking is a confounder in the relationship between coffee drinking and myocardial infarction, then we cannot simply sum the observations in the two contingency tables. If we were to attempt this, the following table would result.

Myocardial Infarction	Coffee		
	Yes	**No**	**Total**
Yes	1,394	147	1,541
No	755	200	955
Total	2,149	347	2,496

Based on these unstratified data, the estimate of the odds of nonfatal myocardial infarction among coffee drinkers relative to those who do not drink coffee is

$$\begin{aligned}\widehat{OR} &= \frac{(1,394)(200)}{(755)(147)} \\ &= 2.51.\end{aligned}$$

This odds ratio is larger than the relative odds in either of the individual strata, which suggests that smoking status is indeed a confounder.

There is another technique–known as the *Mantel-Haenszel method*–that may be used to combine the information in a number of 2 × 2 tables. We begin by determining whether the strength of association is uniform across the tables. Then, if it is appropriate to do so, the method provides a means of calculating both a point estimate and a confidence interval for the overall population relative odds; in addition, it allows us to test the null hypothesis of no association between exposure and disease.

16.2.1 Test of Homogeneity

Before combining contingency tables, we must first verify that the population odds ratios are in fact constant across tables. If they are not, then it is not beneficial to compute a single summary value for the overall relative odds. Instead, it is better to treat the data in the various contingency tables as if they have been drawn from distinct populations and report a different odds ratio for each group.

We determine whether the strength of association between exposure and disease is uniform across a series of g 2×2 tables–where g is an integer greater than or equal to 2–by conducting a test of homogeneity. The *test of homogeneity* evaluates the null hypothesis

H_0 : the population odds ratios for the g tables are identical,

or, equivalently,

$$H_0 : \text{OR}_1 = \text{OR}_2 = \cdots = \text{OR}_i = \cdots = \text{OR}_g.$$

The alternative hypothesis is that not all odds ratios are the same. To perform the test, we calculate the statistic

$$X^2 = \sum_{i=1}^{g} w_i \, (y_i - Y)^2.$$

In this expression, y_i is the logarithm of the estimated odds ratio for the ith table, Y is a weighted average of the g separate log odds ratios, and w_i is the weighting factor for the ith table.

Suppose that the ith 2×2 table has the following format:

	Exposure		
Disease	**Yes**	**No**	**Total**
Yes	a_i	b_i	N_{1i}
No	c_i	d_i	N_{2i}
Total	M_{1i}	M_{2i}	T_i

The estimate of the odds ratio for this table is

$$\widehat{\text{OR}}_i = \frac{a_i d_i}{b_i c_i};$$

the logarithm of the estimated odds ratio is

$$\begin{aligned} y_i &= \ln(\widehat{\text{OR}}_i) \\ &= \ln\left(\frac{a_i d_i}{b_i c_i}\right). \end{aligned}$$

The weighted average Y is calculated using the formula

$$Y = \frac{\sum_{i=1}^{g} w_i \, y_i}{\sum_{i=1}^{g} w_i}$$

for $i = 1, 2, \ldots g$. The weights w_i are computed as

$$\begin{aligned} w_i &= \left[\frac{1}{a_i} + \frac{1}{b_i} + \frac{1}{c_i} + \frac{1}{d_i} \right]^{-1} \\ &= \frac{1}{[(1/a_i) + (1/b_i) + (1/c_i) + (1/d_i)]}. \end{aligned}$$

This quantity is actually the reciprocal of the variance of the log odds ratio. If any one of the cell entries is equal to 0, then w_i is undefined. In this case, we can use the same modification that we applied in Chapter 15. Adding 0.5 to each of the values a_i, b_i, c_i, and d_i, the weighting factor becomes

$$w_i = \left[\frac{1}{(a_i + 0.5)} + \frac{1}{(b_i + 0.5)} + \frac{1}{(c_i + 0.5)} + \frac{1}{(d_i + 0.5)} \right]^{-1}.$$

Under the null hypothesis that the odds ratio is constant across tables, the sum

$$X^2 = \sum_{i=1}^{g} w_i \, (y_i - Y)^2$$

has a distribution that is approximately chi-square with $g - 1$ degrees of freedom. If the p-value associated with this statistic is less than the level of significance of the test, we reject the null hypothesis and report the separate estimates. If p is greater than α, we cannot reject H_0; therefore, we conclude that it is acceptable to combine the information in the g 2×2 tables using the Mantel-Haenszel method.

Recall that the data used to investigate the relationship between coffee consumption and nonfatal myocardial infarction among men under the age of 55 are divided into two distinct strata: a group of smokers and a group of nonsmokers. Since we have only two contingency tables, g is equal to 2. Before attempting to combine the information in the tables, we test the null hypothesis that the population odds ratios for the two groups are identical, or

$$H_0 : \text{OR}_1 = \text{OR}_2,$$

against the alternative

$$H_A : \text{OR}_1 \neq \text{OR}_2.$$

We conduct a two-sided test of homogeneity at the $\alpha = 0.05$ level of significance.

Recall that the estimate of the relative odds for smokers is

$$\begin{aligned}\widehat{\text{OR}}_1 &= \frac{a_1 d_1}{b_1 c_1} \\ &= \frac{(1,011)(77)}{(390)(81)} \\ &= 2.46.\end{aligned}$$

Therefore,

$$\begin{aligned}y_1 &= \ln(\widehat{\text{OR}}_1) \\ &= \ln(2.46) \\ &= 0.900.\end{aligned}$$

The weighting factor for this table is

$$\begin{aligned}w_1 &= \frac{1}{[(1/1,011) + (1/390) + (1/81) + (1/77)]} \\ &= 34.62.\end{aligned}$$

Similarly, the estimate of the relative odds for nonsmokers is

$$\begin{aligned}\widehat{\text{OR}}_2 &= \frac{a_2 d_2}{b_2 c_2} \\ &= \frac{(383)(123)}{(365)(66)} \\ &= 1.96,\end{aligned}$$

and

$$\begin{aligned}y_2 &= \ln(\widehat{\text{OR}}_2) \\ &= \ln(1.96) \\ &= 0.673.\end{aligned}$$

For this table, the weighting factor is

$$\begin{aligned} w_2 &= \frac{1}{[(1/383) + (1/365) + (1/66) + (1/123)]} \\ &= 34.93. \end{aligned}$$

Using these values, the weighted average of y_1 and y_2 is found to be

$$\begin{aligned} Y &= \frac{\sum_{i=1}^{g} w_i y_i}{\sum_{i=1}^{g} w_i} \\ &= \frac{(w_1 y_1 + w_2 y_2)}{(w_1 + w_2)} \\ &= \frac{(34.62)(0.900) + (34.93)(0.673)}{(34.62 + 34.93)} \\ &= 0.786. \end{aligned}$$

Finally, the test statistic is

$$\begin{aligned} X^2 &= \sum_{i=1}^{g} w_i \, (y_i - Y)^2 \\ &= w_1 \, (y_1 - Y)^2 + w_2 \, (y_2 - Y)^2 \\ &= (34.62)(0.900 - 0.786)^2 + (34.93)(0.673 - 0.786)^2 \\ &= 0.896. \end{aligned}$$

Referring to Table A.6, we observe that, for a chi-square distribution with 1 degree of freedom, $p > 0.10$. We cannot reject the null hypothesis; the data do not indicate that the population odds ratio for smokers differs from that for nonsmokers. Consequently, we may assume that the odds ratios for the two strata are in fact estimating the same quantity and proceed with the Mantel-Haenszel method of combining the information.

16.2.2 Summary Odds Ratio

If the odds ratios are found to be uniform across tables, then the next step is to compute an estimate of the strength of the association. The estimate is actually a weighted average of the odds ratios for the separate strata; it is calculated using the formula

$$\widehat{\text{OR}} = \frac{\sum_{i=1}^{g} (a_i d_i / T_i)}{\sum_{i=1}^{g} (b_i c_i / T_i)}$$

where T_i is the total number of observations in the ith table.

For the data relating coffee consumption to nonfatal myocardial infarction,

$$\begin{aligned}\widehat{\text{OR}} &= \frac{\sum_{i=1}^{g}(a_i d_i/T_i)}{\sum_{i=1}^{g}(b_i c_i/T_i)} \\ &= \frac{(a_1 d_1/T_1) + (a_2 d_2/T_2)}{(b_1 c_1/T_1) + (b_2 c_2/T_2)} \\ &= \frac{(1,011)(77)/1,559 + (383)(123)/937}{(390)(81)/1,559 + (365)(66)/937} \\ &= 2.18.\end{aligned}$$

Once differences in smoking status have been taken into account, males under the age of 55 who drink caffeinated coffee have the odds of experiencing nonfatal myocardial infarction that are 2.18 times greater than the odds of males who do not drink coffee.

In addition to computing a point estimate for the summary odds ratio, we may also want to calculate a confidence interval that represents a range of reasonable values for this quantity. When constructing a confidence interval for the odds ratio using data from a single population, we noted that the sampling distribution of the relative odds is skewed to the right. The same is true for the Mantel-Haenszel estimator of the common odds ratio. Since the distribution of the natural logarithm of the odds ratio is more symmetric and approximately normal, we begin by calculating a confidence interval for ln(OR). In addition, to ensure that our strata sample sizes are large enough, we recommend the following restrictions on the expected values of the observations across the g tables:

$$\sum_{i=1}^{g} \frac{M_{1i}N_{1i}}{T_i} \geq 5,$$

$$\sum_{i=1}^{g} \frac{M_{1i}N_{2i}}{T_i} \geq 5,$$

$$\sum_{i=1}^{g} \frac{M_{2i}N_{1i}}{T_i} \geq 5,$$

and

$$\sum_{i=1}^{g} \frac{M_{2i}N_{2i}}{T_i} \geq 5.$$

The quantity Y that we calculated when conducting the test of homogeneity,

$$Y = \frac{\sum_{i=1}^{g} w_i y_i}{\sum_{i=1}^{g} w_i},$$

is the weighted average of g separate log odds ratios and represents an estimator of ln(OR). The estimated standard error of Y is

$$\widehat{\text{se}}(Y) = \frac{1}{\sqrt{\sum_{i=1}^{g} w_i}}.$$

Therefore, a 95% confidence interval for ln(OR) takes the form

$$(Y - 1.96\, \widehat{\text{se}}(Y)\,,\ Y + 1.96\, \widehat{\text{se}}(Y)).$$

If we take the antilogarithm of each limit, a 95% confidence interval for the summary relative odds itself is

$$(e^{Y - 1.96\, \widehat{\text{se}}(Y)}\,,\ e^{Y + 1.96\, \widehat{\text{se}}(Y)}).$$

Before computing a confidence interval for the common odds ratio measuring the strength of association between coffee consumption and myocardial infarction, we first verify the restrictions placed on the expected values of the observations; note that

$$\begin{aligned}
\sum_{i=1}^{2} \frac{M_{1i}N_{1i}}{T_i} &= \frac{M_{11}N_{11}}{T_1} + \frac{M_{12}N_{12}}{T_2} \\
&= \frac{(1,401)(1,092)}{1,559} + \frac{(748)(449)}{937} \\
&= 1,339.8,
\end{aligned}$$

$$\begin{aligned}
\sum_{i=1}^{2} \frac{M_{1i}N_{2i}}{T_i} &= \frac{M_{11}N_{21}}{T_1} + \frac{M_{12}N_{22}}{T_2} \\
&= \frac{(1,401)(467)}{1,559} + \frac{(748)(488)}{937} \\
&= 809.2,
\end{aligned}$$

$$\begin{aligned}
\sum_{i=1}^{2} \frac{M_{2i}N_{1i}}{T_i} &= \frac{M_{21}N_{11}}{T_1} + \frac{M_{22}N_{12}}{T_2} \\
&= \frac{(158)(1,092)}{1,559} + \frac{(189)(449)}{937} \\
&= 201.2,
\end{aligned}$$

and

$$\sum_{i=1}^{2} \frac{M_{2i}N_{2i}}{T_i} = \frac{M_{21}N_{21}}{T_1} + \frac{M_{22}N_{22}}{T_2}$$
$$= \frac{(158)(467)}{1,559} + \frac{(189)(488)}{937}$$
$$= 145.8.$$

Since each of these sums is greater than 5, we proceed with the construction of a confidence interval. We previously found that

$$Y = 0.786.$$

We note that $w_1 = 34.62$ and $w_2 = 34.93$, so

$$\widehat{se}(Y) = \frac{1}{\sqrt{\sum_{i=1}^{2} w_i}}$$
$$= \frac{1}{\sqrt{w_1 + w_2}}$$
$$= \frac{1}{\sqrt{34.62 + 34.93}}$$
$$= 0.120.$$

Therefore, a 95% confidence interval for ln(OR) is

$$(0.786 - 1.96(0.120)\,,\ 0.786 + 1.96(0.120))$$

or

$$(0.551\,,\ 1.021).$$

The 95% confidence interval for the common odds ratio itself is

$$(e^{0.551}\,,\ e^{1.021})$$

or

$$(1.73\,,\ 2.78).$$

We are 95% confident that, after adjusting for the effects of smoking, men who drink caffeinated coffee have the odds of experiencing a nonfatal myocardial infarction that are between 1.73 and 2.78 times greater than the odds of men who do not drink coffee.

16.2.3 Test of Association

The final step in the Mantel-Haenszel method of combining information from 2 × 2 contingency tables is to test whether the summary odds ratio is equal to 1; a relative odds of 1 implies that there is no association between the exposure and the disease. One way to perform this test is to simply refer to the confidence limits calculated above. Because the 95% confidence interval does not contain the value 1, this sample would lead us to reject the null hypothesis at the 0.05 level of significance. Recall, however, that the confidence interval was constructed under the assumption that the sampling distribution of the log of the odds ratio is approximately normal. To evaluate the null hypothesis

$$H_0 : \mathrm{OR} = 1$$

more directly, we could use an alternative method and calculate the test statistic

$$X^2 = \frac{[\sum_{i=1}^{g} a_i - \sum_{i=1}^{g} m_i]^2}{\sum_{i=1}^{g} \sigma_i^2}.$$

In this expression, a_i is the observed number of exposed individuals who develop disease; it is the upper left-hand entry in the ith 2 × 2 table. The term m_i—the expected value of a_i—is computed as

$$m_i = \frac{M_{1i}N_{1i}}{T_i}.$$

Finally, σ_i is the standard deviation of a_i where

$$\sigma_i^2 = \frac{M_{1i}M_{2i}N_{1i}N_{2i}}{T_i^2(T_i - 1)}.$$

Like the chi-square test statistic for a single 2 × 2 table, the quantity

$$X^2 = \frac{[\sum_{i=1}^{g} a_i - \sum_{i=1}^{g} m_i]^2}{\sum_{i=1}^{g} \sigma_i^2}$$

compares the observed frequencies to the corresponding expected frequencies. It has an approximate chi-square distribution with 1 degree of freedom. If the p-value associated with this statistic is less than the level of significance α—suggesting that the deviations between the observed and the expected values are too large to be attributed to chance—then we reject the null hypothesis that the common odds ratio is equal to 1. If p is greater than α, we do not reject H_0.

For the data relating coffee consumption to nonfatal myocardial infarction, we would like to evaluate the null hypothesis

$$H_0 : \mathrm{OR} = 1$$

against the alternative

$$H_A : \text{OR} \neq 1$$

using a two-sided test and setting the significance level at $\alpha = 0.05$. Note that

$$\begin{aligned} a_1 &= 1,011, \\ m_1 &= \frac{M_{11}N_{11}}{T_1} \\ &= \frac{(1,401)(1,092)}{1,559} \\ &= 981.3, \end{aligned}$$

and

$$\begin{aligned} \sigma_1^2 &= \frac{M_{11}M_{21}N_{11}N_{21}}{T_1^2(T_1 - 1)} \\ &= \frac{(1,401)(158)(1,092)(467)}{(1,559)^2(1,559 - 1)} \\ &= 29.81. \end{aligned}$$

Similarly,

$$\begin{aligned} a_2 &= 383, \\ m_2 &= \frac{M_{12}N_{12}}{T_2} \\ &= \frac{(748)(449)}{937} \\ &= 358.4, \end{aligned}$$

and

$$\begin{aligned} \sigma_2^2 &= \frac{M_{12}M_{22}N_{12}N_{22}}{T_2^2(T_2 - 1)} \\ &= \frac{(748)(189)(449)(488)}{(937)^2(937 - 1)} \\ &= 37.69. \end{aligned}$$

Therefore, the test statistic is

$$\begin{aligned} X^2 &= \frac{[\sum_{i=1}^{2} a_i - \sum_{i=1}^{2} m_i]^2}{\sum_{i=1}^{2} \sigma_i^2} \\ &= \frac{[(a_1 + a_2) - (m_1 + m_2)]^2}{\sigma_1^2 + \sigma_2^2} \\ &= \frac{[(1,011 + 383) - (981.3 + 358.4)]^2}{29.81 + 37.69} \\ &= 43.68. \end{aligned}$$

Referring to Table A.6, we observe that the p-value is less than 0.001. Accordingly, we reject the null hypothesis of no association and conclude that the summary odds ratio is not equal to 1. After adjusting for differences in smoking status, we find that adult males under the age of 55 who drink caffeinated coffee face a significantly higher risk of experiencing a nonfatal myocardial infarction than males of the same age who do not drink coffee.

These data represent the results of a single study examining the effects of coffee consumption on human health; other studies have reported conflicting results. At the present time, it appears that, in moderate amounts, coffee is a fairly safe beverage among otherwise healthy individuals [3, 4].

16.3 Further Applications

Suppose that we are presented with a pair of 2×2 contingency tables, each of which provides information about the same dichotomous random variables but originates from different studies. For example, the following data come from two studies conducted in San Francisco that investigate the risk factors for epithelial ovarian cancer [5].

Study 1

Disease Status	Term Pregnancies		
	None	One or more	Total
Cancer	31	80	111
No cancer	93	379	472
Total	124	459	583

Study 2

Disease Status	Term Pregnancies		
	None	One or more	Total
Cancer	39	149	188
No cancer	74	465	539
Total	113	614	727

In the first study, the relative odds of developing ovarian cancer for women who have never had a term pregnancy versus women who have had one or more are estimated by

$$\begin{aligned}\widehat{\text{OR}}_1 &= \frac{(31)(379)}{(80)(93)} \\ &= 1.58.\end{aligned}$$

In the second study, the same quantity is estimated by

$$\begin{aligned}\widehat{\text{OR}}_2 &= \frac{(39)(465)}{(149)(74)} \\ &= 1.64.\end{aligned}$$

In both populations, the odds of developing ovarian cancer are greater for women who have never experienced a term pregnancy. The two ratios are very similar; it is quite possible that they are both estimating the same population value. Are we able to combine the evidence collected in the two different studies to make a single overall statement about the association between the number of term pregnancies a woman has had and the occurrence of epithelial ovarian cancer?

As we have seen, it is not advisable to simply sum the observations in the two contingency tables; instead, we apply the Mantel-Haenszel method for combining the information. This technique consists of four steps: (1) the test of homogeneity, (2) the calculation of a point estimate for the summary odds ratio, (3) the construction of a confidence interval for the odds ratio, and (4) a test of hypothesis to determine whether an association exists between exposure and disease.

The test of homogeneity is conducted to verify that the population odds ratio is uniform across groups; it evaluates the null hypothesis

H_0: the odds ratios of developing ovarian cancer among women who have never had a term pregnancy versus women who have had one or more are identical for the populations represented by the two studies,

or

$$H_0 : \text{OR}_1 = \text{OR}_2.$$

The alternative hypothesis is that the odds ratios are not the same. To carry out the test, we calculate the statistic

$$X^2 = \sum_{i=1}^{2} w_i \, (y_i - Y)^2.$$

Recall that the estimate of the relative odds for the first study is

$$\begin{aligned} \widehat{\text{OR}}_1 &= \frac{(31)(379)}{(80)(93)} \\ &= 1.58; \end{aligned}$$

therefore,

$$\begin{aligned} y_1 &= \ln(\widehat{\text{OR}}_1) \\ &= \ln(1.58) \\ &= 0.457. \end{aligned}$$

The weighting factor for this table is

$$\begin{aligned} w_1 &= \frac{1}{[(1/31) + (1/80) + (1/93) + (1/379)]} \\ &= 17.20. \end{aligned}$$

Similarly, the estimate of the odds ratio for the second study is

$$\widehat{\text{OR}}_2 = \frac{(39)(465)}{(149)(74)}$$
$$= 1.64;$$

thus,

$$y_2 = \ln(\widehat{\text{OR}}_2)$$
$$= \ln(1.64)$$
$$= 0.495,$$

and

$$w_2 = \frac{1}{[(1/39) + (1/149) + (1/74) + (1/465)]}$$
$$= 20.83.$$

The weighted average of y_1 and y_2, or Y, is found to be

$$Y = \frac{\sum_{i=1}^{2} w_i y_i}{\sum_{i=1}^{2} w_i}$$
$$= \frac{(w_1 y_1 + w_2 y_2)}{(w_1 + w_2)}$$
$$= \frac{(17.20)(0.457) + (20.83)(0.495)}{(17.20 + 20.83)}$$
$$= 0.478.$$

Substituting all these values into the formula for the test statistic,

$$X^2 = \sum_{i=1}^{2} w_i (y_i - Y)^2$$
$$= w_1 (y_1 - Y)^2 + w_2 (y_2 - Y)^2$$
$$= (17.20)(0.457 - 0.478)^2 + (20.83)(0.495 - 0.478)^2$$
$$= 0.014.$$

Referring to Table A.6, we note that $p > 0.10$. Since p is greater than the significance level $\alpha = 0.05$, we cannot reject the null hypothesis. The data do not indicate that the population odds ratios for the two studies differ; therefore, we may proceed with the Mantel-Haenszel method for combining the information.

The estimate of the summary odds ratio, which is actually a weighted average of the stratum-specific estimates, is

$$\begin{aligned}
\widehat{\text{OR}} &= \frac{\sum_{i=1}^{2}(a_i d_i / T_i)}{\sum_{i=1}^{2}(b_i c_i / T_i)} \\
&= \frac{(a_1 d_1 / T_1) + (a_2 d_2 / T_2)}{(b_1 c_1 / T_1) + (b_2 c_2 / T_2)} \\
&= \frac{(31)(379)/583 + (39)(465)/727}{(80)(93)/583 + (149)(74)/727} \\
&= 1.61.
\end{aligned}$$

Once we have taken study differences into account, we conclude that women who have never had a term pregnancy have the odds of developing ovarian cancer that are 1.61 times greater than the odds of women who have had one or more term pregnancies.

In addition to computing a point estimate for the odds ratio, we are also able to calculate a confidence interval. Before doing so, however, we should first verify that the sample sizes in the two studies are large enough to justify the assumption of normality in the underlying population. Note that

$$\begin{aligned}
\sum_{i=1}^{2} \frac{M_{1i} N_{1i}}{T_i} &= \frac{(124)(111)}{583} + \frac{(113)(188)}{727} \\
&= 52.8,
\end{aligned}$$

$$\begin{aligned}
\sum_{i=1}^{2} \frac{M_{1i} N_{2i}}{T_i} &= \frac{(124)(472)}{583} + \frac{(113)(539)}{727} \\
&= 184.2,
\end{aligned}$$

$$\begin{aligned}
\sum_{i=1}^{2} \frac{M_{2i} N_{1i}}{T_i} &= \frac{(459)(111)}{583} + \frac{(614)(188)}{727} \\
&= 246.2,
\end{aligned}$$

and

$$\begin{aligned}
\sum_{i=1}^{2} \frac{M_{2i} N_{2i}}{T_i} &= \frac{(459)(472)}{583} + \frac{(614)(539)}{727} \\
&= 826.8.
\end{aligned}$$

Since each of these sums is greater than 5, we proceed with the construction of the confidence interval. Recall that

$$Y = 0.478.$$

The values of the weights are $w_1 = 34.62$ and $w_2 = 34.93$, and the standard error of Y is

$$\begin{aligned}\widehat{\text{se}}(Y) &= \frac{1}{\sqrt{(w_1 + w_2)}}\\ &= \frac{1}{\sqrt{(17.20 + 20.83)}}\\ &= 0.162.\end{aligned}$$

Therefore, a 95% confidence interval for ln(OR) is

$$(0.478 - 1.96(0.162)\,,\ 0.478 + 1.96(0.162))$$

or

$$(0.160\,,\ 0.796),$$

and a 95% confidence interval for the summary odds ratio itself is

$$(e^{0.160}\,,\ e^{0.796})$$

or

$$(1.17\,,\ 2.22).$$

After adjusting for the effects of different studies, we find that women who have never had a term pregnancy have the odds of developing ovarian cancer that are between 1.17 and 2.22 times greater than the odds of women who have had one or more term pregnancies.

The final step in the Mantel-Haenszel method is to test whether the common odds ratio is equal to 1, or

$$H_0 : \text{OR} = 1.$$

The alternative hypothesis is that the odds ratio is not equal to 1. This test allows us to determine whether there is an association between exposure and disease. By noting that the 95% confidence interval does not contain the value 1, we could conclude that the given sample would lead us to reject H_0 at the 0.05 level. To conduct the test more directly, we calculate the statistic

$$X^2 = \frac{[\sum_{i=1}^{2} a_i - \sum_{i=1}^{2} m_i]^2}{\sum_{i=1}^{2} \sigma_i^2}.$$

Note that

$$a_1 = 31,$$

$$m_1 = \frac{M_{11}N_{11}}{T_1} = \frac{(124)(111)}{583} = 23.6,$$

and

$$\sigma_1^2 = \frac{M_{11}M_{21}N_{11}N_{21}}{T_1^2(T_1 - 1)} = \frac{(124)(459)(111)(472)}{(583)^2(583 - 1)} = 15.07.$$

Similarly,

$$a_2 = 39,$$

$$m_2 = \frac{M_{12}N_{12}}{T_2} = \frac{(113)(188)}{727} = 29.2,$$

and

$$\sigma_2^2 = \frac{M_{12}M_{22}N_{12}N_{22}}{T_2^2(T_2 - 1)} = \frac{(113)(614)(188)(539)}{(727)^2(727 - 1)} = 18.32.$$

Therefore,

$$\mathrm{X}^2 = \frac{[\sum_{i=1}^{2} a_i - \sum_{i=1}^{2} m_i]^2}{\sum_{i=1}^{2} \sigma_i^2} = \frac{[(a_1 + a_2) - (m_1 + m_2)]^2}{\sigma_1^2 + \sigma_2^2} = \frac{[(31 + 39) - (23.6 + 29.2)]^2}{(15.07 + 18.32)} = 8.86.$$

Referring to Table A.6, we observe that $0.001 < p < 0.01$. Therefore, we reject H_0 at the 0.05 level of significance and conclude that the summary odds ratio is not equal to 1. After adjusting for study differences, we see that women who have never had a term pregnancy have a significantly higher risk of developing epithelial ovarian cancer than women who have had one or more term pregnancies.

16.4 Review Exercises

1. In the 2 × 2 contingency tables below, the data from Chapter 15 relating smoking to cervical cancer have been stratified by the number of sexual partners that a woman has had [6].

Zero or One Partners

	Smoker		
Cancer	**Yes**	**No**	**Total**
Yes	12	25	37
No	21	118	139
Total	33	143	176

Two or More Partners

	Smoker		
Cancer	**Yes**	**No**	**Total**
Yes	96	92	188
No	142	150	292
Total	238	242	480

a. Estimate the odds ratio for women who have had at most one sexual partner.
b. Estimate the odds ratio for women who have had two or more partners.
c. If possible, you would like to combine the evidence in these two strata to make an overall statement about the relationship between smoking and cervical cancer. What would happen if you were to simply sum the entries in the tables?
d. Conduct a test of homogeneity. Is it appropriate to use the Mantel-Haenszel method to combine the information in these two tables?
e. Compute the Mantel-Haenszel estimate of the summary odds ratio.
f. Construct a 99% confidence interval for the odds ratio.
g. At the 0.01 level of significance, test the null hypothesis that there is no association between smoking status and the presence of cervical cancer.

h. What do you conclude about the relationship between smoking and cervical cancer?

2. In a study investigating the risk factors for heart disease, the relationship between hypertension and coronary artery disease was examined for individuals in two different age groups [7].

35–49 Years of Age

	CAD		
Hypertension	**Yes**	**No**	**Total**
Yes	552	212	764
No	941	495	1,436
Total	1,493	707	2,200

> 65 Years of Age

	CAD		
Hypertension	**Yes**	**No**	**Total**
Yes	1,102	87	1,189
No	1,018	106	1,124
Total	2,120	193	2,313

a. In each category of age, are the odds of suffering from coronary heart disease greater or smaller for individuals with hypertension?
b. Is it appropriate to combine the information in these two tables? Why or why not?
c. Compute an overall point estimate of the strength of association between hypertension and coronary artery disease.
d. Construct a 90% confidence interval for the summary odds ratio.
e. At the $\alpha = 0.10$ level of significance, test the null hypothesis that there is no association between hypertension and coronary artery disease.
f. What do you conclude?

References

[1] Hoagland, P. M., Cook, E. F., Flatley, M., Walker, C., and Goldman, L., "Case-Control Analysis of Risk Factors for the Presence of Aortic Stenosis in Adults (Age 50 Years or Older)," *American Journal of Cardiology*, Volume 55, March 1, 1985, 744–747.

[2] Rosenberg, L., Palmer, J. R., Kelly, J. P., Kaufman, D. W., and Shapiro, S., "Coffee Drinking and Nonfatal Myocardial Infarction in Men Under 55 Years of Age," *American Journal of Epidemiology*, Volume 128, September 1988, 570–578.

[3] Rosemarin, P. C., "Coffee and Coronary Heart Disease: A Review," *Progress in Cardiovascular Diseases*, Volume 32, November–December 1989, 239–245.

[4] Grobbee, D. E., Rimm, E. B., Giovannucci, E., Colditz, G., Stampfer, M., and Willett, W., "Coffee, Caffeine, and Cardiovascular Disease in Men," *The New England Journal of Medicine*, Volume 323, October 11, 1990, 1026–1032.

[5] Wu, M. L., Whittemore, A. S., Paffenbarger, R. S., Sarles, D. L., Kampert, J. B., Grosser, S., Jung, D. L., Ballon, S., Hendrickson, M., and Mohle-Boetani, J., "Personal and Environmental Characteristics Related to Epithelial Ovarian Cancer, Reproductive and Menstrual Events and Oral Contraceptive Use," *American Journal of Epidemiology*, Volume 128, December 1988, 1216–1227.

[6] Nischan, P., Ebeling, K., and Schindler, C., "Smoking and Invasive Cervical Cancer Risk: Results from a Case-Control Study," *American Journal of Epidemiology*, Volume 128, July 1988, 74–77.

[7] Applegate, W. B., Hughes, J. P., and Zwaag, R. V., "Case-Control Study of Coronary Heart Disease Risk Factors in the Elderly," *Journal of Clinical Epidemiology*, Volume 44, Number 4/5, 1991, 409–415.

CHAPTER 17

Correlation

In the preceding chapters, we discussed measures of the strength of association between two dichotomous random variables. We now begin to investigate the relationships that may exist among continuous variables. One statistical technique often employed to measure such an association is known as *correlation analysis*. *Correlation* is defined as the quantification of the degree to which two variables are related, providing that the relationship is linear.

17.1 The Two-Way Scatter Plot

Suppose that we are interested in a pair of continuous random variables, each of which is measured on the same units of study. For example, we might wish to investigate the relationship between the percentage of children who have been immunized against diphtheria, pertussis, and tetanus (DPT) in a given country and the mortality rate for children under 5 years of age. The United Nations Children's Fund considers the under-5 mortality rate to be one of the most important indicators of the level of well-being for a population of children.

The data for a random sample of 20 countries are presented in Table 17.1 [1]. If X represents the percentage of children immunized, and Y represents the under-5 mortality rate, then we have a pair of outcomes (x_i, y_i) for each nation in the sample. The first country on the list, Bolivia, has a DPT immunization percentage of 40% and an under-5 mortality rate of 165 per 1,000 live births; therefore, this nation is represented by the data point (0.40, 0.165).

Before we conduct any type of analysis, we should always create a two-way scatter plot of the data. If we place the possible outcomes of the X variable along the horizontal axis and the outcomes of Y along the vertical axis, then each point on the graph represents a combination of outcomes (x_i, y_i). We can often determine whether a relationship exists between x and y—the outcomes of the

Table 17.1 Percentage of children immunized against DPT and under-5 mortality rate for 20 countries, 1989

Nation	Percentage Immunized	Mortality Rate per 1,000 Live Births
Bolivia	40	165
Brazil	54	85
Canada	85	9
China	95	43
Egypt	81	94
Ethiopia	26	226
Finland	90	7
France	95	9
Greece	83	12
India	83	145
Italy	85	11
Japan	83	6
Mexico	65	51
Poland	98	18
Senegal	47	189
Turkey	74	90
United Kingdom	75	10
United States	97	12
USSR	79	35
Yugoslavia	91	27

random variables X and Y—simply by examining the graph. As an example, the data from Table 17.1 are plotted in Figure 17.1. The percentage of children

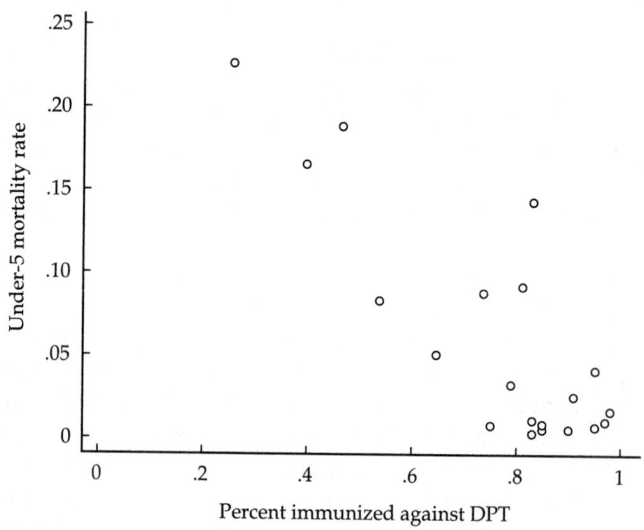

Figure 17.1. Under-5 mortality rate versus percentage of children immunized against DPT for 20 countries, 1989

immunized against DPT appears on the horizontal axis and the under-5 mortality rate on the vertical axis. Not surprisingly, mortality rate tends to diminish as the percentage of children immunized increases.

17.2 Pearson's Correlation Coefficient

In the underlying population from which the sample of points (x_i, y_i) is selected, the correlation between X and Y is denoted by ρ (rho). The correlation quantifies the strength of the linear relationship between the outcomes x and y. It can be thought of as the average of the product of the standard normal deviates of X and Y; in particular,

$$\rho = \text{average}\left[\frac{(X-\mu_x)}{\sigma_x}\frac{(Y-\mu_y)}{\sigma_y}\right].$$

The estimator of the population correlation is known as *Pearson's coefficient of correlation*, or simply the *correlation coefficient*. Pearson's correlation coefficient is denoted by r; it is calculated using the formula

$$\begin{aligned} r &= \frac{1}{(n-1)}\sum_{i=1}^{n}\left(\frac{x_i-\bar{x}}{s_x}\right)\left(\frac{y_i-\bar{y}}{s_y}\right) \\ &= \frac{1}{(n-1)}\frac{\sum_{i=1}^{n}(x_i-\bar{x})(y_i-\bar{y})}{s_x s_y}. \end{aligned}$$

Recall that s_x and s_y are the sample standard deviations of the x and y values, respectively. Therefore, an equivalent formula for r is

$$r = \frac{\sum_{i=1}^{n}(x_i-\bar{x})(y_i-\bar{y})}{\sqrt{[\sum_{i=1}^{n}(x_i-\bar{x})^2][\sum_{i=1}^{n}(y_i-\bar{y})^2]}}.$$

The correlation coefficient is a dimensionless number; it has no associated units of measurement. The maximum value that r can achieve is 1, and its minimum value is -1. Therefore, for any given set of observations, $-1 \leq r \leq 1$. The values $r = 1$ and $r = -1$ occur when there is an exact linear relationship between x and y; if we were to plot all pairs of outcomes (x, y), the points would lie on a straight line. Examples of perfect correlation are illustrated in Figures 17.2 (a) and (b). As the relationship between x and y deviates from perfect linearity, r moves away from 1 or -1 and closer to 0. If y tends to increase in magnitude as x increases, then r is greater than 0 and x and y are said to be *positively correlated*; if y decreases as x increases, then r is less than 0 and the two variables are *negatively correlated*. If $r = 0$, as it does for the samples pictured in Figures 17.2 (c) and (d), then there is no linear relationship between x and y and the variables are uncorrelated. However, a nonlinear relationship may exist.

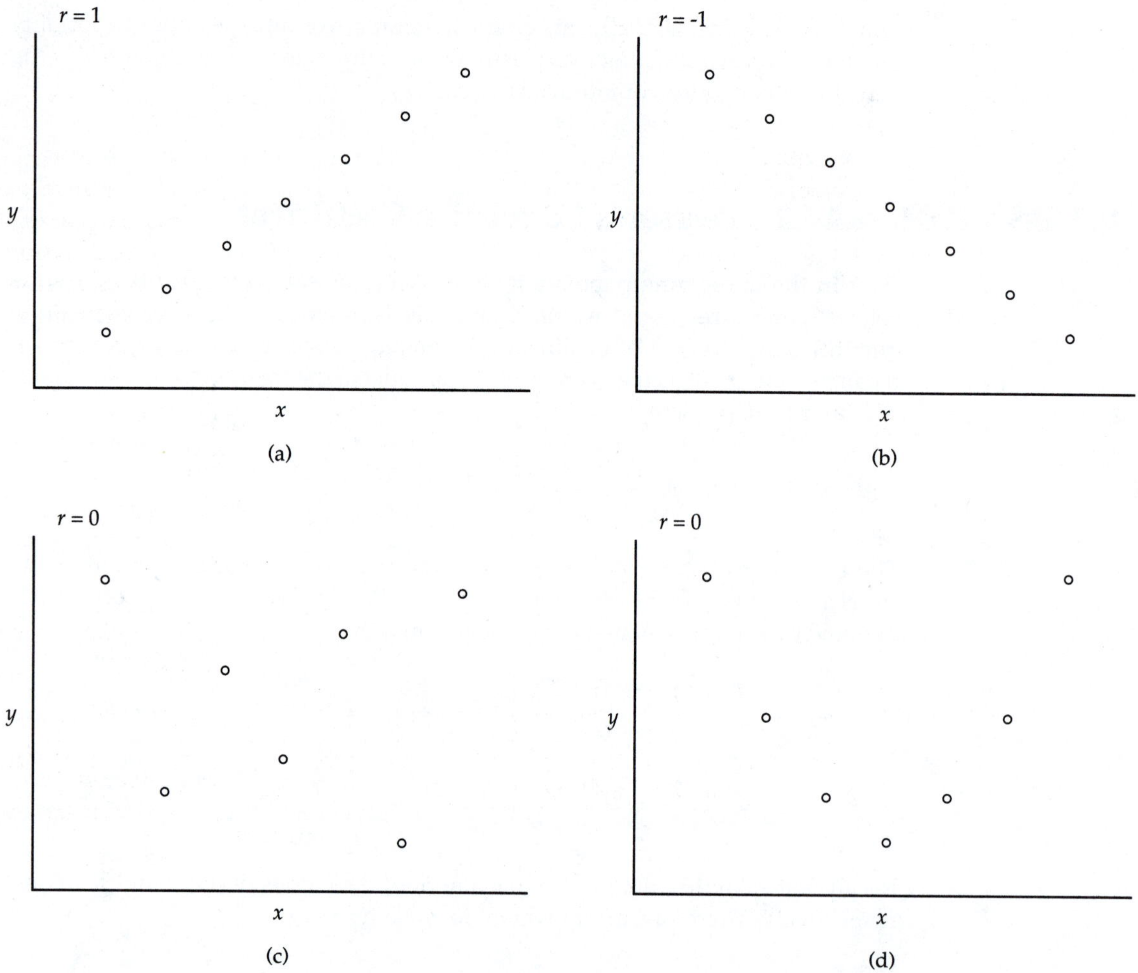

Figure 17.2. Scatter plots showing possible relationships between X and Y

For the data in Table 17.1, the mean percentage of children immunized against DPT is

$$
\begin{aligned}
\bar{x} &= \frac{1}{n}\sum_{i=1}^{n} x_i \\
&= \frac{1}{20}\sum_{i=1}^{20} x_i \\
&= 76.3\%,
\end{aligned}
$$

and the mean value of the under-5 mortality rate is

$$\begin{aligned}
\bar{y} &= \frac{1}{n}\sum_{i=1}^{n} y_i \\
&= \frac{1}{20}\sum_{i=1}^{20} y_i \\
&= 62.2 \text{ per 1,000 live births.}
\end{aligned}$$

In addition,

$$\begin{aligned}
\sum_{i=1}^{n}(x_i - \bar{x})(y_i - \bar{y}) &= \sum_{i=1}^{20}(x_i - 0.763)(y_i - 0.0622) \\
&= -0.2168,
\end{aligned}$$

$$\begin{aligned}
\sum_{i=1}^{n}(x_i - \bar{x})^2 &= \sum_{i=1}^{20}(x_i - 0.763)^2 \\
&= 0.7656,
\end{aligned}$$

and

$$\begin{aligned}
\sum_{i=1}^{n}(y_i - \bar{y})^2 &= \sum_{i=1}^{20}(y_i - 0.0622)^2 \\
&= 0.08932.
\end{aligned}$$

As a result, the coefficient of correlation is

$$\begin{aligned}
r &= \frac{\sum_{i=1}^{n}(x_i - \bar{x})(y_i - \bar{y})}{\sqrt{[\sum_{i=1}^{n}(x_i - \bar{x})^2][\sum_{i=1}^{n}(y_i - \bar{y})^2]}} \\
&= \frac{-0.2168}{\sqrt{(0.7656)(0.08932)}} \\
&= -0.829.
\end{aligned}$$

Based on this sample, there appears to be a fairly strong linear relationship between the percentage of children immunized against DPT in a specified country and its under-5 mortality rate; the Pearson coefficient of correlation is close to its minimum value of −1. Since r is negative, mortality rate decreases in magnitude as percent immunization increases. Care must be taken when interpreting this relationship, however. An effective immunization program might be the primary reason for the decrease in mortality, or it might be a ramification of a successful comprehensive health care system that is itself the cause of the decrease. The correlation coefficient merely tells us that a linear relationship exists; it does not specify the nature of this relationship.

Just as we made inference about a population mean μ based on the sample mean $\bar{x}$, we would also like to be able to draw conclusions about the unknown population correlation ρ using the sample correlation coefficient r. Most frequently, we are interested in determining whether any correlation exists between the random variables X and Y. This may be accomplished by testing the null hypothesis that there is no such correlation in the underlying population, or

$$H_0 : \rho = 0.$$

The procedure is similar to other tests of hypotheses that we have encountered; it involves finding the probability of obtaining a sample correlation coefficient as extreme or more extreme than the observed r given that the null hypothesis is true. Since the estimated standard error of r may be expressed as

$$\widehat{\text{se}}(r) = \sqrt{\frac{1 - r^2}{n - 2}},$$

the test is performed by calculating the statistic

$$\begin{aligned} t &= \frac{r - 0}{\sqrt{(1 - r^2)/(n - 2)}} \\ &= r\sqrt{\frac{n - 2}{1 - r^2}}. \end{aligned}$$

If we assume that the pairs of observations (x_i, y_i) were obtained randomly and that both X and Y are normally distributed, this quantity has a t distribution with $n - 2$ degrees of freedom under the null hypothesis.

Suppose that we want to know whether a linear relationship exists between percent immunization against DPT and the under-5 mortality rate in the population of countries around the world. We conduct a two-sided test of the null hypothesis of no association at the $\alpha = 0.05$ level of significance. Recall that we previously found r to be equal to -0.829. Therefore,

$$\begin{aligned} t &= r\sqrt{\frac{n - 2}{1 - r^2}} \\ &= -0.829\sqrt{\frac{20 - 2}{1 - (-0.829)^2}} \\ &= -6.29. \end{aligned}$$

Referring to Table A.2, we observe that for a t distribution with 18 degrees of freedom, $p < 0.001$. Given that $\rho = 0$, the probability of observing a sample correlation coefficient as large as $r = -0.829$ is very small. Therefore, we reject the null hypothesis at the 0.05 level. Based on this sample, there is evidence that the true population correlation is different from 0. Mortality rate diminishes as the percentage of children immunized increases; therefore, the correlation is negative.

The testing procedure described is valid only for the special case in which the hypothesized value of the population correlation is equal to 0. If ρ is equal to some other value, then the sampling distribution of r is skewed, and the test statistic no longer follows a t distribution. However, methods for testing the more general hypothesis

$$H_0 : \rho = \rho_0$$

are available [2].

The coefficient of correlation has several limitations. First, it quantifies only the strength of the **linear** relationship between two variables; if X and Y have a nonlinear relationship—as they do in Figure 17.2 (d)—then it would not provide a valid measure of this association. Second, care must be taken when the data contain any outliers, or pairs of observations that lie considerably outside the range of the other data points. The sample correlation coefficient is highly sensitive to extreme values and, if one or more are present, often gives misleading results. Third, the estimated correlation should never be extrapolated beyond the observed ranges of the variables; the relationship between X and Y may change outside of this region. Finally, it must be kept in mind that a high correlation between two variables does not in itself imply a cause-and-effect relationship.

17.3 Spearman's Rank Correlation Coefficient

Like other parametric techniques, Pearson's correlation coefficient is very sensitive to outlying values. We may be interested in calculating a measure of association that is more robust. One approach is to rank the two sets of outcomes x and y separately and calculate a coefficient of rank correlation. This procedure—which results in a quantity known as *Spearman's rank correlation coefficient*—may be classified among the nonparametric methods that we studied in Chapter 13.

Spearman's rank correlation coefficient, denoted r_s, is simply Pearson's r calculated for the ranked values of x and y. Therefore,

$$r_s = \frac{\sum_{i=1}^{n}(x_{ri} - \bar{x}_r)(y_{ri} - \bar{y}_r)}{\sqrt{[\sum_{i=1}^{n}(x_{ri} - \bar{x}_r)^2][\sum_{i=1}^{n}(y_{ri} - \bar{y}_r)^2]}}$$

where x_{ri} and y_{ri} are the ranks associated with the ith subject rather than the actual observations. An equivalent method for computing r_s is provided by the formula

$$r_s = 1 - \frac{6 \sum_{i=1}^{n} d_i^2}{n\,(n^2 - 1)};$$

in this case, n is the number of data points in the sample and d_i is the difference between the rank of x_i and the rank of y_i. Like the Pearson correlation coefficient, the Spearman rank correlation ranges in value from −1 to 1. Values of r_s close to the extremes indicate a high degree of correlation between x and y; values near 0 imply a lack of linear association between the two variables.

Suppose that we were to separately rank the percent immunization and under-5 mortality rate data presented in Table 17.1 from the smallest values to the largest, assigning average ranks to tied observations. The results are presented in Table 17.2. The Spearman rank correlation coefficient is

$$\begin{aligned} r_s &= 1 - \frac{6 \sum_{i=1}^{n} d_i^2}{n\,(n^2 - 1)} \\ &= 1 - \frac{6(2,169)}{20(399)} \\ &= -0.631. \end{aligned}$$

This value is somewhat smaller in magnitude than the Pearson correlation coefficient but still suggests a moderately strong relationship between the percentage of children immunized against DPT and the under-5 mortality rate. Again the correlation is negative.

Spearman's rank correlation coefficient may also be thought of as a measure of the concordance of the ranks for the outcomes x and y. If the 20 measurements

Table 17.2 Ranked percentage of children immunized against DPT and under-5 mortality rate for 20 countries, 1989

Nation	Percentage Immunized	Rank	Mortality Rate	Rank	d_i	d_i^2
Ethiopia	0.26	1.0	0.0226	20.0	−19.0	361.00
Bolivia	0.40	2.0	0.0165	18.0	−16.0	256.00
Senegal	0.47	3.0	0.0189	19.0	−16.0	256.00
Brazil	0.54	4.0	0.0085	14.0	−10.0	100.00
Mexico	0.65	5.0	0.0051	13.0	−8.0	64.00
Turkey	0.74	6.0	0.0090	15.0	−9.0	81.00
United Kingdom	0.75	7.0	0.0010	5.0	2.0	4.00
USSR	0.79	8.0	0.0035	11.0	−3.0	9.00
Egypt	0.81	9.0	0.0094	16.0	−7.0	49.00
Japan	0.83	11.0	0.0006	1.0	10.0	100.00
Greece	0.83	11.0	0.0012	7.5	3.5	12.25
India	0.83	11.0	0.0145	17.0	−6.0	36.00
Canada	0.85	13.5	0.0009	3.5	10.0	100.00
Italy	0.85	13.5	0.0011	6.0	7.5	56.25
Finland	0.90	15.0	0.0007	2.0	13.0	169.00
Yugoslavia	0.91	16.0	0.0027	10.0	6.0	36.00
France	0.95	17.5	0.0009	3.5	14.0	196.00
China	0.95	17.5	0.0043	12.0	5.5	30.25
United States	0.97	19.0	0.0012	7.5	11.5	132.25
Poland	0.98	20.0	0.0018	9.0	11.0	121.00
						2,169.00

of percent immunization and under-5 mortality rate in Table 17.2 happened to be ranked in the same order for each variable—meaning that the country with the ith largest percent immunization also has the ith largest mortality rate for all values of i—then each difference d_i would be equal to 0, and

$$\begin{aligned} r_s &= 1 - \frac{6(0)}{n\,(n^2 - 1)} \\ &= 1. \end{aligned}$$

If the ranking of the first variable is the inverse of the ranking of the second—so that the country with the largest percentage of children immunized against DPT has the smallest under-5 mortality rate, and so forth—it can be shown that

$$r_s = -1.$$

When there is no linear correspondence between the two sets of ranks, then

$$r_s = 0.$$

If the sample size n is not too small—in particular, if it is greater than or equal to ten—and if we assume that pairs of ranks (x_{ri}, y_{ri}) are chosen randomly, then we may test the null hypothesis that the unknown population correlation is equal to 0,

$$H_0 : \rho = 0,$$

using the same procedure that we used for Pearson's r. If n is less than ten, then tables of critical values must be used to evaluate H_0 [2]. For the ranked data of Table 17.2, the test statistic is

$$\begin{aligned} t_s &= r_s \sqrt{\frac{n-2}{1-r_s^2}} \\ &= -0.631 \sqrt{\frac{18}{1-(-0.631)^2}} \\ &= -3.45. \end{aligned}$$

For a t distribution with 18 degrees of freedom, $0.001 < p < 0.01$. Therefore, we reject H_0 at the 0.05 level and conclude that the true population correlation is smaller than 0.

Like other nonparametric techniques, Spearman's rank correlation coefficient has advantages and disadvantages. It is much less sensitive to outlying values than is Pearson's correlation coefficient. In addition, it can be used when one or both of the relevant variables are ordinal. Because they rely on ranks rather than actual observations, however, nonparametric methods do not use all the information that is known about a distribution.

17.4 Further Applications

Suppose that we wish to examine the relationship between the percentage of births attended by trained health care personnel—including physicians, nurses, midwives, and other health care workers—and the maternal mortality rate per 100,000 live births. The data for a random sample of 20 countries appear in Table 17.3 [1]. We would like to determine whether there is any evidence of a linear relationship between these two variables.

We begin our analysis by constructing a two-way scatter plot of the data, placing percentage of births attended on the horizontal axis and maternal mortality rate on the vertical axis. The graph is displayed in Figure 17.3. Note that higher values of percentage of births attended appear to be associated with lower values of maternal mortality rate.

Let x represent the observed outcomes of percentage of births attended and y the observed maternal mortality rates. The mean percentage of births attended is

$$\begin{aligned}\bar{x} &= \frac{1}{20}\sum_{i=1}^{20} x_i \\ &= 66\%,\end{aligned}$$

Table 17.3 Percentage of births attended by trained health care personnel and maternal mortality rate for 20 countries, 1980–1988

Nation	Percentage Attended	Maternal Mortality Rate per 100,000 Live Births
Bangladesh	5	600
Belgium	100	9
Chile	98	47
Ecuador	27	190
Hong Kong	92	5
Hungary	99	26
Iran	82	120
Kenya	28	170
Morocco	29	300
Nepal	6	830
Nigeria	40	800
Norway	100	2
Pakistan	24	500
Philippines	57	93
Portugal	87	12
South Korea	70	26
Spain	96	11
Switzerland	99	5
United States	99	8
Venezuela	82	59

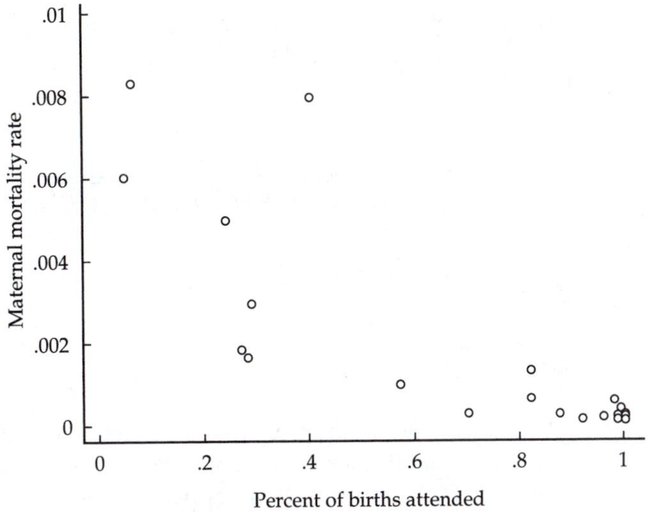

Figure 17.3. Maternal mortality rate versus percentage of births attended by trained health care personnel for 20 countries, 1980–1988

and the mean value of maternal mortality rate is

$$\begin{aligned}\bar{y} &= \frac{1}{20}\sum_{i=1}^{20} y_i \\ &= 190.65 \text{ per } 100{,}000 \text{ live births.}\end{aligned}$$

Also,

$$\begin{aligned}\sum_{i=1}^{n}(x_i - \bar{x})(y_i - \bar{y}) &= \sum_{i=1}^{20}(x_i - 0.66)(y_i - 0.0019065) \\ &= -0.014716,\end{aligned}$$

$$\begin{aligned}\sum_{i=1}^{n}(x_i - \bar{x})^2 &= \sum_{i=1}^{20}(x_i - 0.66)^2 \\ &= 2.3324,\end{aligned}$$

and

$$\begin{aligned}\sum_{i=1}^{n}(y_i - \bar{y})^2 &= \sum_{i=1}^{20}(y_i - 0.0019065)^2 \\ &= 0.00013975.\end{aligned}$$

The Pearson coefficient of correlation, which quantifies the degree to which these outcomes are linearly related, is

$$
\begin{aligned}
r &= \frac{\sum_{i=1}^{n}(x_i - \bar{x})(y_i - \bar{y})}{\sqrt{[\sum_{i=1}^{n}(x_i - \bar{x})^2][\sum_{i=1}^{n}(y_i - \bar{y})^2]}} \\
&= \frac{-0.014716}{\sqrt{(2.3324)(0.00013975)}} \\
&= -0.815.
\end{aligned}
$$

A value of -1 would imply that an exact negative linear relationship exists; if r were to equal 0, then there would be no linear relationship at all. Based on the sample of 20 countries, there appears to be a fairly strong relationship between percentage of births attended and maternal mortality rate. Since r is negative, mortality rate diminishes as percentage of births attended increases.

We might wish to test

$$H_0 : \rho = 0,$$

the null hypothesis that there is no linear relationship between percentage of births attended and the maternal mortality rate in the underlying population of countries around the world. We use a two-sided test conducted at the 0.05 level. Since $r = -0.815$,

$$
\begin{aligned}
\mathrm{t} &= r\sqrt{\frac{n-2}{1-r^2}} \\
&= -0.815\sqrt{\frac{20-2}{1-(-0.815)^2}} \\
&= -5.97.
\end{aligned}
$$

For a t distribution with 18 degrees of freedom, $p < 0.001$. We reject H_0 and conclude that the true population correlation is not equal to 0; since r is negative, maternal mortality rate decreases in magnitude as percentage of births attended increases.

If we are interested in computing a more robust measure of association, we could order the two sets of outcomes x and y and calculate the rank correlation instead. Spearman's rank correlation is simply Pearson's r calculated using ranks rather than actual outcomes. The ranked percentage of births attended and maternal mortality rate data are displayed in Table 17.4. The Spearman rank correlation for these data is

$$\begin{aligned} r_s &= 1 - \frac{6\sum_{i=1}^{n} d_i^2}{n\,(n^2-1)} \\ &= 1 - \frac{6(2,506)}{20(399)} \\ &= -0.884. \end{aligned}$$

This value is somewhat larger in magnitude than the Pearson correlation coefficient; it also signifies a fairly strong relationship between percentage of births attended and maternal mortality rate.

To test the null hypothesis that the unknown population correlation is equal to 0, or

$$H_0 : \rho = 0,$$

Table 17.4 Ranked percentage of births attended by trained health care personnel and maternal mortality rate for 20 countries, 1980–1988

Nation	Percentage Attended	Rank	Mortality Rate	Rank	d_i	d_i^2
Bangladesh	0.05	1.0	0.00600	18.0	−17.0	289.00
Nepal	0.06	2.0	0.00830	20.0	−18.0	324.00
Pakistan	0.24	3.0	0.00500	17.0	−14.0	196.00
Ecuador	0.27	4.0	0.00190	15.0	−11.0	121.00
Kenya	0.28	5.0	0.00170	14.0	−9.0	81.00
Morocco	0.29	6.0	0.00300	16.0	−10.0	100.00
Nigeria	0.40	7.0	0.00800	19.0	−12.0	144.00
Philippines	0.57	8.0	0.00093	12.0	−4.0	16.00
South Korea	0.70	9.0	0.00026	8.5	0.5	0.25
Venezuela	0.82	10.5	0.00059	11.0	−0.5	0.25
Iran	0.82	10.5	0.00120	13.0	−2.5	6.25
Portugal	0.87	12.0	0.00012	7.0	5.0	25.00
Hong Kong	0.92	13.0	0.00005	2.5	10.5	110.25
Spain	0.96	14.0	0.00011	6.0	8.0	64.00
Chile	0.98	15.0	0.00047	10.0	5.0	25.00
Switzerland	0.99	17.0	0.00005	2.5	14.5	210.25
United States	0.99	17.0	0.00008	4.0	13.0	169.00
Hungary	0.99	17.0	0.00026	8.5	8.5	72.25
Norway	1.00	19.5	0.00002	1.0	18.5	342.25
Belgium	1.00	19.5	0.00009	5.0	14.5	210.25
						2,506.00

Table 17.5 Stata output displaying the Pearson coefficient of correlation and the Spearman rank correlation

(obs=20)

	attend	maternal
attend	1.0000	
maternal	-0.8151	1.0000

Spearman Rank Correlation

(obs=20)

	attend	maternal
attend	1.0000	
maternal	-0.8842	1.0000

we calculate

$$
\begin{aligned}
t_s &= r_s \sqrt{\frac{n-2}{1-r_s^2}} \\
&= -0.884 \sqrt{\frac{20-2}{1-(-0.884)^2}} \\
&= -8.02.
\end{aligned}
$$

For a t distribution with 18 degrees of freedom, $p < 0.001$. Using this nonparametric procedure, we again reject the null hypothesis that the population correlation is equal to 0.

Finding the coefficient of correlation for two random variables involves quite a bit of computation; however, most computer packages will calculate both the Pearson and Spearman correlation coefficients for us. Table 17.5 shows the appropriate output from Stata if the 20 measurements of percentage of births attended are saved under the variable name `attend` and the corresponding measurements of maternal mortality rate are saved under the name `maternal`. Since a perfect relationship exists, the correlation of any variable with itself is equal to 1.

17.5 Review Exercises

1. In a study conducted in Italy, ten patients with hypertriglyceridemia were placed on a low-fat, high-carbohydrate diet. Before the the start of the diet, cholesterol and triglyceride measurements were recorded for each subject [3].

Patient	Cholesterol Level (mmol/l)	Triglyceride Level (mmol/l)
1	5.12	2.30
2	6.18	2.54
3	6.77	2.95
4	6.65	3.77
5	6.36	4.18
6	5.90	5.31
7	5.48	5.53
8	6.02	8.83
9	10.34	9.48
10	8.51	14.20

a. Construct a two-way scatter plot for these data.
b. Does there appear to be any evidence of a relationship between cholesterol and triglyceride levels?
c. Compute r, the Pearson correlation coefficient.
d. At the 0.05 level of significance, test the null hypothesis that the population correlation ρ is equal to 0.
e. Calculate r_s, the Spearman rank correlation.
f. How does the value of r_s compare to r?
g. Using r_s, again test the null hypothesis that the population correlation is equal to 0.
h. Do you reach the same conclusion for each test?

2. Thirty-five patients with ischemic heart disease took part in a series of tests designed to evaluate the perception of pain. In one part of the study, the patients exercised until they experienced angina; time until the onset of angina and the duration of the attack were recorded. The data are saved on your disk in the file `ischemic`. Time to angina in seconds is saved under the variable name `time`; the duration of angina, also in seconds, is saved under the name `duration` [4].

a. Create a two-way scatter plot for these data.
b. In the population of patients with ischemic heart disease, does there appear to be any evidence of a linear relationship between time to angina and the duration of the attack?
c. Calculate Pearson's correlation coefficient.
d. Does the duration of angina tend to increase or decrease as time to angina increases?
e. Test the null hypothesis

$$H_0: \rho = 0.$$

f. Compute Spearman's rank correlation.
g. Using r_s, again test the null hypothesis that the population correlation is equal to 0.

3. Suppose that you are interested in determining whether a relationship exists between the fluoride content in a public water supply and the dental caries

experience of children using this water. Data from a study examining 7,257 children in 21 cities are saved on your disk in the file `water`. The fluoride content of the public water supply in each city, measured in parts per million, is saved under the variable name `fluoride`; the number of dental caries per 100 children examined is saved under the name `caries` [5]. The total dental caries experience is obtained by summing the numbers of filled teeth, teeth with untreated dental caries, teeth requiring extraction, and missing teeth.

a. Construct a two-way scatter plot.
b. What is the correlation between the number of dental caries per 100 children and the fluoride content of the water?
c. Is this correlation significantly different from 0?
d. For the 21 cities in the study, the highest fluoride content in a given water supply is 2.6 ppm. If you were to increase the fluoride content of the water to more than 4 ppm, do you believe that the number of dental caries per 100 children would decrease?

References

[1] United Nations Children's Fund, *The State of the World's Children 1991*, New York: Oxford University Press.

[2] Snedecor, G. W., and Cochran, W. G., *Statistical Methods*, Ames, Iowa: The Iowa State University Press, 1980.

[3] Cominacini, L., Zocca, I., Garbin, U., Davoli, A., Compri, R., Brunetti, L., and Bosello, O., "Long-Term Effect of a Low-Fat, High-Carbohydrate Diet on Plasma Lipids of Patients Affected by Familial Endogenous Hypertriglyceridemia," *American Journal of Clinical Nutrition*, Volume 48, July 1988, 57–65.

[4] Miller, P. F., Sheps, D. S., Bragdon, E. E., Herbst, M. C., Dalton, J. L., Hinderliter, A. L., Koch, G. G., Maixner, W., and Ekelund, L. G., "Aging and Pain Perception in Ischemic Heart Disease," *American Heart Journal*, Volume 120, July 1990, 22–30.

[5] Dean, H. T., Arnold, F. A., and Elvove, E., "Domestic Water and Dental Caries," *Public Health Reports*, Volume 57, August 7, 1942, 1155–1179.

CHAPTER

18

Simple Linear Regression

Like correlation analysis, *simple linear regression* is a technique that is used to explore the nature of the relationship between two continuous variables. The primary difference between these two analytical methods is that regression enables us to investigate the change in one variable, called the *response*, which corresponds to a given change in the other, known as the *explanatory variable*. Correlation analysis makes no such distinction; the two variables involved are treated symmetrically. The ultimate objective of regression analysis is to predict or estimate the value of the response that is associated with a fixed outcome of the explanatory variable.

An example of a situation in which regression analysis might be preferred to correlation is illustrated by the pediatric growth charts in Figures 18.1 and 18.2. Among children of both sexes, head circumference appears to increase linearly between the ages of 2 and 18 years. Rather than quantifying the strength of this association, we might be interested in predicting the change in head circumference that corresponds to a given change in age. In this case, head circumference is the response, and age is the explanatory variable. An understanding of their relationship helps parents and pediatricians to monitor growth and detect possible cases of macrocephaly or microcephaly.

18.1 Regression Concepts

Suppose that we are interested in the distribution of a continuous random variable Y. The outcomes of Y, denoted y, are the head circumference measurements in centimeters for the population of low birth weight infants—those

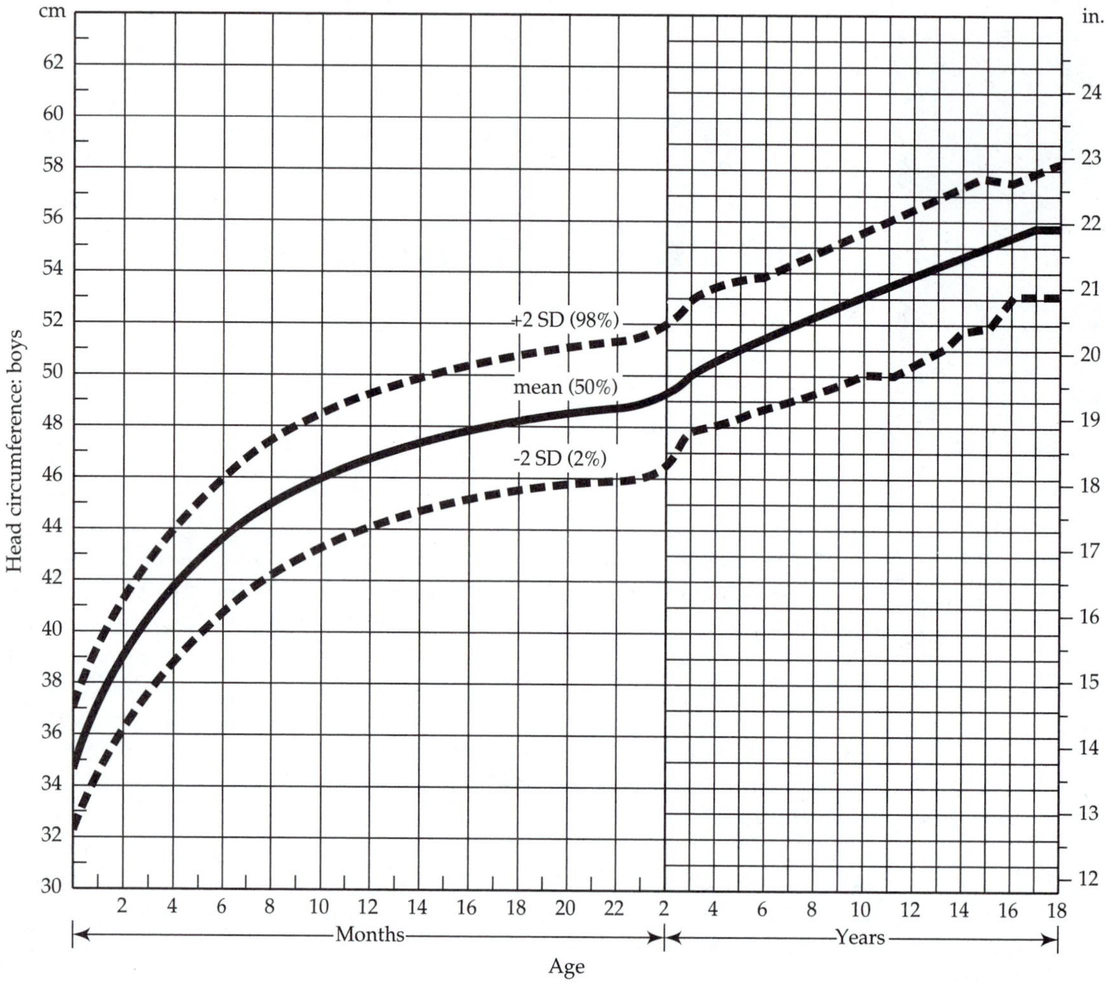

Figure 18.1. Head circumference versus age for boys (*Source*: From G. Nellhaus, *Pediatrics*, 1968, 41:106. Copyright 1968 American Academy of Pediatrics)

weighing less than 1,500 grams–born in two teaching hospitals in Boston, Massachusetts [1]. We are told that the mean head circumference for the infants in this population is

$$\mu_y = 27 \text{ cm}$$

and that the standard deviation is

$$\sigma_y = 2.5 \text{ cm}.$$

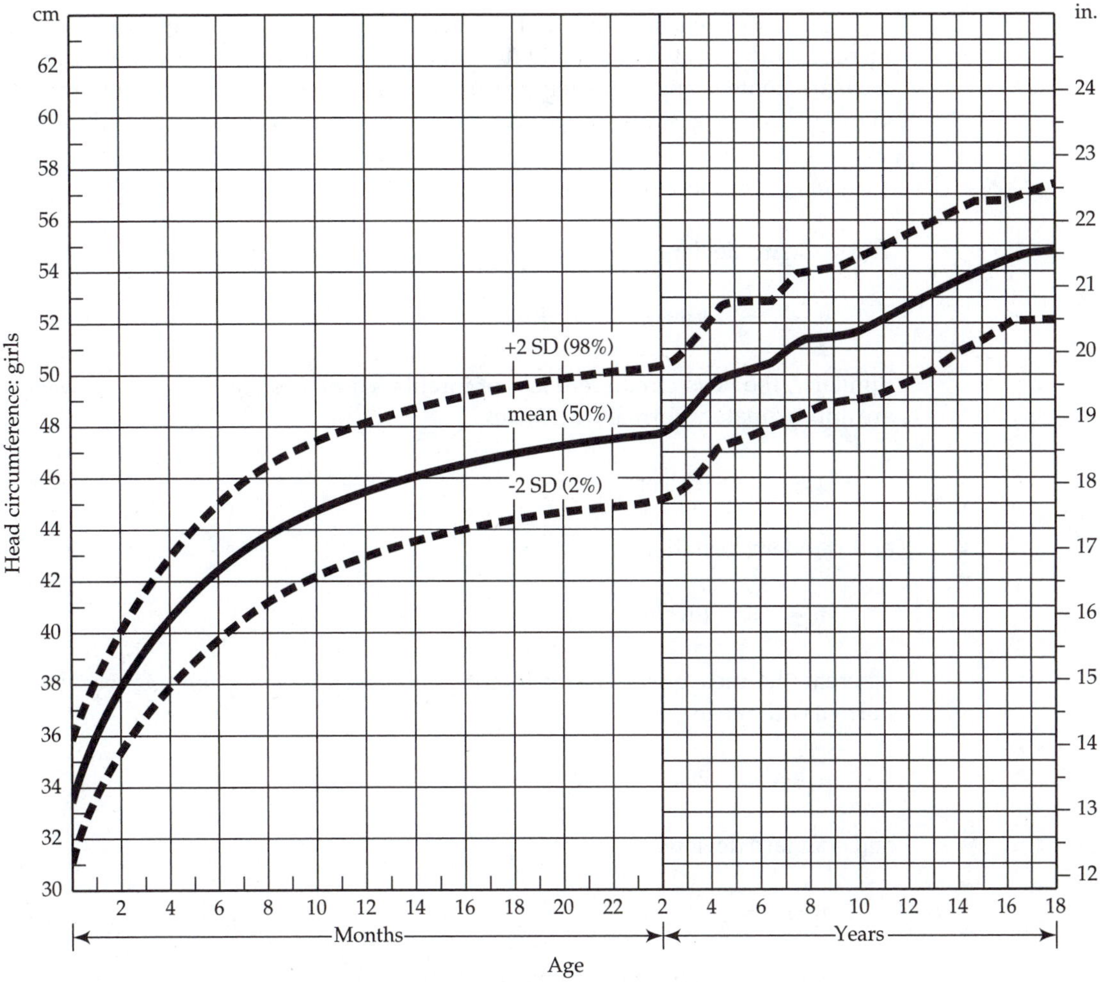

Figure 18.2. Head circumference versus age for girls (*Source*: From G. Nellhaus, *Pediatrics*, 1968, 41:106. Copyright 1968 American Academy of Pediatrics)

Since the distribution of measurements is roughly normal, we are able to say that approximately 95% of the infants have head circumferences that measure between

$$\begin{aligned} \mu_y - 1.96\,\sigma_y &= 27 - (1.96)(2.5) \\ &= 22.1 \text{ cm} \end{aligned}$$

and

$$\begin{aligned} \mu_y + 1.96\,\sigma_y &= 27 + (1.96)(2.5) \\ &= 31.9 \text{ cm.} \end{aligned}$$

Suppose we also know that the head circumferences of newborn infants increase with gestational age and that for each specified age x the distribution of measurements is approximately normal. For example, the head circumferences of infants whose gestational age is 26 weeks are normally distributed with mean

$$\mu_{y|26} = 24 \text{ cm}$$

and standard deviation

$$\sigma_{y|26} = 1.6 \text{ cm}.$$

Similarly, the head circumferences of infants whose gestational age is 29 weeks are approximately normal with mean

$$\mu_{y|29} = 26.5 \text{ cm}$$

and standard deviation

$$\sigma_{y|29} = 1.6 \text{ cm},$$

whereas the measurements for infants whose gestational age is 32 weeks are normal with mean

$$\mu_{y|32} = 29 \text{ cm}$$

and standard deviation

$$\sigma_{y|32} = 1.6 \text{ cm}.$$

For each value of gestational age x, the standard deviation $\sigma_{y|x}$ is constant and is less than σ_y. In fact, it can be shown that, in general,

$$\sigma_{y|x}^2 = (1 - \rho^2)\,\sigma_y^2$$

where ρ is the correlation between X and Y in the underlying population [2]. If X and Y have no linear relationship, then $\rho = 0$ and

$$\begin{aligned}\sigma_{y|x}^2 &= (1 - 0)\,\sigma_y^2 \\ &= \sigma_y^2.\end{aligned}$$

For the head circumference and gestational age data, $\sigma_y = 2.5$ cm and $\sigma_{y|x} = 1.6$ cm; therefore,

$$(1.6)^2 = (1 - \rho^2)\,(2.5)^2,$$

and

$$\begin{aligned}\rho &= \sqrt{1 - \frac{(1.6)^2}{(2.5)^2}} \\ &= \sqrt{0.5904} \\ &= \pm 0.768.\end{aligned}$$

There is a fairly strong correlation between head circumference and gestational age in the underlying population of low birth weight infants; using this method of calculation, however, we cannot tell whether the correlation is positive or negative.

Because the standard deviation of the distribution of head circumference measurements for infants of a specified gestational age ($\sigma_{y|x} = 1.6$ cm) is smaller than the standard deviation for infants of all ages combined ($\sigma_y = 2.5$ cm), working with a single value of gestational age allows us to be more precise in our descriptions. For example, we are able to say that approximately 95% of the values of head circumference for the population of infants whose gestational age is 26 weeks lie between

$$\begin{aligned}\mu_{y|26} - 1.96\,\sigma_{y|26} &= 24 - (1.96)(1.6) \\ &= 20.9 \text{ cm}\end{aligned}$$

and

$$\begin{aligned}\mu_{y|26} + 1.96\,\sigma_{y|26} &= 24 + (1.96)(1.6) \\ &= 27.1 \text{ cm.}\end{aligned}$$

In addition, roughly 95% of the infants whose gestational age is 29 weeks have head circumferences between

$$\begin{aligned}\mu_{y|29} - 1.96\,\sigma_{y|29} &= 26.5 - (1.96)(1.6) \\ &= 23.4 \text{ cm}\end{aligned}$$

and

$$\begin{aligned}\mu_{y|29} + 1.96\,\sigma_{y|29} &= 26.5 + (1.96)(1.6) \\ &= 29.6 \text{ cm,}\end{aligned}$$

whereas 95% of the infants whose gestational age is 32 weeks have measurements that lie between

$$\begin{aligned}\mu_{y|32} - 1.96\,\sigma_{y|32} &= 29 - (1.96)(1.6) \\ &= 25.9 \text{ cm}\end{aligned}$$

and

$$\mu_{y|32} + 1.96\,\sigma_{y|32} = 29 + (1.96)(1.6) = 32.1 \text{ cm.}$$

In summary, the respective intervals are as follows:

Gestational Age (weeks)	Interval Containing 95% of the Observations
26	(20.9 , 27.1)
29	(23.4 , 29.6)
32	(25.9 , 32.1)

Each of these intervals is constructed to enclose 95% of the population head circumference values for infants of a particular gestational age. None is as wide as (22.1 , 31.9), the interval computed for the entire population of low birth weight infants. In addition, the intervals shift to the right as gestational age increases.

18.2 The Model

18.2.1 The Population Regression Line

As noted above, mean head circumference tends to become larger as gestational age increases. Based on the data plotted in Figure 18.3, the relationship is linear. We might wish to quantify this relationship by fitting a model of the form

$$\mu_{y|x} = \alpha + \beta x$$

where $\mu_{y|x}$ is the mean head circumference of low birth weight infants whose gestational age is x weeks. This model–the *population regression line*–is the equation of a straight line. The parameters α and β are constants and are called the *coefficients* of the equation; α is the y-intercept of the line and β is its slope. The *y-intercept* is defined as the mean value of the response y when x is equal to 0. The *slope* is interpreted as the change in y that corresponds to a one unit change in x. If β is positive, then $\mu_{y|x}$ increases in magnitude as x increases; if β is negative, $\mu_{y|x}$ decreases as x increases.

Although the relationship between mean head circumference and gestational age might be a perfect straight line, the relationship between individual values of head circumference and age is not. As previously mentioned, the distribution of head circumference measurements for all low birth weight infants of a particular gestational age x is approximately normal with mean $\mu_{y|x}$ and standard

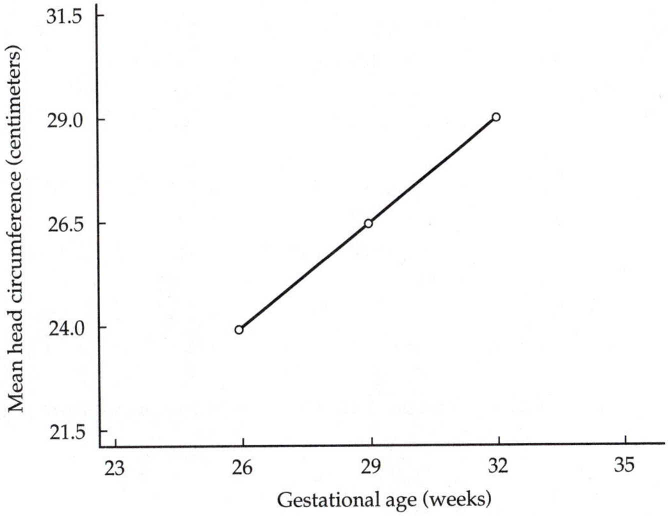

Figure 18.3. Population regression line of mean head circumference versus gestational age for low birth weight infants, $\mu_{y|x} = 2.3 + 0.83x$

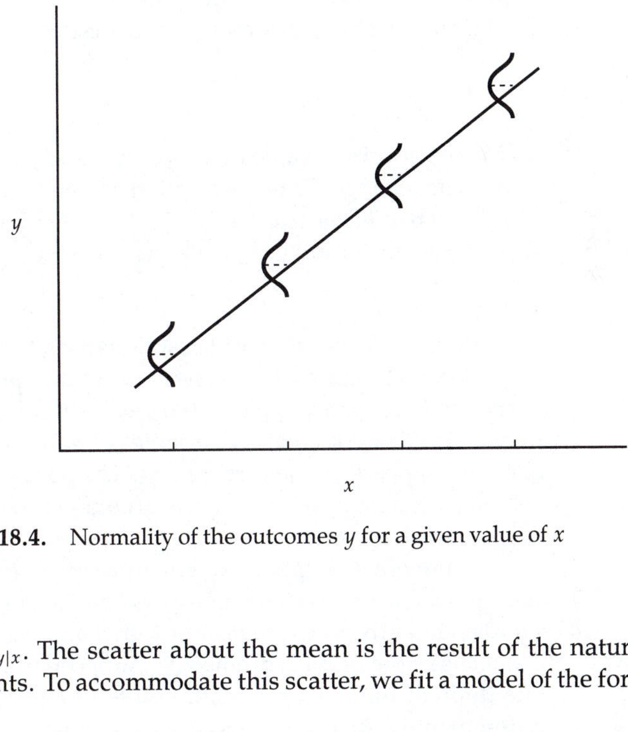

Figure 18.4. Normality of the outcomes y for a given value of x

deviation $\sigma_{y|x}$. The scatter about the mean is the result of the natural variation among infants. To accommodate this scatter, we fit a model of the form

$$y = \alpha + \beta x + \epsilon$$

where ϵ, known as the *error*, is the distance that a particular outcome y lies from the population regression line

$$\mu_{y|x} = \alpha + \beta x.$$

If the deviation ϵ is positive, then the point (x, y) lies above the line; if ϵ is negative, then (x, y) lies below it.

In simple linear regression, the straight line that we fit to the scatter of points takes the form

$$\hat{y} = a + bx.$$

This line serves as an estimate of the population regression line

$$\mu_{y|x} = \alpha + \beta x.$$

Before we attempt to fit such a line, we make a few assumptions:

1. For a specified value of x, which is considered to be measured without error, the distribution of the y values is normal with mean $\mu_{y|x}$ and standard deviation $\sigma_{y|x}$. This concept is illustrated in Figure 18.4.
2. The relationship between $\mu_{y|x}$ and x is described by the straight line

   $$\mu_{y|x} = \alpha + \beta x.$$

3. For any specified value of x, $\sigma_{y|x}$—the standard deviation of the outcomes y—does not change. This assumption of constant variability across all values of x is known as *homoscedasticity*.
4. The outcomes y are independent.

18.2.2 The Method of Least Squares

Consider Figure 18.5, the two-way scatter plot of head circumference versus gestational age for a sample of 100 low birth weight infants born in Boston, Massachusetts. The explanatory variable is measured on the horizontal axis and the response appears on the vertical axis. The points themselves vary widely, but the overall pattern suggests that head circumference tends to increase in magnitude as gestational age increases.

In previous chapters we attempted to estimate a population parameter—such as a mean or an odds ratio—based on the observations in a randomly chosen sample; similarly, we estimate the coefficients of a population regression line using a single sample of measurements. Suppose that we were to draw an arbitrary line through the scatter of data points in Figure 18.5. Lines sketched by two different individuals are unlikely to be identical, even though both persons might be attempting to depict the same trend. The question then arises as to which line best describes the relationship between head circumference and gestational

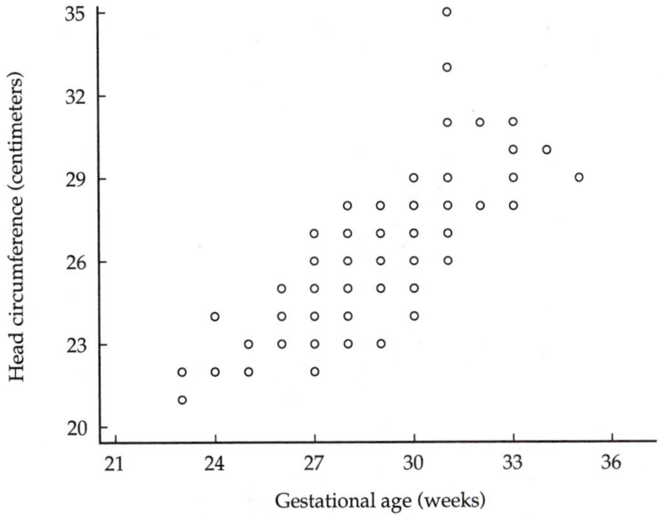

Figure 18.5. Head circumference versus gestational age for a sample of 100 low birth weight infants

age. What is needed is a procedure for obtaining the line that is somewhat more objective. (A word of explanation regarding Figure 18.5 is necessary: although the graph contains information for 100 infants, there appear to be far fewer data points. Since the outcomes of both variables are rounded to the nearest integer, many infants have identical values of head circumference and gestational age; consequently, some points are plotted on top of others.)

One mathematical technique for fitting a straight line to a set of points is known as the *method of least squares*. Observe that each data point (x_i, y_i) lies some vertical distance from the arbitrary line; we label this distance e_i. If y_i is the observed outcome of Y for a particular value of x_i, and $\hat{y}_i$—or y_i-hat—is the corresponding point on the fitted line, then

$$e_i = y_i - \hat{y}_i.$$

The distance e_i is known as the *residual*. Ideally, we would like all of the residuals to be equal to 0; this would imply that each point (x_i, y_i) lies on the population regression line. Since this is impossible, however, we must settle on some other criterion. The sum of the squares of the residuals,

$$\sum_{i=1}^{n} e_i^2 = \sum_{i=1}^{n} (y_i - \hat{y}_i)^2,$$

is often called the *error sum of squares*, or the *residual sum of squares*. The least squares line is constructed so that the error sum of squares is minimized.

The process of fitting the least squares line

$$\hat{y} = a + bx$$

involves finding a and b, the estimates of the population values α and β. Using calculus to minimize

$$\sum_{i=1}^{n} e_i^2 = \sum_{i=1}^{n} (y_i - \hat{y}_i)^2 = \sum_{i=1}^{n} (y_i - a - bx_i)^2,$$

we find that

$$b = \frac{\sum_{i=1}^{n} (x_i - \bar{x})(y_i - \bar{y})}{\sum_{i=1}^{n} (x_i - \bar{x})^2}$$

and

$$a = \bar{y} - b\bar{x}.$$

These equations yield the slope and the y-intercept for the fitted least squares line. In the expression for b, the numerator is the sum of cross-products of deviations about the mean for x and y; the denominator is the sum of squared deviations about the mean for x alone. The equation for a is expressed in terms of the estimated slope b. Once we know a and b, we are able to substitute various values of x into the equation for the line, solve for the corresponding values of $\hat{y}$, and plot these points.

The least squares regression line fitted to the data on head circumference and gestational age is

$$\hat{y} = 3.9143 + 0.7801x.$$

This line–which is plotted in Figure 18.6–has an error sum of squares that is smaller than the sum for any other line that could be drawn through the scatter of points. The y-intercept of the fitted line is 3.9143. Theoretically, this is the predicted value of head circumference that corresponds to a gestational age of 0 weeks. In this example, an age of 0 weeks has no real meaning. The slope of the line is 0.7801, implying that for each one-week increase in gestational age, an infant's head circumference increases by 0.7801 centimeters on average.

18.2.3 Inference for Regression Coefficients

We would like to be able to use the least squares regression line

$$\hat{y} = a + bx$$

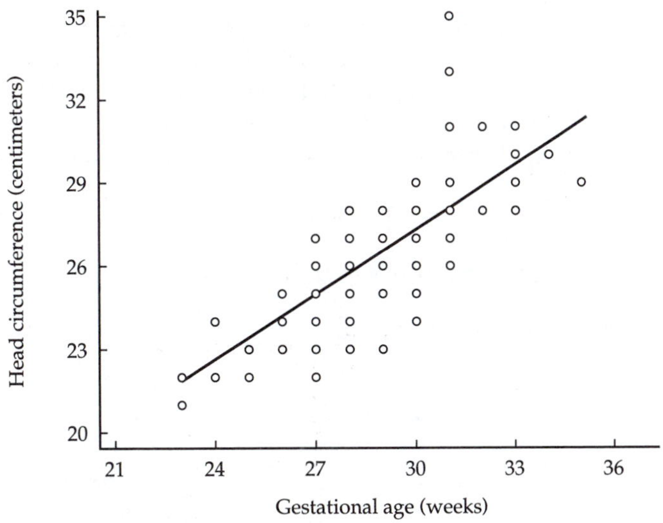

Figure 18.6. Least squares regression of head circumference on gestational age, $\hat{y} = 3.9143 + 0.7801x$

to make inference about the population regression line

$$\mu_{y|x} = \alpha + \beta x.$$

Recall that a is an estimate of the population intercept α and b is an estimate of the slope β. If we were to select repeated samples of size n from the underlying population of paired outcomes (x, y) and calculate a least squares line for each such set of observations, the computed values of a and b would vary from sample to sample. We need the standard errors of these quantities–just as we needed $\sigma/\sqrt{n}$, the standard error of the sample mean $\overline{X}$–to construct confidence intervals and conduct tests of hypotheses.

It can be shown that

$$\text{se}(b) = \frac{\sigma_{y|x}}{\sqrt{\sum_{i=1}^{n}(x_i - \bar{x})^2}}$$

and

$$\text{se}(a) = \sigma_{y|x}\sqrt{\frac{1}{n} + \frac{\bar{x}^2}{\sum_{i=1}^{n}(x_i - \bar{x})^2}}.$$

The standard errors of the estimated coefficients a and b depend on $\sigma_{y|x}$, the standard deviation of the y values for a given x. In practice, this quantity is usually unknown. As a result, we must estimate $\sigma_{y|x}$ by the sample standard deviation $s_{y|x}$, where

$$s_{y|x} = \sqrt{\frac{\sum_{i=1}^{n}(y_i - \hat{y}_i)^2}{n-2}}.$$

This formula involves the sum of the squared deviations of the actual observations y_i from the fitted values $\hat{y}_i$; the sum of squared residuals is the quantity that was minimized when we fit the least squares line. The estimate $s_{y|x}$ is often called the *standard deviation from regression.*

For the least squares regression of head circumference on gestational age,

$$s_{y|x} = 1.5904.$$

This estimate may be used to compute

$$\begin{aligned}\widehat{\text{se}}(b) &= \frac{s_{y|x}}{\sqrt{\sum_{i=1}^{n}(x_i - \bar{x})^2}} \\ &= 0.0631\end{aligned}$$

and

$$\begin{aligned}\widehat{\text{se}}(a) &= s_{y|x}\sqrt{\frac{1}{n} + \frac{\bar{x}^2}{\sum_{i=1}^{n}(x_i - \bar{x})^2}} \\ &= 1.8291.\end{aligned}$$

The slope is usually the more important coefficient in the regression equation; it quantifies the average change in y that corresponds to each unit change in x. We can test the null hypothesis that the population slope is equal to β_0, or

$$H_0 : \beta = \beta_0,$$

by finding p, the probability of observing an estimated slope as extreme or more extreme than the sample slope b given that the true population value is β_0. The test is carried out by calculating the statistic

$$\text{t} = \frac{b - \beta_0}{\widehat{\text{se}}(b)}.$$

Under the null hypothesis, this ratio has a t distribution with $n - 2$ degrees of freedom. Using Table A.2, we find the probability p. We then compare p to α–the significance level of the test–to determine whether we should reject or not reject H_0.

Most frequently we consider the case in which $\beta_0 = 0$. If the population slope is equal to 0, then there is no linear relationship between x and y; in this sense, x is of no value in predicting or estimating y. Suppose that we wish to test

$$H_0 : \beta = 0$$

for the head circumference and gestational age data using a two-sided test conducted at the $\alpha = 0.05$ level. This test is mathematically equivalent to the test of

$$H_0 : \rho = 0$$

where ρ is the correlation between head circumference and gestational age in the underlying population of low birth weight infants. In fact, it can be shown algebraically that

$$b = r\left(\frac{s_y}{s_x}\right)$$

where s_x and s_y are the standard deviations of the x and y values respectively [2]. Both null hypotheses claim that head circumference does not change as gestational age increases.

To conduct the test, we calculate

$$\begin{aligned} \text{t} &= \frac{b - \beta_0}{\widehat{\text{se}}(b)} \\ &= \frac{0.7801 - 0}{0.0631} \\ &= 12.36. \end{aligned}$$

For a t distribution with $100 - 2 = 98$ degrees of freedom, $p < 0.001$. We reject the null hypothesis that the population slope is equal to 0. In the underlying population of low birth weight infants, there is a statistically significant linear relationship between head circumference and gestational age; head circumference increases in magnitude as gestational age increases.

In addition to conducting a test of hypothesis, we are also able to calculate confidence limits for the population slope β. For a t distribution with 98 degrees of freedom, approximately 95% of the observations fall between -1.98 and 1.98. Therefore,

$$(b - 1.98\,\widehat{\text{se}}(b)\,,\; b + 1.98\,\widehat{\text{se}}(b))$$

is a 95% confidence interval for β. Given the head circumference and gestational age data, we find the interval to be

$$(0.7801 - 1.98(0.0631)\,,\; 0.7801 + 1.98(0.0631))$$

or

$$(0.6564\,,\; 0.9038).$$

We are 95% confident that these limits cover the true population slope β.

If we are interested in testing whether the population intercept is equal to a specified value α_0, we use calculations that are analogous to those for the slope. We compute the statistic

$$\text{t} = \frac{a - \alpha_0}{\widehat{\text{se}}(a)}$$

and compare this value to the t distribution with $n - 2$ degrees of freedom. We are also able to construct a confidence interval for the true population intercept α just as we calculated an interval for the slope β. However, if the observed data points tend to be far from the intercept–as they are for the head circumference and gestational age data–there is very little practical value in making inference about the intercept. In fact, it is dangerous to extrapolate the fitted line beyond the range of the observed values x; the relationship between X and Y might be quite different outside this range.

18.2.4 Inference for Predicted Values

In addition to inference regarding the population slope and intercept, we might also be interested in putting confidence limits around the mean value of y for a given x. If we have a sample of 100 observations, the confidence interval would be of the form

$$(\hat{y} - 1.98\,\widehat{\text{se}}(\hat{y})\,,\ \hat{y} + 1.98\,\widehat{\text{se}}(\hat{y}))$$

where $\hat{y}$ is the predicted mean of the normally distributed outcomes, and the standard error of $\hat{y}$ is estimated by

$$\widehat{\text{se}}(\hat{y}) = \sqrt{s_{y|x}^2 \left[\frac{1}{n} + \frac{(x - \bar{x})^2}{\sum_{i=1}^{n} (x_i - \bar{x})^2} \right]}.$$

Note the term $(x - \bar{x})^2$ in the preceding expression. This quantity assumes the value 0 when x is equal to $\bar{x}$ and grows larger as x moves farther and farther away. As a result, if x is near $\bar{x}$, the confidence interval is relatively narrow. It grows wider as x moves away from $\bar{x}$. In a sense, we are most confident about the mean value of y when we are close to the mean value of the explanatory variable.

Return once again to the head circumference and gestational age data. When $x = 29$ weeks, for instance,

$$\begin{aligned} \hat{y} &= a + bx \\ &= 3.9143 + (0.7801)(29) \\ &= 26.54 \text{ cm.} \end{aligned}$$

The estimated standard error of $\hat{y}$ is

$$\widehat{\text{se}}(\hat{y}) = 0.159 \text{ cm.}$$

Therefore, a 95% confidence interval for the mean value of y is

$$(26.54 - 1.98(0.159)\,,\ 26.54 + 1.98(0.159))$$

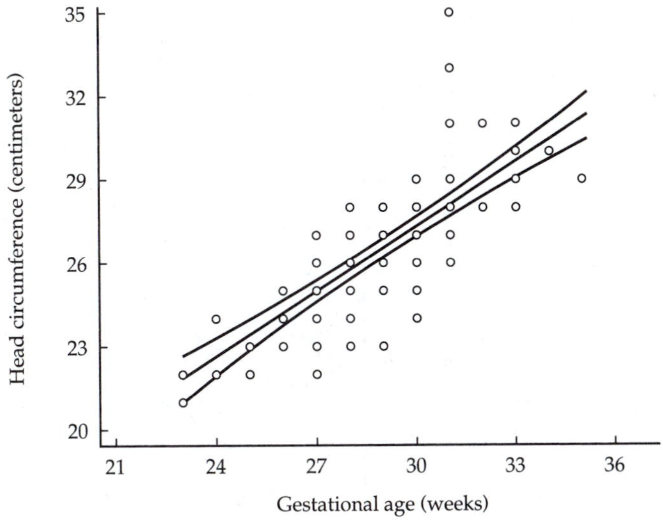

Figure 18.7. The 95% confidence limits on the predicted mean of y for a given value of x

or

$$(26.23\,,\, 26.85).$$

The curved lines in Figure 18.7 represent the 95% confidence limits on the mean value of y for each observed x. As we move away from $x = 29$ weeks, which is very close to $\bar{x}$, the confidence limits gradually become wider.

Sometimes, instead of predicting the mean value of y for a given x, we would prefer to predict an individual y for a new member of the population. This individual outcome y is denoted by $\tilde{y}$, or y-tilde. The predicted individual value $\tilde{y}$ is identical to the predicted mean $\hat{y}$; in particular,

$$\tilde{y} \quad = \quad a + bx.$$

The standard error of $\tilde{y}$, however, is not the same. When computing $\widehat{\text{se}}(\hat{y})$, we were interested only in the variability of the mean of the y values. When considering an individual y, we have an extra source of variability to account for: the dispersion of the y values about their mean $\mu_{y|x}$. Recall that for a given value of x, the outcomes y are normally distributed with standard deviation $\sigma_{y|x}$. Therefore, we would expect the expression for the standard error of $\tilde{y}$ to incorporate an extra term involving $\sigma_{y|x}$—or $s_{y|x}$—which is not included in the expression for the standard error of $\hat{y}$. In fact,

$$\begin{aligned} \widehat{\text{se}}(\tilde{y}) &= \sqrt{s_{y|x}^2 + \widehat{\text{se}}(\hat{y})^2} \\ &= \sqrt{s_{y|x}^2 \left[1 + \frac{1}{n} + \frac{(x - \bar{x})^2}{\sum_{i=1}^{n}(x_i - \bar{x})^2}\right]}. \end{aligned}$$

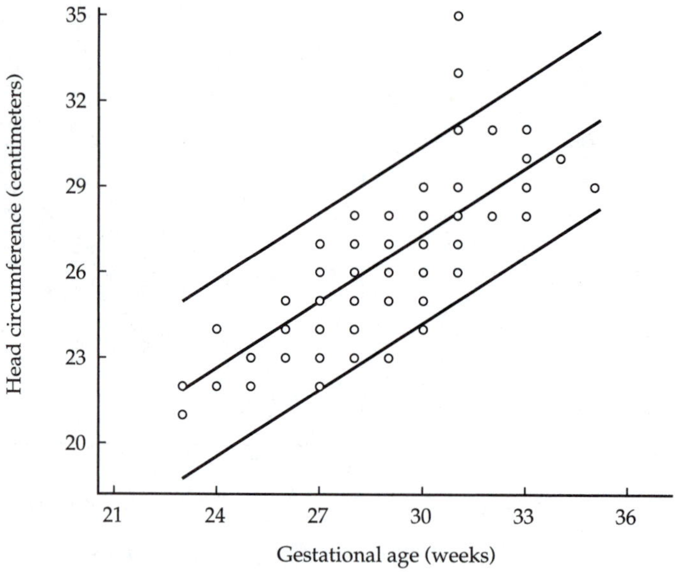

Figure 18.8. The 95% confidence limits on an individual predicted y for a given value of x

If we have a sample of 100 observations, a 95% confidence interval for an individual outcome y takes the form

$$(\tilde{y} - 1.98\,\widehat{\text{se}}(\tilde{y})\,,\ \tilde{y} + 1.98\,\widehat{\text{se}}(\tilde{y})).$$

Because of the extra source of variability, the limits on a predicted **individual** value of y are wider than the limits on a predicted **mean** y.

Suppose that a new child is selected from the underlying population of low birth weight infants. If this newborn has a gestational age of 29 weeks, then

$$\begin{aligned}\tilde{y} &= a + bx \\ &= 3.9143 + (0.7801)(29) \\ &= 26.54 \text{ cm.}\end{aligned}$$

The standard error of $\tilde{y}$ is estimated as

$$\begin{aligned}\widehat{\text{se}}(\tilde{y}) &= \sqrt{s_{y|x}^2 + \widehat{\text{se}}(\hat{y})^2} \\ &= \sqrt{(1.5904)^2 + (0.159)^2} \\ &= 1.598 \text{ cm.}\end{aligned}$$

Therefore, a 95% confidence interval for an individual new value of head circumference is

$$(26.54 - 1.98(1.598)\,,\ 26.54 + 1.98(1.598))$$

or

$$(23.38\,,\,29.70).$$

The curved lines in Figure 18.8 are the 95% confidence limits around the individual value of y for each observed x. Note that these bands are farther apart than the 95% confidence limits around the mean value of y.

18.3 Evaluation of the Model

18.3.1 The Coefficient of Determination

After constructing a least squares regression line

$$\hat{y} = a + bx,$$

we might wonder how well this model actually fits the observed data. One way to get a sense of the fit is to compute the coefficient of determination. The *coefficient of determination* is represented by R^2 and is the square of the Pearson correlation coefficient r; consequently,

$$r^2 = R^2.$$

Since r may assume any value in the range -1 to 1, R^2 must lie between 0 and 1. The coefficient of determination can be interpreted as the proportion of the variability among the observed values of y that is explained by the linear regression of y on x.

For the regression of head circumference on gestational age, the coefficient of determination is

$$R^2 = 0.6095.$$

This implies that 60.95% of the variation among the observed values of head circumference is explained by its linear relationship with gestational age. The remaining

$$100 - 60.95 = 39.05\%$$

of the variation is unexplained.

18.3.2 Residual Plots

Another strategy for evaluating how well the least squares line fits the observed data is to plot the residuals against the fitted or predicted values of the response variable. For example, the first child in the sample of 100 low birth weight infants has a gestational age of 29 weeks and a head circumference of 27 centimeters. The child's predicted head circumference, given that $x_1 = 29$ weeks, is $\hat{y}_1 = 26.536$ centimeters. The residual for this first observation is

$$\begin{aligned} e_1 &= y_1 - \hat{y}_1 \\ &= 27 - 26.536 \\ &= 0.464; \end{aligned}$$

therefore, the point (26.536, 0.464) would be included in the graph. Figure 18.9 is a plot of the points $(\hat{y}_i, e_i)$ for all 100 observations in the sample.

A plot of the residuals serves three purposes. First, it can help us to detect outlying observations in the sample. In Figure 18.9, two residuals are somewhat larger than the others; the bigger of the two is associated with a child whose gestational age is 31 weeks and whose head circumference is 35 centimeters. We would predict the infant's head circumference to be only

$$\begin{aligned} \hat{y} &= 3.914 + 0.7801(31) \\ &= 28.10 \text{ cm}. \end{aligned}$$

The method of least squares is very sensitive to such outliers in the data. When it is believed that an outlier is the result of an error in measuring or recording a particular observation, removal of this point improves the fit of the regression line. However, care must be taken not to throw away unusual data points that are in fact valid; these might be the most interesting observations in the data set.

A plot of the residuals might also suggest a failure in the assumption of homoscedasticity. If the range of the magnitude of the residuals either increases or decreases as $\hat{y}$ becomes larger—producing a fan-shaped scatter such as the

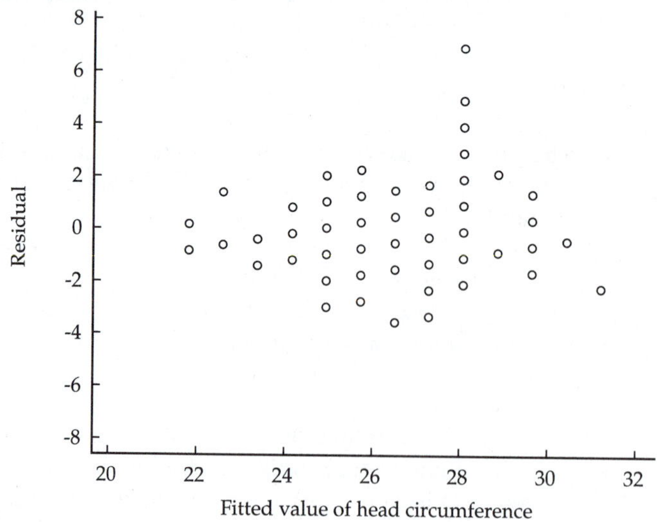

Figure 18.9. Residuals versus fitted values of head circumference

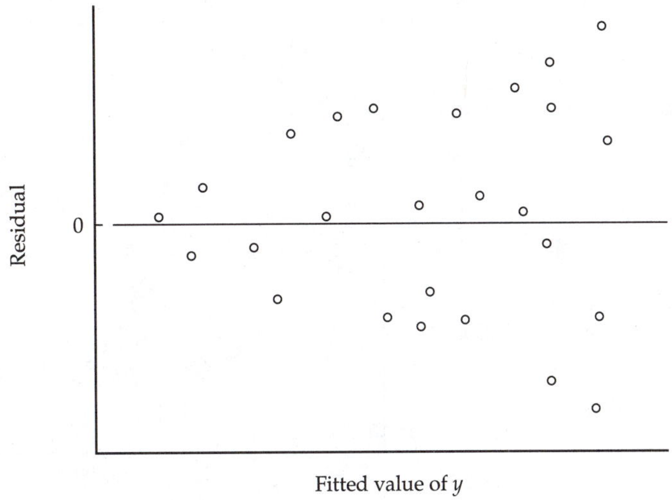

Figure 18.10. Violation of the assumption of homoscedasticity

one in Figure 18.10—then $\sigma_{y|x}$ is not constant across the values of x. In this situation, simple linear regression is not the appropriate technique for modeling the relationship between x and y. No such pattern is evident in Figure 18.9, the residual plot for the head circumference and gestational age data. Thus, the assumption of homoscedasticity does not appear to have been violated.

Finally, if the residuals do not exhibit a random scatter but instead follow a distinct trend, this suggests that the true relationship between x and y might not be linear. Under these circumstances, a *transformation* of either x or y or both could be in order. When transforming a variable, we simply measure it on a different scale. In many situations, a curvilinear relationship can be transformed to a more straightforward linear one; if this is the case, then we may use simple linear regression to fit a model to the transformed data.

18.3.3 Transformations

Consider Figure 18.11. This graph is a scatter plot of crude birth rate per 1,000 population versus gross national product per capita for 127 countries [3]. Note that birth rate tends to decrease as gross national product increases. This relationship, however, is not a linear one. Instead, birth rate drops off rapidly at first; when the GNP per capita reaches approximately \$5,000, it begins to level off. Consequently, if we wish to describe the relationship between birth rate and gross national product using linear regression analysis, we see that some type of transformation is necessary.

When the relationship between x and y is not linear, we begin by looking at transformations of the form x^p or y^p where

$$p = \ldots -3, -2, -1, -\frac{1}{2}, \ln, \frac{1}{2}, 1, 2, 3, \ldots .$$

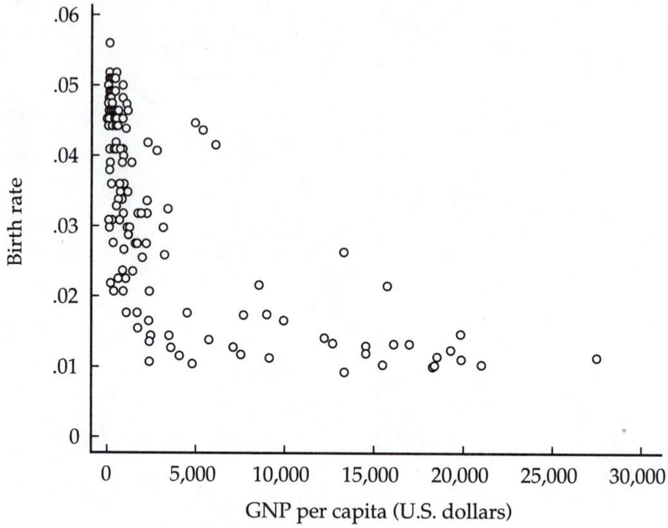

Figure 18.11. Birth rate per 1,000 population versus gross national product per capita for 127 countries, 1988–1989

Note that "ln" refers to the natural logarithm of x or y rather than an exponent. In general, we prefer to transform x whenever possible.

The *circle of powers*—or the *ladder of powers*, as it is sometimes called—provides a general guideline for choosing a transformation. The strategy is illustrated in Figure 18.12. If the plotted data resemble the pattern in Quadrant I, for instance, then an appropriate transformation would involve raising either x or y to a power

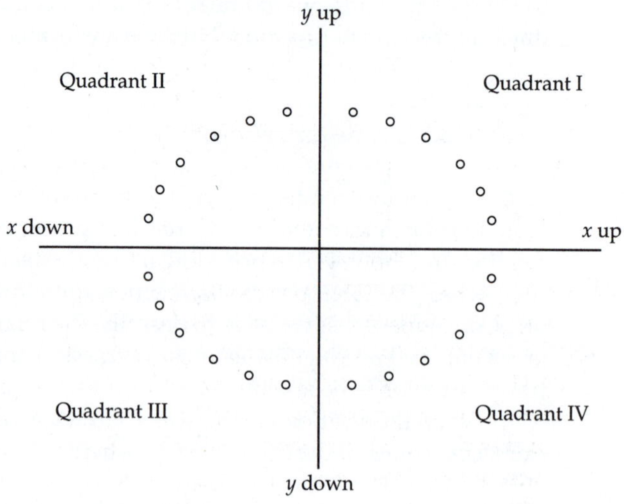

Figure 18.12. The circle of powers

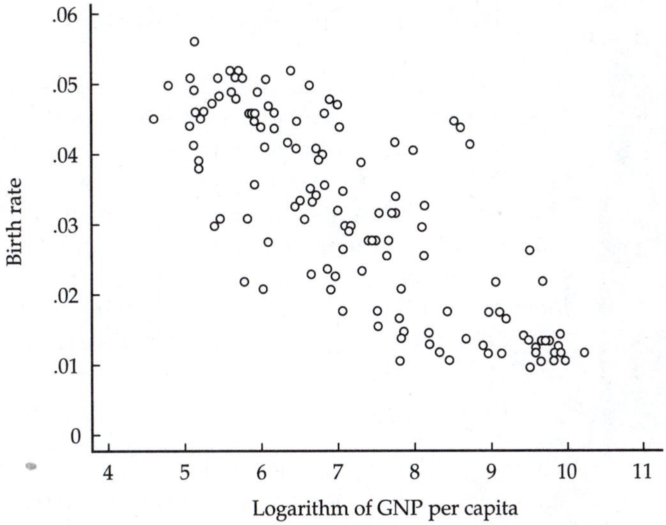

Figure 18.13. Birth rate per 1,000 population versus the natural logarithm of gross national product per capita for 127 countries, 1988–1989

greater than $p = 1$; we might try replacing x by x^2, for example. In this case, we would fit a model of the form

$$\hat{y} = a + bx^2$$

instead of the usual

$$\hat{y} = a + bx.$$

If the data follow the trend in Quadrant II, we would want either to raise x to a power that is lower than 1 or to raise y to a power that is greater than 1. Therefore, we might try replacing x by $\sqrt{x}$ or $\ln(x)$. Whichever transformation is chosen, we must always verify that the assumption of homoscedasticity remains valid.

The data in Figure 18.11 most closely resemble the pattern in Quadrant III; therefore, we would want to raise either x or y to a power that is less than 1. We might try replacing gross national product by its natural logarithm, for instance. This transformation is illustrated in Figure 18.13; note that the relationship between birth rate and the logarithm of GNP appears much more linear than the relationship between birth rate and GNP itself.

18.4 Further Applications

Suppose that we are interested in the relationship between length and gestational age for the population of low birth weight infants. In particular, we want to know what happens to length as gestational age increases.

We begin our analysis by constructing a two-way scatter plot of these variables for the 100 infants in the previously mentioned sample. The plot is displayed in Figure 18.14. The points in the graph exhibit a great deal of scatter; however, there is a distinct tendency for length to increase in magnitude as gestational age increases. It is possible that this relationship is a linear one.

To estimate the true population regression line

$$\mu_{y|x} = \alpha + \beta x$$

where $\mu_{y|x}$ is the mean length for low birth weight infants of a specified gestational age x, we use the method of least squares to fit the model

$$\hat{y} = a + bx.$$

Rather than attempt to fit this equation by hand, we rely on the computer to do the calculations for us. Suppose that the 100 measurements of length are saved under the variable name `length`, and the measurements of gestational age are saved under `gestage`. Table 18.1 shows the relevant output from Stata.

The top portion of Table 18.1 contains an ANOVA table and other descriptive measures. The bottom part displays information relating to the estimated coefficients of the least squares equation. The column on the left lists the names of the

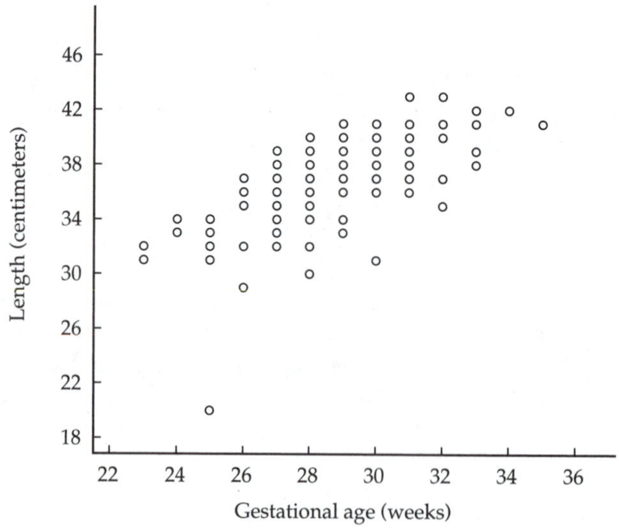

Figure 18.14. Length versus gestational age for a sample of 100 low birth weight infants

Table 18.1 Stata output displaying the simple linear regression of length on gestational age

Source	SS	df	MS
Model	575.73916	1	575.73916
Residual	687.02084	98	7.01041674
Total	1262.76	99	12.7551515

Number of obs	=	100
F(1,98)	=	82.13
Prob > F	=	0.0000
R-square	=	0.4559
Adj R-square	=	0.4504
Root MSE	=	2.6477

Variable	Coefficient	Std. Error	t	P>|t|	[95% Conf.	Interval]
length						
gestage	0.9516035	0.1050062	9.062	0.000	.7432221	1.159985
_cons	9.3281740	3.0451630	3.063	0.003	3.285149	15.3712

response and the explanatory variables; the label "_cons" refers to the y-intercept or constant term in the equation. The estimated values a and b appear in the second column. Rounding these values to four decimal places, we see that the least squares regression line for this sample of data is

$$\hat{y} = 9.3282 + 0.9516x.$$

The y-intercept of 9.3282 is the predicted value of length that corresponds to an x value of 0; in this example, however, a gestational age of 0 weeks does not make sense. The slope of the line indicates that for each one-week increase in gestational age, an infant's length increases by 0.9516 centimeters on average.

The third column in the bottom section of Table 18.1 displays the estimated standard errors of a and b. Suppose that we wish to test the null hypothesis that the population slope is equal to 0, or

$$H_0 : \beta = 0.$$

The appropriate test statistic is

$$\begin{aligned} t &= \frac{b - \beta_0}{\widehat{\text{se}}(b)} \\ &= \frac{0.9516 - 0}{0.1050} \\ &= 9.062; \end{aligned}$$

this statistic is provided in the fourth column of the table. The adjacent column contains the p-value that corresponds to a two-sided test. Since p is less than 0.001, we reject the null hypothesis that the population slope is equal to 0. Based on this sample of low birth weight infants, length increases in magnitude as gestational age increases.

Since the data set contains a total of 100 observations, a 95% confidence interval for the population slope β takes the form

$$(b - 1.98\,\widehat{\text{se}}(b)\,,\; b + 1.98\,\widehat{\text{se}}(b)).$$

If we substitute the values of b and $\widehat{\text{se}}(b)$, the interval is

$$(0.7432\,,\; 1.1600).$$

These confidence limits are provided in the sixth and seventh columns of Table 18.1.

We might also be interested in putting confidence limits around the predicted mean y for a specified value of x. For each of the 100 infants in the sample, we can use the computer to obtain the predicted value of length as well as its standard error. The data for the first ten infants are shown in Table 18.2. When x is equal to 29 weeks, we observe that

$$\begin{aligned} \hat{y} &= 9.3282 + 0.9516(29) \\ &= 36.93 \text{ cm.} \end{aligned}$$

The estimated standard error of $\hat{y}$ is 0.265. Therefore, a 95% confidence interval for the mean value of length is

$$(36.93 - 1.98(0.265)\,,\; 36.93 + 1.98(0.265))$$

or

$$(36.41\,,\; 37.45).$$

Table 18.2 Predicted values of length and standard errors for the first ten infants in the sample

Gestational Age	Fitted Length $\hat{y}$	Standard Error of $\hat{y}$
29	36.92467	0.2650237
31	38.82788	0.3452454
33	40.73109	0.5063217
31	38.82788	0.3452454
30	37.87628	0.2892917
25	33.11826	0.4867806
27	35.02147	0.3308946
29	36.92467	0.2650237
28	35.97307	0.2807812
29	36.92467	0.2650237

The predicted value of y for an individual member of the population is identical to the predicted mean of y; for an infant whose gestational age is 29 weeks,

$$\begin{aligned} \tilde{y} &= 9.3282 + 0.9516(29) \\ &= 36.93 \text{ cm.} \end{aligned}$$

Recall that $s_{y|x}$ is the standard deviation from regression; its value is contained in the output in Table 18.1 under the heading Root MSE. Therefore,

$$s_{y|x} = 2.6477,$$

and the estimated standard error of $\tilde{y}$ is

$$\begin{aligned} \widehat{\text{se}}(\tilde{y}) &= \sqrt{s_{y|x}^2 + \widehat{\text{se}}(\hat{y})^2} \\ &= \sqrt{(2.6477)^2 + (0.265)^2} \\ &= 2.661. \end{aligned}$$

A 95% confidence interval for the individual value of y is

$$(36.93 - 1.98(2.661)\,,\ 36.93 + 1.98(2.661))$$

or

$$(31.66\,,\ 42.20).$$

This interval is quite a bit wider than the corresponding 95% confidence interval for the predicted mean value of y.

After obtaining the least squares regression line, we might also wish to have some idea of how well this model fits the observed data. One way in which the fit can be evaluated is by examining the coefficient of determination. In Table 18.1, the value of R^2 is displayed in the top portion of the output in the column on the right. For the regression of length on gestational age, $R^2 = 0.4559$; this indicates that approximately 45.59% of the variability in length is explained by its linear relationship with gestational age. The remaining 54.41% is left unexplained. The line of output directly below R^2, labeled adjusted R^2 (Adj R-square), will be discussed in the following chapter.

A second technique for evaluating the fit of the least squares regression line to the observed data involves looking at a plot of the residuals versus the predicted values of length. The residuals are obtained by subtracting the fitted values $\hat{y}_i$ from the actual observations y_i; the calculations may be performed using a computer package. Table 18.3 shows the relevant data for the first ten infants in the sample. Figure 18.15 is a two-way scatter plot of the points $(\hat{y}_i\,,\ e_i)$ for all 100 observations in the sample.

Table 18.3 Residuals for the first ten infants in the sample

Length	Fitted Length $\hat{y}$	Residual
41	36.92467	4.075324
40	38.82788	1.172117
38	40.73109	−2.731091
38	38.82788	−0.827883
38	37.87628	0.123720
32	33.11826	−1.118262
33	35.02147	−2.021469
38	36.92467	1.075324
30	35.97307	−5.973073
34	36.92467	−2.924676

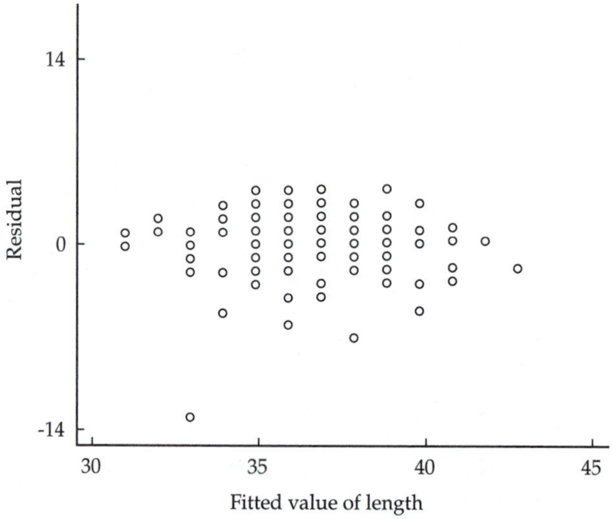

Figure 18.15. Residuals versus fitted values of length for a sample of 100 low birth weight infants

Looking at the plot, we see that there is one point that appears to be an outlier. We might try removing this point, fitting a new line, and then comparing the two models. However, there is no evidence that the assumption of homoscedasticity has been violated or that a transformation of variables is necessary.

18.5 Review Exercises

1. The file `lowbwt` contains data for the sample of 100 low birth weight infants discussed in the text. Measurements of systolic blood pressure are saved

under the variable name `sbp`; the corresponding gestational ages are saved under `gestage`.

a. Construct a two-way scatter plot of systolic blood pressure versus gestational age.
b. Does the graph suggest anything about the relationship between these variables?
c. Using systolic blood pressure as the response variable, compute the least squares regression line. Interpret the slope and the y-intercept of the line.
d. At the 0.05 level of significance, test the null hypothesis that the population slope is equal to 0.
e. What is the estimated mean systolic blood pressure for the population of infants whose gestational age is 31 weeks?
f. Construct a 95% confidence interval for the true mean value of systolic blood pressure.
g. Suppose that you randomly select a new infant from the population and find that his or her gestational age is 31 weeks. What is the predicted systolic blood pressure for this child?
h. Construct a 95% confidence interval for this new value of systolic blood pressure.
i. Does the regression model seem to fit the observed data? Comment on the coefficient of determination and a plot of the residuals versus the fitted values of systolic blood pressure.

2. The measurements of length and weight for a sample of 20 low birth weight infants are contained in the data set `twenty`. The length measurements are saved under the variable name `length`, and the corresponding birth weights are saved under `weight`.

a. Construct a graph of weight versus length for the 20 infants in the sample.
b. Without doing any calculations, sketch your best guess for the least squares regression line directly on the scatter plot.
c. Compute the least squares regression line.
d. Draw this line on the scatter plot. Does the actual least squares line concur with your guess?

From the two-way plot, it is clear that one point lies outside the range of the rest of the data. This point corresponds to the ninth infant in the sample. To illustrate the effect that the outlier has on the model, remove this point from the data set.

e. Compute the new least squares regression line.
f. Sketch this line on the original scatter plot. How did the line change? In particular, comment on the slope and the intercept.
g. Compare the coefficients of determination (R^2) and the standard deviations from regression ($s_{y|x}$) for the two lines. Explain how and why these values changed when you removed the outlier from the original data set.

3. In the 11 years before the passage of the Federal Coal Mine Health and Safety Act of 1969, the fatality rates for underground miners varied little. After

the implementation of the Act, however, fatality rates decreased steadily until 1979. The fatality rates for the years 1970 to 1981 are provided below [4]; for computational purposes, calendar years have been converted to a scale beginning at 1. The data are in the file `miner`. The response variable, fatality rate, is saved under the name `rate`; the explanatory variable, converted calendar year, is saved under `year`.

Calendar Year	Year	Fatality Rate per 100 Employees
1970	1	0.2419
1971	2	0.1732
1972	3	0.1361
1973	4	0.1108
1974	5	0.0996
1975	6	0.0952
1976	7	0.0904
1977	8	0.0792
1978	9	0.0701
1979	10	0.0890
1980	11	0.0799
1981	12	0.1084

a. Construct a two-way scatter plot of fatality rate versus year.

b. To model the trend in fatality rates, fit the least squares regression line

$$\hat{y} = a + bx$$

where x represents year. Using both the coefficient of determination and a plot of the residuals, comment on the fit of the model.

c. Transform x to $\ln(x)$. Create a scatter plot of fatality rate versus the natural logarithm of year.

d. Fit the model

$$\hat{y} = a + b\ \ln(x).$$

Again using the coefficient of determination and a plot of the residuals, compare the fit of this model to the model in part (b).

e. Transform x to $1/x$. Construct a two-way scatter plot of fatality rate versus the reciprocal of year.

f. Fit the model

$$\hat{y} = a + b\left(\frac{1}{x}\right).$$

Comment on the fit of this model and compare it to the previous models.

g. Which of the three models appears to best fit the data? Defend your selection.

4. Statistics that summarize personal health care expenditures by state for the years 1966 to 1982 have been examined in an attempt to understand issues

relating to rising health care costs. Suppose that you are interested in focusing on the relationship between expense per admission into a community hospital and average length of stay. The file `hospital` contains data for each state in the United States (including the District of Columbia) for the year 1982 [6]. The measures of average expense per admission are saved under the variable name `expadm`; the corresponding lengths of stay are saved under `los`.

a. Summarize the variables expense per admission and length of stay both graphically and numerically.
b. Construct a scatter plot of expense per admission versus length of stay. What does the graph suggest about the relationship between these variables?
c. Using `expadm` as the response variable, compute the least squares regression line. Interpret the estimated slope.
d. Is there a significant linear relationship between expense per admission and length of stay?
e. Evaluate the fit of the model to the data. In particular, discuss the coefficient of determination and a plot of the residuals.

5. Figure 18.16 displays a two-way scatter plot of CBVR–the response of cerebral blood volume to changes in arterial carbon dioxide tension–versus gestational age for a sample of 17 newborn infants [5]. The graph also contains the fitted regression line. The investigators found the coefficient of gestational age to be significantly larger than 0.

a. Suppose that you are interested only in infants who are born prematurely. If you were to eliminate the four data points corresponding to infants

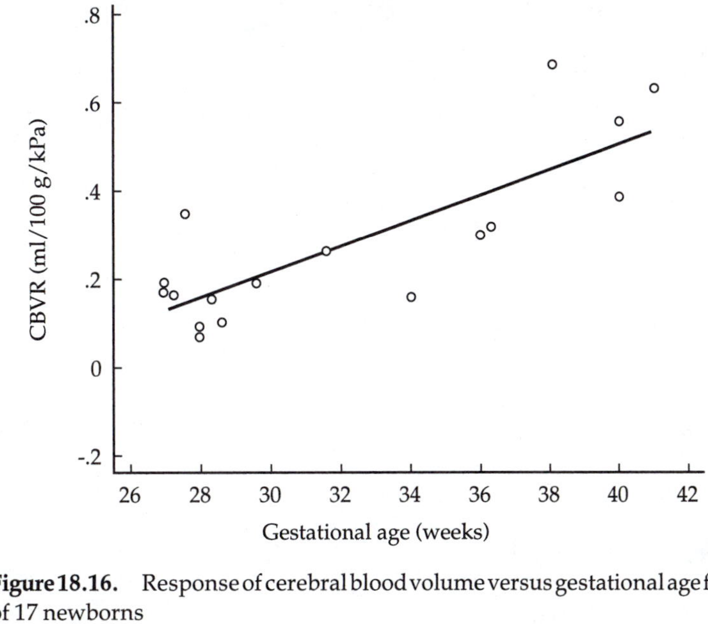

Figure 18.16. Response of cerebral blood volume versus gestational age for a sample of 17 newborns

whose gestational age is 38 weeks or greater, would you still believe that CBVR increases significantly as gestational age increases?

b. In an earlier study, the same investigators found no obvious relationship to exist between CBVR in newborn infants and gestational age. Would this information cause you to modify your answer above?

References

[1] Leviton, A., Fenton, T., Kuban, K. C. K., and Pagano, M., "Labor and Delivery Characteristics and the Risk of Germinal Matrix Hemorrhage in Low Birth Weight Infants," *Journal of Child Neurology*, Volume 6, October 1991, 35–40.

[2] Kleinbaum, D. G., Kupper, L. L., and Muller, K. E., *Applied Regression Analysis and Other Multivariable Methods*, Boston: PWS-Kent, 1988.

[3] United Nations Children's Fund, *The State of the World's Children 1991*, New York: Oxford University Press.

[4] Weeks, J. L., and Fox, M., "Fatality Rates and Regulatory Policies in Bituminous Coal Mining, United States, 1959–1981," *American Journal of Public Health*, Volume 73, November 1983, 1278–1280.

[5] Wyatt, J. S., Edwards, A. D., Cope, M., Delpy, D. T., McCormick, D. C., Potter, A., and Reynolds, E. O. R., "Response of Cerebral Blood Volume to Changes in Arterial Carbon Dioxide Tension in Preterm and Term Infants," *Pediatric Research*, Volume 29, June 1991, 553–557.

[6] Levit, K. R., "Personal Health Care Expenditures, by State: 1966–1982," *Health Care Financing Review*, Volume 6, Summer 1985, 1–25.

CHAPTER

19

Multiple Regression

In the previous chapter, we saw how simple linear regression can be used to explore the nature of the relationship between two continuous random variables. If knowing the value of a single explanatory variable improves our ability to predict the outcome of the response, however, then we might suspect that additional variables could be used to our advantage. To investigate the more complicated relationship among a number of different variables, we use a natural extension of simple linear regression known as *multiple regression* analysis.

19.1 The Model

When studying multiple regression, we consider models of the form

$$y = \alpha + \beta_1 x_1 + \beta_2 x_2 + \cdots + \beta_q x_q + \epsilon$$

where x_1, $x_2, \ldots,$ and x_q are the outcomes of q distinct explanatory variables. As in the simple case, ϵ is the random error associated with y. The population intercept α is the mean value of the response when all the explanatory variables take the value 0; the population slope β_i is the estimated change in y that corresponds to a one-unit increase in x_i given that all other explanatory variables remain constant.

Just as we had to make a number of assumptions before fitting a regression model involving a single explanatory variable, we must make a set of analogous assumptions before fitting a more complex multiple regression model. These assumptions are as follows:

1. For specified values of x_1, $x_2, \ldots,$ and x_q that are considered to be measured without error, the outcomes y are normally distributed with mean $\mu_{y|x_1,\ldots,x_q}$ and standard deviation $\sigma_{y|x_1,\ldots,x_q}$.

2. The relationship between $\mu_{y|x_1,\ldots,x_q}$ and $x_1, x_2, \ldots,$ and x_q is represented by the equation

$$\mu_{y|x_1,\ldots,x_q} = \alpha + \beta_1 x_1 + \beta_2 x_2 + \cdots + \beta_q x_q.$$

3. For any specified set of values $x_1, x_2, \ldots$ and x_q, $\sigma_{y|x_1,\ldots,x_q}$ is constant.
4. The outcomes y are independent.

19.1.1 The Least Squares Regression Equation

To estimate the population regression equation

$$\mu_{y|x_1,\ldots,x_q} = \alpha + \beta_1 x_1 + \beta_2 x_2 + \cdots + \beta_q x_q,$$

we use the method of least squares to fit the model

$$\hat{y} = a + b_1 x_1 + b_2 x_2 + \cdots + b_q x_q.$$

This technique requires that we minimize the sum of the squared residuals, or, in this case,

$$\sum_{i=1}^{n}(y_i - \hat{y}_i)^2 = \sum_{i=1}^{n}(y_i - a - b_1 x_{1i} - b_2 x_{2i} - \cdots - b_q x_{qi})^2.$$

When a single explanatory variable was involved, the fitted model was simply a straight line. With two explanatory variables the model represents a plane in three-dimensional space; with three or more variables it becomes a hyperplane in higher dimensional space. Although the calculations are more complicated than they were for a single explanatory variable, they do not present a major problem as long as a computer is available.

In the previous chapter, we found a significant linear relationship to exist between the variables head circumference and gestational age for the population of low birth weight infants [1]; the fitted least squares regression line was

$$\hat{y} = 3.9143 + 0.7801x.$$

We might suspect that head circumference also depends on the birth weight of an infant. Consider Figure 19.1; the plotted data suggest that head circumference increases in magnitude as weight increases. Given that we have already accounted for gestational age, does birth weight further improve our ability to predict head circumference?

Suppose that we let x_1 represent gestational age and x_2 designate birth weight. If we fit a model of the form

$$\hat{y} = a + b_1 x_1 + b_2 x_2,$$

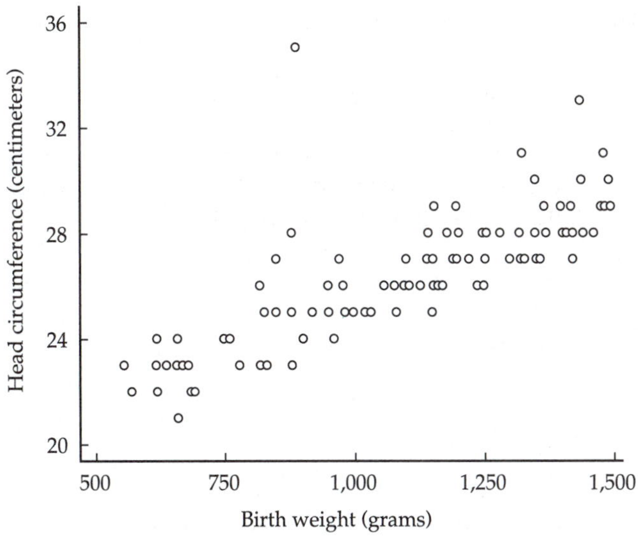

Figure 19.1. Head circumference versus weight for a sample of 100 low birth weight infants

then a is an estimate of the population intercept α, b_1 is an estimate of the slope β_1, and b_2 is an estimate of the slope β_2. For the sample of size 100, the least squares regression equation is

$$\hat{y} = 8.3080 + 0.4487x_1 + 0.0047x_2.$$

Note that the addition of birth weight causes the coefficient of gestational age to decrease. The new estimate is 0.4487; given two infants with the same weight, if the gestational age of the first child is one week greater than the gestational age of the second, then the first child has a head circumference that is 0.4487 centimeters larger on average. Similarly, the coefficient of weight indicates that if two infants have the same gestational age but one weighs one gram more than the other, then the heavier child would have a head circumference that is approximately 0.0047 centimeters larger.

19.1.2 Inference for Regression Coefficients

Since the regression coefficients a, b_1, and b_2 were estimated from a sample of data rather than the entire population, each has a corresponding standard error. For the 100 infants mentioned above,

$$\widehat{\text{se}}(a) = 1.5789,$$

$$\widehat{\text{se}}(b_1) = 0.0672,$$

and

$$\widehat{\text{se}}(b_2) = 0.0006.$$

Tests of hypotheses for the population parameters can be carried out just as they were for the model containing a single explanatory variable. There is one difference; under the null hypothesis

$$H_0 : \beta_i = \beta_{i0},$$

the test statistic

$$\text{t} = \frac{b_i - \beta_{i0}}{\widehat{\text{se}}(b_i)}$$

does not follow a t distribution with $n - 2$ degrees of freedom. Instead, it has a t distribution with $n - q - 1$ degrees of freedom where q is the number of explanatory variables in the model. For the model containing gestational age and birth weight, for instance, q is equal to 2, and the appropriate t distribution has $100 - 2 - 1 = 97$ degrees of freedom.

To test the null hypothesis that β_1 is equal to 0 using the sample of 100 infants, we calculate the statistic

$$\begin{aligned} \text{t} &= \frac{b_1 - \beta_{10}}{\widehat{\text{se}}(b_1)} \\ &= \frac{0.4487 - 0}{0.0672} \\ &= 6.68. \end{aligned}$$

For a t distribution with 97 degrees of freedom, $p < 0.001$; therefore, we reject the null hypothesis and conclude that β_1 is greater than 0. Conducting a similar test, we would find that β_2 is significantly larger than 0 as well. We must bear in mind, however, that multiple tests of hypotheses based on the same set of data are not independent; if each individual test is conducted at the α level of significance, then the overall probability of committing a type I error—or rejecting a null hypothesis that is true—is larger than α.

In addition to conducting tests of hypotheses, we are also able to calculate confidence limits for the population regression coefficients. Furthermore, we can construct confidence intervals for the predicted mean y and the predicted individual y given the values of the explanatory variables. In all cases, the procedures are analogous to those used when a single explanatory variable was involved.

19.1.3 Evaluation of the Model

Using descriptors such as the coefficient of determination and a plot of the residuals, we are able to assess how well a particular model fits the observed data. For example, the model that contains gestational age and weight explains 75.20% of the variation in the observed head circumference measurements; the model that contains gestational age alone explains about 60.95%. This increase in R^2 suggests that adding the explanatory variable weight to the model improves our ability to predict head circumference for the population of low birth weight infants.

The inclusion of an additional variable in a model can never cause the coefficient of determination to decrease. To get around this, we may use a second measure–called the *adjusted* R^2–that compensates for the added complexity of a model. Unlike R^2, the adjusted R^2 increases when the inclusion of a variable improves our ability to predict the response and decreases when it does not. Consequently, the adjusted R^2 allows us to make a more valid comparison among models that contain unequal numbers of explanatory variables.

19.1.4 Indicator Variables

All the explanatory variables that we have considered up to this point have been measured on a continuous scale. However, regression analysis can be generalized to incorporate discrete or nominal explanatory variables as well. For example, we might wonder whether an expectant mother's diagnosis of toxemia affects the head circumference of her child. The presence of toxemia is a dichotomous random variable; a woman either has it or she does not. We would like to be able to quantify the effect of toxemia on head circumference by comparing infants whose mothers suffered from this condition to infants whose mothers did not.

Since the explanatory variables in a regression analysis must assume numerical values, we designate the presence of toxemia by 1 and its absence by 0. These numbers do not represent any actual measurements; they simply identify the categories of a nominal random variable. Because its values do not have any quantitative meaning, an explanatory variable of this sort is called an *indicator variable*, or a *dummy variable*.

Suppose that we add the indicator variable toxemia to the equation that already contains gestational age. For the moment, we will ignore birth weight. The fitted least squares regression model is

$$\hat{y} = 1.4956 + 0.8740x_1 - 1.4123x_3$$

where x_1 represents gestational age and x_3 represents toxemia. The coefficient of toxemia is negative; a test of the null hypothesis

$$H_0: \beta_3 = 0$$

indicates that it is significantly different from 0. Given two infants with identical gestational ages, the measure of head circumference would tend to be smaller for the child whose mother had experienced toxemia than for the child whose mother had not.

From this single regression equation, we are able to obtain two different models that correspond to the two possible values of the dichotomous random variable toxemia. When $x_3 = 1$,

$$\begin{aligned}\hat{y} &= 1.4956 + 0.8740x_1 - 1.4123(1) \\ &= 0.0833 + 0.8740x_1.\end{aligned}$$

When $x_3 = 0$,

$$\begin{aligned}\hat{y} &= 1.4956 + 0.8740x_1 - 1.4123(0) \\ &= 1.4956 + 0.8740x_1.\end{aligned}$$

These two lines are plotted in Figure 19.2. The equations for the distinct groups of infants have identical slopes; in either group, therefore, a one-week increase in gestational age is associated with a 0.8740 centimeter increase in head circumference. Since one line lies entirely above the other, the models also suggest that, across all values of gestational age, children whose mothers did not have toxemia tend to have larger head circumference measurements than children whose mothers did have it.

19.1.5 Interaction Terms

In some situations, one explanatory variable has a different effect on the predicted response y depending on the value of a second explanatory variable. To model an effect of this type, we create what is known as an *interaction term*.

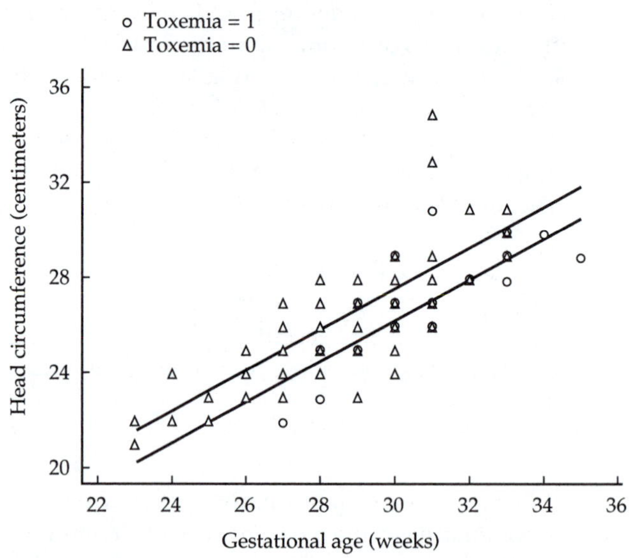

Figure 19.2. Fitted regression lines for different levels of toxemia

An interaction term is generated by multiplying together the outcomes of two variables x_i and x_j to create a new variable $x_i x_j$. An interaction that is the product of gestational age and toxemia, for example, would suggest that a one-week increase in gestational age has a different effect on head circumference depending on whether an infant's mother had experienced toxemia or not. If we add such a term to the equation that already contains gestational age and toxemia, the fitted least squares model is

$$\hat{y} = 1.7629 + 0.8646x_1 - 2.8150x_3 + 0.0462x_1x_3.$$

A test of hypothesis indicates that the coefficient of the interaction term is not significantly different from 0 at the 0.05 level; therefore, we are unable to reject the null hypothesis that gestational age has the same effect on head circumference regardless of a mother's toxemia status.

If the coefficient of the interaction term had achieved significance, we would again want to evaluate the models that correspond to the two possible values of the dichotomous random variable toxemia. When $x_3 = 1$, the least squares equation would reduce to

$$\begin{aligned} \hat{y} &= 1.7629 + 0.8646x_1 - 2.8150(1) + 0.0462x_1(1) \\ &= -1.0521 + 0.9108x_1. \end{aligned}$$

When $x_3 = 0$,

$$\begin{aligned} \hat{y} &= 1.7629 + 0.8646x_1 - 2.8150(0) + 0.0462x_1(0) \\ &= 1.7629 + 0.8646x_1. \end{aligned}$$

These two lines are plotted in Figure 19.3; note that they have different intercepts **and** different slopes. In the range of interest, however, one line still lies completely above the other. This implies that, across all relevant values of gestational age, infants whose mothers did not suffer from toxemia tend to have larger head circumference measurements than infants whose mothers did.

19.2 Model Selection

If a particular regression coefficient β_i is not significantly different from 0, then we are able to simplify the model under study by dropping the explanatory variable x_i. In general, if we are presented with a number of potential explanatory variables, how do we decide which ones to include in the model and which to leave out? This decision is usually made based on a combination of statistical and nonstatistical considerations. Initially, we should have some prior knowledge as to which variables might be relevant. To study the full effect of each of these explanatory variables, however, it would be necessary to perform a regression

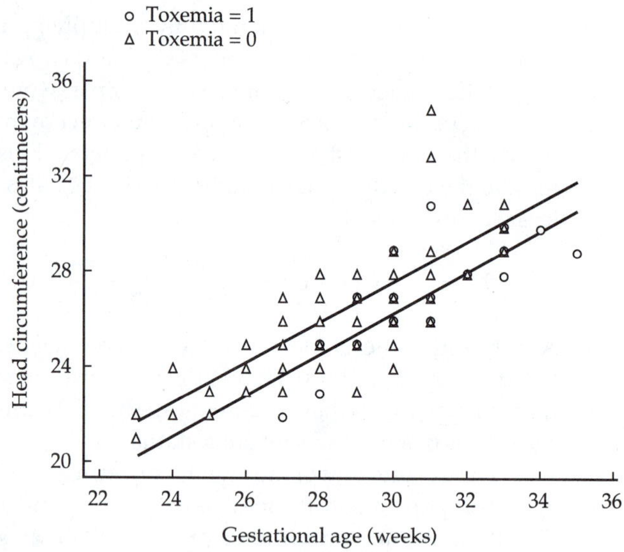

Figure 19.3. Fitted regression lines for different levels of toxemia, interaction term included

analysis for all possible combinations of the variables. The resulting models could then be evaluated according to some statistical criteria. This strategy for finding the best regression equation is known as the *all possible models* approach. It is the most thorough method, but it is also extremely time consuming. If we have a large number of potential explanatory variables, the procedure may not be feasible. As a result, we frequently resort to a stepwise approach to choose a regression model. There are two commonly used stepwise procedures: the first is known as forward selection and the second as backward elimination.

Forward selection proceeds by introducing variables into the model one at a time. The model is evaluated at each step, and the process continues until some specified statistical criterion is achieved. For example, we might begin by including the single explanatory variable that yields the largest coefficient of determination. This is the quantity that explains the greatest proportion of the variability in y. We next put into the model the variable that increases R^2 the most. The procedure continues until we reach a point where none of the remaining variables explains a significant amount of the additional variability in y.

Backward elimination begins by including all the explanatory variables in the model. Variables are dropped one at a time, and the equation is assessed at each step. The procedure is repeated until a satisfactory model is achieved. In this case, we might begin by dropping the variable that explains the smallest proportion of the variability in y, given the other variables in the model. We would continue this process until each of the remaining variables explains a significant portion of the variation. (Note that it is possible that we could end up with different final models, depending on which stepwise method is applied.)

Regardless of the strategy that we choose to fit a model, we should always check for the presence of collinearity. *Collinearity* occurs when two or more of the explanatory variables are correlated to the extent that they convey essentially the same information about the variation in y. One symptom of collinearity is the instability of the estimated coefficients and their standard errors. In particular, the standard error often becomes very large; this implies that there is a great deal of sampling variability in the estimated coefficient.

Refer back to the model that includes gestational age, toxemia, and the interaction between the two. Toxemia and the gestational age–toxemia interaction are highly correlated; in fact, their Pearson correlation coefficient is equal to 0.997. This model and the model that did not include the interaction term are contrasted below.

	Interaction Term Not Included	**Interaction Term Included**
Coefficient	−1.412	−2.815
Standard error	0.406	4.985
Test statistic	−3.477	−0.565
p-value	0.001	0.574
R^2	0.653	0.653
Adjusted R^2	0.646	0.642

When the interaction term is included, the estimated coefficient of toxemia doubles in magnitude. In addition, its standard error increases by a factor of 12. In the model without the interaction term, the coefficient of toxemia is significantly different from 0 at the 0.05 level; when the interaction term is present, it no longer achieves significance. Also, the coefficient of determination does not change; it remains 65.3%. Furthermore, the adjusted R^2 decreases slightly. These facts indicate that the inclusion of the interaction term in the model does not explain any additional variability in head circumference beyond that which is explained by gestational age and toxemia alone. The information supplied by this term is redundant.

19.3 Further Applications

In the Further Applications section in the previous chapter, we used the variable gestational age to explain a portion of the variability in the length measurements of a sample of low birth weight infants. We found that a significant linear relationship exists between these two variables; in particular, length tends to increase in magnitude as gestational age increases.

We now wish to determine whether the length of an infant also depends on the age of its mother. To begin the analysis, we create a two-way scatter plot of length versus mother's age for the 100 infants in the sample. The graph is displayed in Figure 19.4. There does not appear to be much of a relationship between these two variables.

To estimate the true population regression line of length on gestational age and mother's age,

$$\mu_{y|x_1,x_2} = \alpha + \beta_1 x_1 + \beta_2 x_2,$$

we fit a least squares model of the form

$$\hat{y} \quad = \quad a + b_1 x_1 + b_2 x_2.$$

Suppose that the 100 measurements of length are saved under the variable name `length`, the measurements of gestational age under the name `gestage`, and the values of mother's age under `momage`. Table 19.1 shows the relevant output from Stata. The bottom portion of the table contains an extra row for the additional

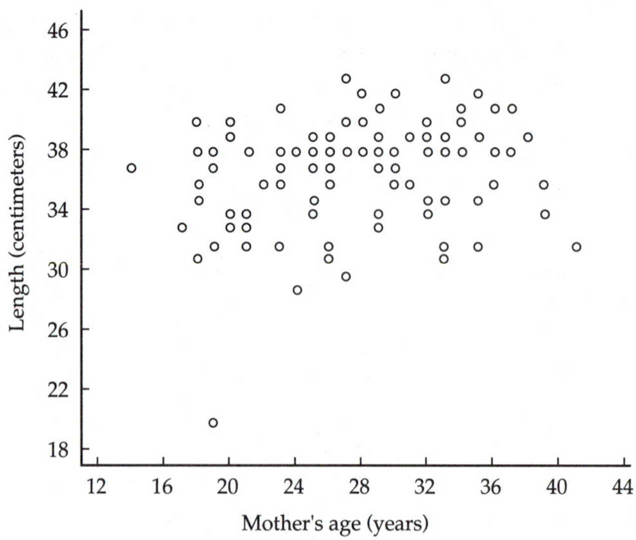

Figure 19.4. Length versus mother's age for a sample of 100 low birth weight infants

Table 19.1 Stata output displaying the regression of length on gestational age and mother's age

Source	SS	df	MS			
				Number of obs	=	100
				F(2,97)	=	40.91
Model	577.752198	2	288.876099	Prob > F	=	0.0000
Residual	685.007802	97	7.06193611	R-square	=	0.4575
Total	1262.76	99	12.7551515	Adj R-square	=	0.4463
				Root MSE	=	2.6574

length	Coef.	Std. Err.	t	P> |t|	[95% Conf.	Interval]
gestage	.9360867	.1093252	8.562	0.000	.7191064	1.153067
momage	.0247236	.0463071	0.534	0.595	−.0671832	.1166305
_cons	9.090871	3.088481	2.943	0.004	2.961091	15.22065

variable. If we round the estimated coefficients to four decimal places, the fitted least squares equation is

$$\hat{y} = 9.0909 + 0.9361x_1 + 0.0247x_2.$$

The coefficient of gestational age is 0.9361; if a child's mother's age remains constant while his or her gestational age increases by one week, then the infant's length would increase by 0.9361 centimeters on average. Similarly, the coefficient of mother's age suggests that, if gestational age were to remain constant and mother's age increase by one year, then a child's length would increase by approximately 0.0247 centimeters. Tests of the null hypotheses

$$H_0 : \beta_1 = 0$$

and

$$H_0 : \beta_2 = 0$$

indicate that the population parameter β_1 is significantly different from 0 at the 0.05 level, whereas β_2 is not.

Recall that gestational age alone explains about 45.59% of the variability in the observed values of length; gestational age and mother's age together explain 45.75%. The adjusted coefficient of determination–which appears directly below R^2 in Table 19.1–has actually decreased slightly. It is obvious that adding the explanatory variable mother's age to the model does not improve our ability to predict length for the infants in this population. Consequently, we drop this variable from the model.

We now wish to investigate whether an expectant mother's diagnosis of toxemia affects the length of her child. To do this, we add the indicator variable `tox` to the model that already contains `gestage`. If `tox` takes the value 1, then a

Table 19.2 Stata output displaying the regression of length on gestational age and toxemia

Source	SS	df	MS			
				Number of obs	=	100
Model	619.253622	2	309.626811	F(2,97)	=	46.67
Residual	643.506378	97	6.63408638	Prob > F	=	0.0000
Total	1262.76	99	12.7551515	R-square	=	0.4904
				Adj R-square	=	0.4799
				Root MSE	=	2.5757

length	Coef.	Std. Err.	t	P>|t|	[95% Conf.	Interval]
gestage	1.069883	.1121039	9.544	0.000	.8473879	1.292378
tox	−1.777381	.6939918	−2.561	0.012	−3.154763	−.3999997
_cons	6.284326	3.191824	1.969	0.052	−.050561	12.61921

mother suffered from toxemia; if it takes the value 0, she did not. As shown in Table 19.2, the least squares equation for this model is

$$\hat{y} = 6.2843 + 1.0699x_1 - 1.7774x_3$$

where x_1 represents gestational age, and x_3 represents toxemia. The coefficient of toxemia is negative and is significantly different from 0 at the 0.05 level; given two infants with identical gestational ages, therefore, the child whose mother had suffered from toxemia would tend to be shorter than the child whose mother had not.

From this single fitted model, we are able to obtain two different lines that correspond to the two possible values of the dichotomous random variable toxemia. When $x_3 = 1$,

$$\begin{aligned}\hat{y} &= 6.2843 + 1.0699x_1 - 1.7774(1)\\ &= 4.5069 + 1.0699x_1.\end{aligned}$$

When $x_3 = 0$,

$$\begin{aligned}\hat{y} &= 6.2843 + 1.0699x_1 - 1.7774(0)\\ &= 6.2843 + 1.0699x_1.\end{aligned}$$

The two lines are plotted in Figure 19.5. The lines have identical slopes; for either group, a one-week increase in gestational age corresponds to a 1.0699 centimeter increase in length, on average.

To determine whether an increase in gestational age has a different effect on length for infants whose mothers had experienced toxemia versus infants whose mothers had not, we could add the gestational age–toxemia interaction to the model. The interaction `gesttox` is obtained by multiplying the outcomes of the two random variables `gestage` and `tox`. The output corresponding to this model is presented in Table 19.3. The coefficient of the interaction term is not significantly different from 0 at the 0.05 level. In addition, the high correlation between `tox` and `gesttox`—the Pearson correlation coefficient is equal to 0.997—has introduced

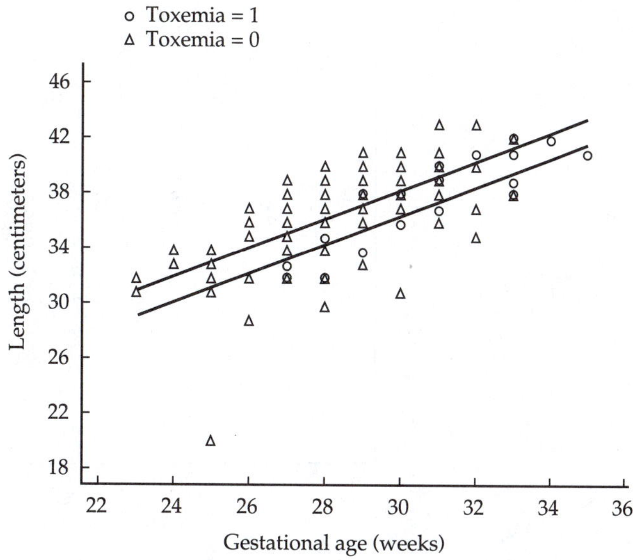

Figure 19.5. Fitted regression lines for different levels of toxemia

Table 19.3 Stata output displaying the linear regression of length on gestational age, toxemia, and their interaction

Source	SS	df	MS
Model	619.522097	3	206.507366
Residual	643.237903	96	6.70039483
Total	1262.76	99	12.7551515

Number of obs	=	100
F(3,96)	=	30.82
Prob > F	=	0.0000
R-square	=	0.4906
Adj R-square	=	0.4747
Root MSE	=	2.5885

length	Coef.	Std. Err.	t	P>|t|	[95% Conf.	Interval]
gestage	1.058458	.1262952	8.381	0.000	.8077647	1.309152
tox	−3.477085	8.5198381	−0.408	0.684	−20.38883	13.43466
gesttox	.0559409	.2794651	0.200	0.842	−4.4987929	.6106747
_cons	6.608269	3.592847	1.893	0.069	−.5234754	13.74001

collinearity into the model. Note the huge increase in the standard error of the estimated coefficient of toxemia.

If we drop the interaction term from the equation, the model we are left with is

$$\hat{y} = 6.2843 + 1.0699x_1 - 1.7774x_3$$

where x_1 represents gestational age, and x_3 represents toxemia. The two variables together explain 49.06 % of the observed variability in length; the remaining 50.94 % is unexplained.

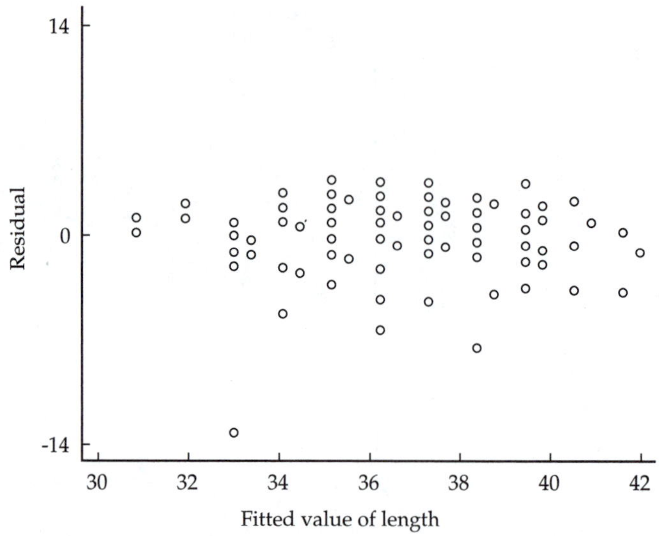

Figure 19.6. Residuals from the model containing gestational age and toxemia

A plot of the residuals from this model is displayed in Figure 19.6. There appears to be one outlier in the data set. We might consider dropping this observation, refitting the least squares equation, and comparing the two models. However, the assumption of homoscedasticity has not been violated, and a transformation of variables does not appear to be necessary.

19.4 Review Exercises

1. In a study designed to examine the effects of adding oats to the typical American diet, individuals were randomly divided into two different groups. Twice a day, the first group substituted oats for other foods containing carbohydrates; the members of the second group did not make any changes to their diet. One outcome of interest is the serum cholesterol level of each individual eight weeks after the onset of the study. Explanatory variables that might affect this response include diet group, serum cholesterol level at the start of the study, body mass index, and gender. The estimated coefficients and standard errors from the multiple regression model containing these four variables are displayed below [2].

Variable	Coefficient	Standard Error
Diet group	−11.25	4.33
Baseline cholesterol	0.85	0.07
Body mass index	0.23	0.65
Gender	−3.02	4.42

a. At the 0.05 level of significance, which of the four explanatory variables has an effect on serum cholesterol level eight weeks after the onset of the study?
b. If an individual's body mass index were to increase by 10 kg/m^2 while the values of all other explanatory variables remain constant, what would happen to his or her serum cholesterol level?
c. The indicator variable gender is coded so that 1 represents a male and 0 designates a female. Who is more likely to have a higher serum cholesterol level, a man or a woman?

2. For the population of low birth weight infants, a significant linear relationship was observed to exist between systolic blood pressure and gestational age. Recall that the relevant data are in the file `lowbwt`. The measurements of systolic blood pressure are saved under the variable name `sbp`, and the corresponding gestational ages are saved under `gestage`. Also included in the data set is `apgar5`, the five-minute apgar score for each infant. (The apgar score is an indicator of a child's general state of health five minutes after it is born; although it is actually an ordinal measurement, it is often treated as if it were continuous.)

a. Construct a scatter plot of systolic blood pressure versus the five-minute apgar score.
b. Does there appear to be any relationship between the two variables?
c. Fit the least squares model

$$\hat{y} = a + b_1 x_1 + b_2 x_2$$

where x_1 represents gestational age and x_2 represents five-minute apgar score. Interpret b_2, the estimated coefficient of `apgar5`.
d. What is the estimated mean systolic blood pressure for the population of infants whose gestational age is 31 weeks and whose five-minute apgar score is 7?
e. Construct a 95% confidence interval for the true mean value of systolic blood pressure.
f. Test the null hypothesis

$$H_0 : \beta_2 = 0$$

at the 0.05 level of significance.
g. Comment on the magnitude of R^2. Does the inclusion of the five-minute apgar score in the model improve your ability to predict systolic blood pressure?
h. Construct a plot of the residuals versus the fitted values of systolic blood pressure.

3. The data set `lowbwt` also contains `sex`, a dichotomous random variable designating the gender of each infant.

a. Add the indicator variable `sex`—where 1 represents a male and 0 a female—to the model that contains gestational age. Given two infants with identical gestational ages, one male and the other female, which would tend to have the higher systolic blood pressure?
b. Construct a two-way scatter plot of systolic blood pressure versus gestational age. On the graph, draw the two regression lines that correspond to the cases where `sex` = 1 and `sex` = 0.
c. Does there appear to be a different relationship between systolic blood pressure and gestational age for males as opposed to females?
d. Add to the model a new variable that is the interaction between gestational age and sex. Does gestational age have a different effect on systolic blood pressure depending on the gender of the infant?
e. Would you choose to include both sex and the gestational age–sex interaction in the model? Why or why not?

4. In the previous chapter, the relationship between expense per admission into a community hospital and average length of stay was examined for each state in the United States in 1982. The relevant data are in the file `hospital` [3]; average expense per admission is saved under the variable name `expadm` and length of stay under the name `los`. Also included in the data set is `salary`, the average salary per employee in 1982.

a. Summarize `salary` both graphically and numerically.
b. Construct a scatter plot of expense per admission versus salary. What does the graph suggest about the relationship between these two variables?
c. Fit the least squares model containing both length of stay and salary. Interpret the estimated regression coefficients.
d. What happens to the coefficient of `los` when `salary` is added to the model?
e. Does the inclusion of salary improve your ability to predict expense per admission? Explain.
f. Examine a plot of the residual versus the fitted values of expense per admission. Do any of the assumptions of the model appear to have been violated?

References

[1] Leviton, A., Fenton, T., Kuban, K. C. K., and Pagano, M., "Labor and Delivery Characteristics and the Risk of Germinal Matrix Hemorrhage in Low Birth Weight Infants," *Journal of Child Neurology*, Volume 6, October 1991, 35–40.

[2] Van Horn, L., Moag-Stahlberg, A., Liu, K., Ballew, C., Ruth, K., Hughes, R., and Stamler, J., "Effects on Serum Lipids of Adding Instant Oats to Usual

American Diets," *American Journal of Public Health*, Volume 81, February 1991, 183–188.

[3] Levit, K. R., "Personal Health Care Expenditures, by State: 1966–1982," *Health Care Financing Review*, Volume 6, Summer 1985, 1–25.

CHAPTER

20

Logistic Regression

When studying linear regression, we investigated models of the form

$$y = \alpha + \beta_1 x_1 + \beta_2 x_2 + \cdots + \beta_q x_q + \epsilon.$$

The response Y was continuous, and we were interested in identifying a set of explanatory variables that helped us to predict its mean value while explaining the observed variability of the outcomes.

In many situations, however, we are concerned with a response variable Y which is dichotomous rather than continuous. The outcome of Y is able to assume only two possible values; in general, the value 1 represents a "success," and 0 represents a "failure." The mean of the dichotomous random variable, designated p, is the proportion of times that it takes the value 1. Equivalently,

$$\begin{aligned} p &= \mathrm{P}(Y = 1) \\ &= \mathrm{P}(\text{"success"}). \end{aligned}$$

In this case, we would like to be able to estimate the probability p and determine the factors or explanatory variables that influence its value. To do this, we use a technique known as *logistic regression*.

20.1 The Model

Consider the population of low birth weight infants–in this case, those weighing less than 1,750 grams–who satisfy the following criteria: they are confined to a neonatal intensive care unit, they require intubation during the first 12 hours of life, and they survive for at least 28 days. In a sample of 223 such infants,

76 developed bronchopulmonary dysplasia (BPD), a chronic lung disease [1]. The remaining 147 did not. Let Y be a dichotomous random variable such that the outcome 1 represents the presence of BPD and 0 indicates its absence. Based on the sample, we would estimate the probability that a child in this population develops bronchopulmonary dysplasia by

$$\begin{aligned}\tilde{p} &= P(Y = 1) \\ &= \frac{76}{223} \\ &= 0.341.\end{aligned}$$

We might suspect that there are certain factors—both maternal and neonatal—that affect the likelihood that a particular infant will develop BPD. If we are able to classify a child according to these characteristics, then it is possible that we could estimate his or her probability of developing BPD with greater precision than that afforded by $\tilde{p}$ and subsequently take measures to decrease its value.

One factor of interest might be the birth weight of an infant. If the response variable Y was continuous, we would begin the analysis by constructing a two-way scatter plot of the data. A graph of bronchopulmonary dysplasia versus birth weight is displayed in Figure 20.1. Note that all the points lie on one of two parallel lines, depending on whether Y takes the value 1 or 0. There does appear to be a tendency for infants who do not develop BPD to have larger birth weights. However, the nature of this relationship is not clear.

To explore whether an association exists between susceptibility to bronchopulmonary dysplasia and birth weight, we begin by subdividing the population of infants into three categories: those weighing 950 grams or less, those

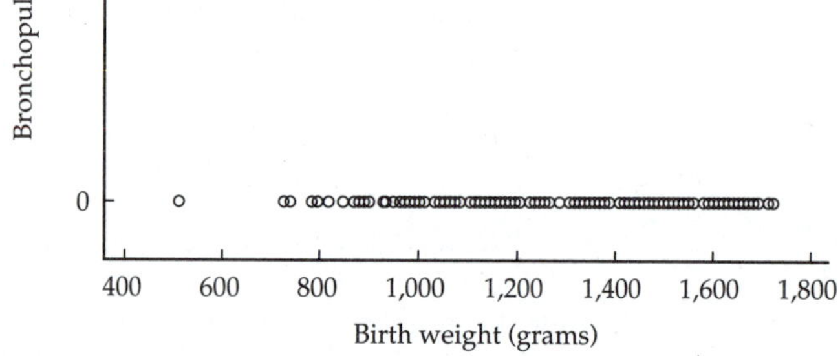

Figure 20.1. Bronchopulmonary dysplasia versus birth weight for a sample of 223 low birth weight infants

weighing between 951 and 1,350 grams, and those weighing 1,351 grams or more. We then estimate the probability that a child will develop BPD in each of these subgroups individually.

Birth Weight (grams)	Sample Size	Number with BPD	$\tilde{p}$
0–950	68	49	0.721
951–1,350	80	18	0.225
1,351–1,750	75	9	0.120
	223	76	0.341

The estimated probability of developing bronchopulmonary dysplasia decreases as the birth weight of an infant increases. Since there does appear to be a relationship between these two variables, we would like to be able to model p as a function of birth weight.

20.1.1 The Logistic Function

Our first strategy might be to fit a model of the form

$$p = \alpha + \beta_1 x_1$$

where x_1 represents weight. This is simply the standard linear regression model in which y—the outcome of a continuous random variable—has been replaced by p. On inspection, however, this model is not feasible. Since p is a probability, it is restricted to taking on values between 0 and 1. The term $\alpha + \beta_1 x_1$, in contrast, could easily yield a value that lies outside this range.

We might try to solve this problem by fitting the model

$$p = e^{\alpha + \beta_1 x_1}$$

instead. This equation guarantees that the estimate of p is positive. We would soon realize, however, that this model is also unsuitable. Although the term $e^{\alpha + \beta_1 x_1}$ cannot produce a negative estimate of p, it can result in a value that is greater than 1.

To accommodate this final constraint, we fit a model of the form

$$p = \frac{e^{\alpha + \beta_1 x_1}}{1 + e^{\alpha + \beta_1 x_1}}.$$

The expression on the right, called a *logistic function*, cannot yield a value that is either negative or greater than 1; consequently, it restricts the estimated value of p to the required range.

If

$$p = \frac{e^{\alpha+\beta_1 x_1}}{1 + e^{\alpha+\beta_1 x_1}},$$

then

$$\begin{aligned} \frac{p}{1-p} &= \frac{e^{\alpha+\beta_1 x_1}/(1 + e^{\alpha+\beta_1 x_1})}{1/(1 + e^{\alpha+\beta_1 x_1})} \\ &= e^{\alpha+\beta_1 x_1}. \end{aligned}$$

Taking the natural logarithm of each side of the equation,

$$\ln\left[\frac{p}{1-p}\right] = \alpha + \beta_1 x_1.$$

Recall that $p/(1-p)$ is defined as the odds of success. Therefore, modeling p with the logistic function is equivalent to fitting a linear regression model where the continuous outcome y has been replaced by the logarithm of the odds of success. Instead of assuming that the relationship between p and x_1 is linear, we assume that the relationship between $\ln[p/(1-p)]$ and x_1 is linear. The technique of fitting a model of this form is known as *logistic regression.*

20.1.2 The Fitted Equation

To model p—the probability that an infant develops bronchopulmonary dysplasia—as a function of the explanatory variable birth weight, we fit the equation

$$\ln\left[\frac{\hat{p}}{1-\hat{p}}\right] = a + b_1 x_1.$$

Although we categorized birth weight into three different intervals when we were exploring its relationship with BPD, we use the original continuous random variable for the logistic regression analysis. We cannot apply the method of least squares to fit a logistic model; instead, we use maximum likelihood estimation [2]. As with the linear regression model, a is an estimate of the population intercept α, and b_1 is an estimate of the slope β_1. These estimates are the values of the parameters that are most likely to have produced the observed data.

For the 223 infants in the sample, the estimated equation is

$$\ln\left[\frac{\hat{p}}{1-\hat{p}}\right] = 4.0343 - 0.0042 x_1.$$

The coefficient of weight implies that, for each one-gram increase in birth weight, the log odds that the infant develops BPD decrease by 0.0042. When the log odds decrease, the probability p decreases as well. A test of the null hypothesis

$$H_0: \beta_1 = 0$$

indicates that the coefficient of birth weight is significantly different from 0 at the 0.05 level.

To estimate the probability that an infant weighing 750 grams at birth develops BPD, we simply substitute the value $x_1 = 750$ into the preceding equation; therefore,

$$\begin{aligned} \ln\left[\frac{\hat{p}}{1-\hat{p}}\right] &= 4.0343 - 0.0042(750) \\ &= 0.8843. \end{aligned}$$

Taking the antilogarithm of each side,

$$\begin{aligned} \frac{\hat{p}}{1-\hat{p}} &= e^{0.8843} \\ &= 2.4213. \end{aligned}$$

Finally, solving for $\hat{p}$,

$$\begin{aligned} \hat{p} &= \frac{2.4213}{1+2.4213} \\ &= \frac{2.4213}{3.4213} \\ &= 0.708. \end{aligned}$$

Using similar calculations, we find that the estimated probability of developing BPD for an infant weighing 1,150 grams at birth is

$$\hat{p} = 0.311,$$

whereas the probability for a child weighing 1,550 grams is

$$\hat{p} = 0.078.$$

If we were to calculate the estimated probability $\hat{p}$ for each observed value of birth weight and plot these data, the result would be the curve in Figure 20.2. According to the logistic regression model, the estimated value of p decreases as birth weight increases. As previously mentioned, however, the relationship between p and x_1 is not linear.

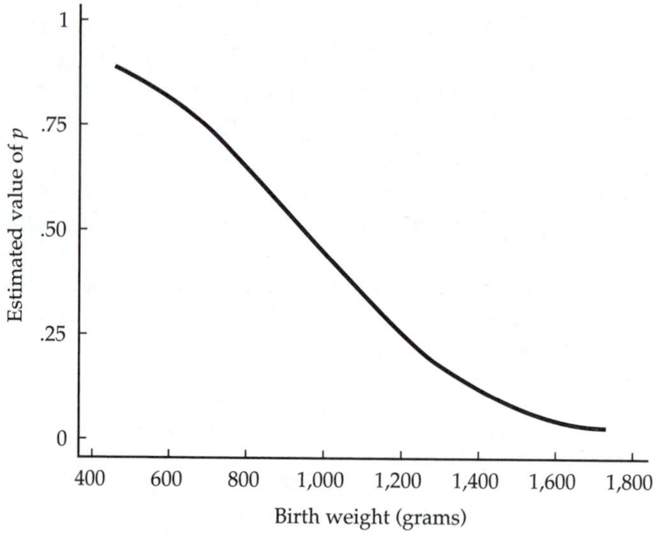

Figure 20.2. Logistic regression of bronchopulmonary dysplasia on birth weight, $\ln[\hat{p}/(1-\hat{p})] = 4.0343 - 0.0042x_1$

20.2 Multiple Logistic Regression

Now that we have determined that birth weight influences the probability that an infant will develop BPD, we might also wonder whether the variable gestational age would help to predict p. To explore this possibility, we again subdivide the population of infants into three categories: those whose gestational age is less than or equal to 28 weeks, those whose gestational age is between 29 and 30 weeks, and those whose gestational age is 31 weeks or more.

Gestational Age (weeks)	Sample Size	Number with BPD	$\tilde{p}$
≤ 28	58	40	0.690
29–30	73	26	0.356
≥ 31	92	10	0.109
	223	76	0.341

The estimated probability of developing bronchopulmonary dysplasia decreases as gestational age increases. We would like to be able to use this information to further improve our ability to predict p.

To determine whether this is feasible, we first cross-classify the sample of 223 infants according to birth weight and gestational age and estimate the probability of developing BPD in each category. The numbers in parentheses are the sample sizes in each subgroup.

Birth Weight (grams)	Gestational Age (weeks)		
	≤ 28	29–30	≥ 31
0–950	0.805 (41)	0.714 (21)	0.167 (6)
951–1,350	0.412 (17)	0.194 (36)	0.148 (27)
1,351–1,750	0.000 (0)	0.250 (16)	0.085 (59)

There appears to be a trend in the estimates of p. For a given category of birth weight, $\tilde{p}$ decreases as gestational age increases. In addition, for a specified category of gestational age, $\tilde{p}$ decreases as weight increases. There are slight discrepancies in these trends; they occur when the number of infants in a particular subgroup is very small. For example, there are no infants who weigh more than 1,350 grams and have a gestational age of less than 29 weeks, and only 16 who weigh more than 1,350 grams and have a gestational age between 29 and 30 weeks. In instances such as these, $\tilde{p}$ is subject to a great deal of fluctuation. Ignoring such aberrations, however, we are able to estimate the probability that a child develops bronchopulmonary dysplasia with greater precision when we know his or her birth weight and gestational age.

Since a relationship appears to exist among the three variables, we would now like to model p as a function of both birth weight and gestational age. To do this, we fit a model of the form

$$\ln\left[\frac{\hat{p}}{1-\hat{p}}\right] = a + b_1x_1 + b_2x_2$$

where x_1 designates weight, and x_2 represents age. Like birth weight, gestational age is now treated as a continuous random variable.

The fitted logistic regression equation is

$$\ln\left[\frac{\hat{p}}{1-\hat{p}}\right] = 13.8273 - 0.0024x_1 - 0.3983x_2.$$

The estimated coefficient of birth weight has decreased somewhat now that another variable has been added to the model; however, it is still significantly different from 0 at the 0.05 level. The coefficient of gestational age implies that, if birth weight remains constant while gestational age increases by one week, then the log odds of developing BPD decrease by 0.3983. A test of the null hypothesis

$$H_0: \beta_2 = 0$$

indicates that this coefficient is also different from 0 at the 0.05 level of significance. Therefore, given that we already know the birth weight of an infant, knowledge of his or her gestational age will further improve our ability to predict p.

To estimate the probability that an infant weighing 750 grams at birth with gestational age 27 weeks will develop BPD, we substitute the values $x_1 = 750$ and $x_2 = 27$ into the fitted equation;

$$\begin{aligned} \ln\left[\frac{\hat{p}}{1-\hat{p}}\right] &= 13.8273 - 0.0024(750) - 0.3983(27) \\ &= 1.2732. \end{aligned}$$

Taking the antilogarithm of each side results in

$$\begin{aligned} \frac{\hat{p}}{1-\hat{p}} &= e^{1.2732} \\ &= 3.5723, \end{aligned}$$

and solving for $\hat{p}$ we have

$$\begin{aligned} \hat{p} &= \frac{3.5723}{4.5723} \\ &= 0.781. \end{aligned}$$

For an infant who weighs 1,150 grams and has gestational age 32 weeks,

$$\begin{aligned} \ln\left[\frac{\hat{p}}{1-\hat{p}}\right] &= 13.8273 - 0.0024(1,150) - 0.3983(32) \\ &= -1.6783. \end{aligned}$$

Therefore,

$$\begin{aligned} \frac{\hat{p}}{1-\hat{p}} &= e^{-1.6783} \\ &= 0.1867, \end{aligned}$$

and

$$\begin{aligned} \hat{p} &= \frac{0.1867}{1.1867} \\ &= 0.157. \end{aligned}$$

In general, as either birth weight or gestational age increases, the estimated probability that an infant will develop bronchopulmonary dysplasia decreases.

20.3 Indicator Variables

Like the multiple regression model, the logistic regression model can be generalized to include explanatory variables that are not continuous. In fact, if an explanatory variable x_i is dichotomous, then the estimated coefficient b_i has a special interpretation. In this case, the antilogarithm of b_i—or e^{b_i}—is the estimated odds ratio of the response for the two possible levels of x_i. For example, suppose that we fit a model of the form

$$\ln\left[\frac{\hat{p}}{1-\hat{p}}\right] = a + b_3 x_3$$

where $\hat{p}$ is the estimated probability of developing BPD, and x_3 is the outcome of a dichotomous random variable that indicates whether a mother was diagnosed with toxemia during her pregnancy. If the presence of toxemia is represented by 1 and its absence by 0, the fitted model is

$$\ln\left[\frac{\hat{p}}{1-\hat{p}}\right] = -0.5718 - 0.7719 x_3.$$

Note that the estimated coefficient of toxemia is -0.7719. As a result, the relative odds of developing bronchopulmonary dysplasia for children whose mothers suffered from toxemia versus children whose mothers did not is

$$\begin{aligned} \widehat{\text{OR}} &= e^{b_3} \\ &= e^{-0.7719} \\ &= 0.46. \end{aligned}$$

Children whose mothers experienced toxemia during pregnancy appear to be less likely to develop chronic lung disease.

The same results could have been obtained by examining a 2×2 contingency table. Of the 76 infants who developed bronchopulmonary dysplasia, 6 had mothers who were diagnosed with toxemia during pregnancy; of the 147 children who did not, 23 had mothers who suffered from toxemia.

	Toxemia		
BPD	**Yes**	**No**	**Total**
Yes	6	70	76
No	23	124	147
Total	29	194	223

Note that

$$\widehat{\text{OR}} = \frac{(6)(124)}{(70)(23)} = 0.46.$$

Confidence limits for the odds ratio may be obtained from the logistic regression model by computing a confidence interval for β_3 and taking the antilogarithm of its upper and lower limits. Recall that a 95% confidence interval takes the form

$$(b_3 - 1.96\,\text{se}(b_3)\,,\; b_3 + 1.96\,\text{se}(b_3)).$$

If $\widehat{\text{se}}(b_3) = 0.4822$, then an approximate confidence interval for β_3 is

$$(-0.7719 - 1.96(0.4822)\,,\; -0.7719 + 1.96(0.4822))$$

or

$$(-1.7170\,,\; 0.1732).$$

Furthermore, a 95% confidence interval for e^{β_3}, the relative odds of developing bronchopulmonary dysplasia, is

$$(e^{-1.7170}\,,\; e^{0.1732})$$

or

$$(0.18\,,\; 1.19).$$

We are 95% confident that these limits cover the true population odds ratio for infants whose mothers had experienced toxemia versus infants whose mothers had not. Note that the interval contains the value 1; therefore, we would not have rejected the null hypothesis that β_3 is equal to 0 at the 0.05 level of significance.

Suppose that we add a second dichotomous explanatory variable to the model that already contains toxemia and fit an equation of the form

$$\ln\left[\frac{\hat{p}}{1-\hat{p}}\right] = a + b_3x_3 + b_4x_4.$$

The outcome x_4 indicates whether a mother was administered steroids during her pregnancy. The fitted model for the sample of 223 infants is

$$\ln\left[\frac{\hat{p}}{1-\hat{p}}\right] = -0.7172 - 0.7883x_3 + 0.3000x_4.$$

Including a second dichotomous random variable in the logistic regression model is equivalent to performing a stratified analysis; the coefficient of toxemia provides an estimate of the relative odds of developing BPD that has been adjusted for the effect of steroid use. In this case, the adjusted relative odds for children whose mothers experienced toxemia versus children whose mothers did not is

$$\begin{aligned} \widehat{\text{OR}} &= e^{-0.7883} \\ &= 0.45. \end{aligned}$$

A test of the null hypothesis

$$H_0: \beta_4 = 0$$

indicates that the coefficient of steroid use is not significantly different from 0 at the 0.05 level. This in turn implies that steroid use is not a confounder in the relationship between bronchopulmonary dysplasia and toxemia. Also note that the adjusted estimate of the odds ratio, 0.45, is nearly identical to the crude estimate, 0.46. This confirms that steroid use does not distort the relationship between the two variables.

20.4 Further Applications

Suppose that we wish to identify the factors that influence the probability that a low birth weight infant will experience a germinal matrix hemorrhage. The occurrence of a hemorrhage is a dichotomous random variable; a child either experiences one or does not. We use a logistic regression model and the sample of 100 infants born in Boston, Massachusetts, to estimate the probability that this event occurs [3].

To begin, we attempt to determine whether the head circumference of an infant affects the probability that he or she will suffer a hemorrhage by fitting a model of the form

$$\ln\left[\frac{\hat{p}}{1-\hat{p}}\right] = a + b_1 x_1$$

where x_1 represents head circumference. Suppose that the 100 indicators of hemorrhage are saved under the variable name `grmhem`; 1 designates the presence of a hemorrhage and 0 its absence. In addition, the measurements of head circumference are saved under the variable name `headcirc`. We use a computer package to perform the analysis; Table 20.1 shows the relevant output from Stata.

The top portion of the table displays details of the process of maximum likelihood estimation. Like the output for a linear regression analysis, the bottom portion contains information relating to the estimated regression coefficients. Rounding the coefficients to four decimal places, the fitted equation is

$$\ln\left[\frac{\hat{p}}{1-\hat{p}}\right] = 1.1929 - 0.1117 x_1.$$

The coefficient of head circumference implies that, for each one-centimeter increase in head circumference, the log odds of experiencing a hemorrhage decrease by 0.1117. Since the log odds decrease, the probability p decreases as well.

Table 20.1 Stata output displaying the logistic regression of germinal matrix hemorrhage on head circumference

```
Iteration 0: Log Likelihood = -42.270909
Iteration 1: Log Likelihood = -41.792731
Iteration 2: Log Likelihood = -41.788107
Iteration 3: Log Likelihood = -41.788106

Logit Estimates                                  Number of obs  =      100
                                                 chi2(1)        =     0.97
                                                 Prob > chi2    =   0.3258
Log Likelihood = -41.788106                      Pseudo R2      =   0.0114
```

grmhem	Coef.	Std. Err.	t	P>|t|	[95% Conf.	Interval]
headcirc	−.1117081	.1152569	−0.969	0.335	−.3404317	.1170155
_cons	1.192855	3.00632	0.397	0.692	−4.77309	7.158799

Suppose we wish to estimate the probability that an infant whose head circumference is 28 centimeters will suffer a germinal matrix hemorrhage. If we substitute the value $x_1 = 28$ into the fitted equation,

$$\ln\left[\frac{\hat{p}}{1-\hat{p}}\right] = 1.1929 - 0.1117(28)$$
$$= -1.9347.$$

Therefore,

$$\frac{\hat{p}}{1-\hat{p}} = e^{-1.9347}$$
$$= 0.1445,$$

and

$$\hat{p} = \frac{0.1445}{1.1445}$$
$$= 0.126.$$

The computer can be used to calculate the predicted value of p for each child in the data set. The predicted probabilities of experiencing a hemorrhage for the first ten infants in the sample are listed in Table 20.2.

A test of the null hypothesis

$$H_0 : \beta_1 = 0$$

results in a p-value of 0.335; therefore, we are unable to reject H_0 at the 0.05 level. Based on this sample, head circumference does not appear to have a significant effect on the probability that a hemorrhage occurs.

Table 20.2 Predicted probabilities of experiencing a germinal matrix hemorrhage for the first ten infants in the sample

Head Circumference (cm)	Hemorrhage	Predicted $\hat{p}$
29	0	0.114
23	1	0.202
28	0	0.126
27	0	0.139
26	0	0.153
26	1	0.153
27	0	0.139
28	0	0.126
28	0	0.126
26	0	0.153

We now attempt to determine whether the gender of an infant affects the probability that he or she experiences a germinal matrix hemorrhage. To do this, we fit a model of the form

$$\ln\left[\frac{\hat{p}}{1-\hat{p}}\right] = a + bx_2$$

where x_2 represents sex. Table 20.3 shows the relevant output from Stata if the 100 indicators of gender are saved under the variable name sex. The fitted logistic regression equation is

$$\ln\left[\frac{\hat{p}}{1-\hat{p}}\right] = -1.4088 - 0.8938x_2.$$

If males are designated by 1 and females by 0, then the estimated odds ratio for boys relative to girls is

$$\begin{aligned}\widehat{\text{OR}} &= e^{-0.8938}\\ &= 0.41.\end{aligned}$$

Males appear to be less likely to suffer a hemorrhage than females; however, a test of the null hypothesis

$$H_0\colon \beta_2 = 0$$

indicates that this coefficient is not significantly different from 0.

Table 20.3 Stata output displaying the logistic regression of germinal matrix hemorrhage on gender

```
Iteration 0: Log Likelihood = -42.270909
Iteration 1: Log Likelihood = -41.174001
Iteration 2: Log Likelihood = -41.147057
Iteration 3: Log Likelihood = -41.147023

Logit Estimates                              Number of obs  =     100
                                             chi2(1)        =    2.25
                                             Prob > chi2    =  0.1338
Log Likelihood = -41.147023                  Pseudo R2      =  0.0266
```

grmhem	Coef.	Std. Err.	t	P>|t|	[95% Conf.	Interval]
sex	−.8938179	.6230019	−1.435	0.155	−2.130142	.3425063
_cons	−1.408767	.33635	−4.188	0.000	−2.076243	−.7412917

20.5 Review Exercises

1. Suppose that you are interested in studying intravenous drug use among high school students in the United States. Drug use is characterized as a dichotomous random variable where 1 indicates that an individual has injected drugs within the past year and 0 that he or she has not. Factors that might be related to this outcome are instruction about the human immunodeficiency virus (HIV) in school, age, gender, and general knowledge about HIV, including the various modes of transmission and ways of reducing risk. The estimated coefficients and standard errors from a logistic regression model containing each of these variables and the interaction between instruction and gender are displayed below [4].

Variable	Coefficient	Standard Error
Intercept	−1.183	0.859
HIV instruction	0.039	0.421
Age	−0.164	0.092
Gender	1.212	0.423
HIV knowledge	−0.187	0.048
Instruction/gender	−0.663	0.512

a. At the 0.05 level of significance, which of the explanatory variables influence intravenous drug use in the past year?
b. As a student becomes older, does the probability that he or she has used intravenous drugs in the past year increase or decrease?
c. The dichotomous random variable gender is coded so that 1 represents a male and 0 a female. What are the odds of injecting drugs for males relative to females?
d. Does HIV instruction have a different relationship with intravenous drug use for males as opposed to females?

2. In a study investigating maternal risk factors for congenital syphilis, syphilis is treated as a dichotomous outcome variable where 1 represents the presence of disease in a newborn and 0 its absence. The estimated coefficients from a logistic regression model containing cocaine or crack use, marital status, number of prenatal visits to a doctor, alcohol use, and level of education are listed below [5].

Variable	Coefficient
Cocaine/crack use	1.354
Marital status	0.779
Number of prenatal visits	−0.098
Alcohol use	0.723
Level of education	0.298

a. As an expectant mother's number of prenatal visits to the doctor increases, what happens to the probability that her child will be born with congenital syphilis?
b. Marital status is coded as a dichotomous random variable where 1 indicates that a woman is unmarried and 0 that she is married. What are the relative odds that a newborn will suffer from syphilis for unmarried versus married mothers?
c. If a woman uses either cocaine or crack, what are the relative odds that her child will be born with congenital syphilis?
d. The estimated coefficient of cocaine use has standard error 0.162. Construct a 95% confidence interval for the population odds ratio.

3. In the data set `lowbwt`, the variable `grmhem` is a dichotomous random variable that indicates whether an infant experienced a germinal matrix hemorrhage. The value 1 represents the presence of a hemorrhage and 0 its absence. Outcomes of the infants' five-minute apgar scores are saved under the name `apgar5`, and indicators of toxemia— where 1 designates a diagnosis of toxemia for the child's mother—under the variable name `tox`.

a. Fit a model of the form

$$\ln\left[\frac{\hat{p}}{1-\hat{p}}\right] = a + b_1 x_1$$

where $\hat{p}$ is the estimated probability of experiencing a germinal matrix hemorrhage, and x_1 represents the five-minute apgar score.
b. If a particular child has a five-minute apgar score of 3, what is the estimated probability that this child will experience a hemorrhage?
c. At the 0.05 level of significance, test the null hypothesis that the population parameter β_1 is equal to 0.
d. Fit the model

$$\ln\left[\frac{\hat{p}}{1-\hat{p}}\right] = a + b_2 x_2$$

where x_2 represents toxemia.
e. What is the estimated probability that a child whose mother suffered from toxemia during pregnancy will experience a germinal matrix hemorrhage?
f. What are the estimated odds of suffering a germinal matrix hemorrhage for children whose mothers experienced toxemia relative to children whose mothers did not?
g. Construct a 95% confidence interval for the population odds ratio.

References

[1] Van Marter, L. J., Leviton, A., Kuban, K. C. K., Pagano, M., and Allred, E. N., "Maternal Glucocorticoid Therapy and Reduced Risk of Bronchopulmonary Dysplasia," *Pediatrics*, Volume 86, September 1990, 331–336.

[2] Hosmer, D. W., and Lemeshow, S., *Applied Logistic Regression*, New York: Wiley, 1989.

[3] Leviton, A., Fenton, T., Kuban, K. C. K., and Pagano, M., "Labor and Delivery Characteristics and the Risk of Germinal Matrix Hemorrhage in Low Birth Weight Infants," *Journal of Child Neurology*, Volume 6, October 1991, 35–40.

[4] Holtzman, D., Anderson, J. E., Kann, L., Arday, S. L., Truman, B. I., and Kohbe, L. J., "HIV Instruction, HIV Knowledge, and Drug Injection Among High School Students in the United States," *American Journal of Public Health*, Volume 81, December 1991, 1596–1601.

[5] Zweig, M. S., Singh, T., Htoo, M., and Schultz, S., "The Association Between Congenital Syphilis and Cocaine/Crack Use in New York City: A Case-Control Study," *American Journal of Public Health*, Volume 81, October 1991, 1316–1318.

CHAPTER 21

Survival Analysis

In some studies, the response variable of interest is the amount of time from an initial observation–for example, the time of diagnosis or the start of a new treatment–until the occurrence of a subsequent event–such as infection, the relapse of disease, or death. This time until failure is called the *survival time*. Although the measurements are continuous, the distributions of survival times are rarely normal; instead, they tend to be skewed to the right. The analysis of this type of data generally focuses on estimating the probability that an individual will survive for a given length of time.

One common occurrence in working with a sample of survival data is that not all the individuals are observed until their respective times of failure. The data may be analyzed before the event of interest has occurred in all patients, for instance. Other patients who either move away before the study is complete or who refuse to participate any longer are said to be *lost to follow-up*. The incomplete observation of a time to failure is known as *censoring*; the presence of censored observations distinguishes the analysis of survival data from other types of analyses.

A distribution of survival times can be characterized by a *survival function* $S(t)$. $S(t)$ is defined as the probability that an individual survives beyond time t. Equivalently, for a given t, $S(t)$ specifies the proportion of individuals who have not yet failed. If T is a continuous random variable representing survival time, then

$$S(t) = P(T > t).$$

The graph of $S(t)$ versus t is called a *survival curve*.

Survival curves have been in use for many years. For example, a study published in 1938 investigated the effects of tobacco on human longevity among white males over the age of 30 years [1]. Three categories of individuals were considered: nonusers of tobacco, moderate smokers, and heavy smokers. The results of the study are presented in Figure 21.1. As is evident from the graph,

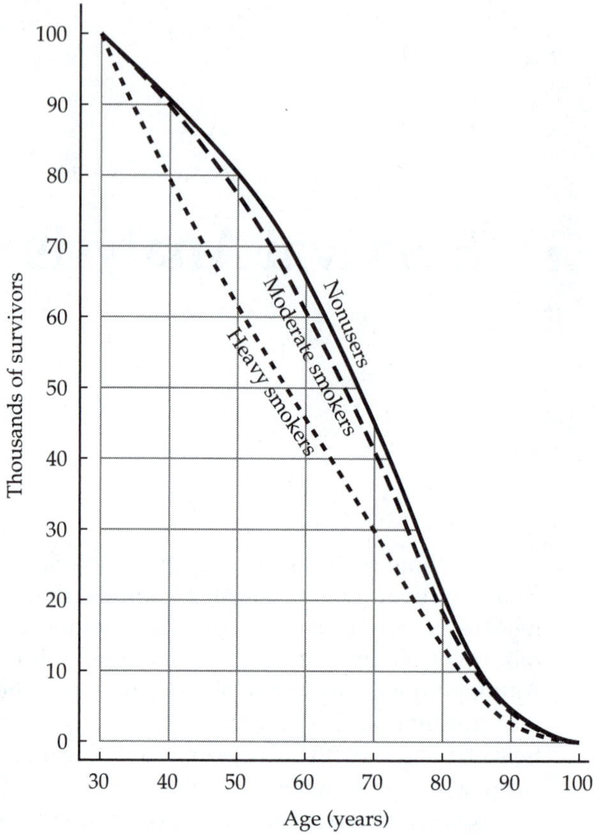

Figure 21.1. Survival curves for three categories of white males: nonusers of tobacco, moderate smokers, and heavy smokers, 1938

the smoking of tobacco is associated with a shorter duration of life; in addition, longevity is affected by the amount of tobacco used. It seems that these results were ignored not only at the time of publication but in subsequent years as well.

21.1 The Life Table Method

The classical method of estimating a survival curve is known as the *life table* method. This technique parallels our previous study of life tables. Whereas in Chapter 5 we examined descriptive statistics for an entire population, however, in this chapter we explore methods of statistical inference that are based on a sample of observations. Survival times are grouped within intervals of fixed length. The first columns of the table enumerate the time interval t to $t + n$, the proportion of individuals alive at time t who fail prior to $t + n$, and the number of individuals still alive at time t. The proportion of individuals who have not yet failed at time t can be calculated as

$$S(t) = \frac{l_t}{l_0}$$

where l_0 is the number of persons in the initial cohort, and l_t is the number who have not yet failed at time t.

Table 21.1 reproduces a portion of the first three columns of the complete U.S. life table for 1979–1981; this table was provided at the end of Chapter 5. The fourth column is the estimate of the survival function at time t where

$$S(t) = \frac{l_t}{100,000}.$$

Table 21.1 Complete U.S. life table for individuals less than 30 years of age, 1979–1981

$[t, t+n)$	${}_nq_t$	l_t	S(t)
0–1	0.01260	100,000	1.0000
1–2	0.00093	98,740	0.9874
2–3	0.00065	98,648	0.9865
3–4	0.00050	98,584	0.9858
4–5	0.00040	98,535	0.9854
5–6	0.00037	98,495	0.9850
6–7	0.00033	98,459	0.9846
7–8	0.00030	98,426	0.9843
8–9	0.00027	98,396	0.9840
9–10	0.00023	98,370	0.9837
10–11	0.00020	98,347	0.9835
11–12	0.00019	98,328	0.9833
12–13	0.00025	98,309	0.9831
13–14	0.00037	98,285	0.9829
14–15	0.00053	98,248	0.9825
15–16	0.00069	98,196	0.9820
16–17	0.00083	98,129	0.9813
17–18	0.00095	98,047	0.9805
18–19	0.00105	97,953	0.9795
19–20	0.00112	97,851	0.9785
20–21	0.00120	97,741	0.9774
21–22	0.00127	97,623	0.9762
22–23	0.00132	97,499	0.9750
23–24	0.00134	97,370	0.9737
24–25	0.00133	97,240	0.9724
25–26	0.00132	97,110	0.9711
26–27	0.00131	96,982	0.9698
27–28	0.00130	96,856	0.9686
28–29	0.00130	96,730	0.9673
29–30	0.00131	96,604	0.9660

The corresponding survival curve for the entire population is plotted in Figure 21.2. Table 21.1 is an example of a *current*, or *cross-sectional, life table*. It is constructed from data gathered over a relatively short period of time; the persons represented in the different intervals are not the same group of individuals throughout. For Table 21.1, it is assumed that the entire population is represented and hence that there is no sampling variability in the estimate of $S(t)$. However, the life table method can also be applied to a sample of individuals drawn from a population.

Ideally, we would prefer to work with a *longitudinal life table*, which tracks an actual cohort of individuals over their entire lifetimes. This method is not practical for large population studies; it would involve following a sizable group of individuals for over 100 years. However, it is often used in smaller clinical studies in which patients are enrolled sequentially and followed for shorter periods of time.

Consider the data presented in Table 21.2. A total of 12 hemophiliacs, all under 41 years of age at the time of HIV seroconversion, were followed from primary AIDS diagnosis until death [2]. We would prefer that our starting point be the time at which an individual contracts AIDS rather than the time at which he or she is diagnosed, but this information is unavailable. Nevertheless, what are we able to infer about the survival of the population of hemophiliacs on the basis of these 12 individuals?

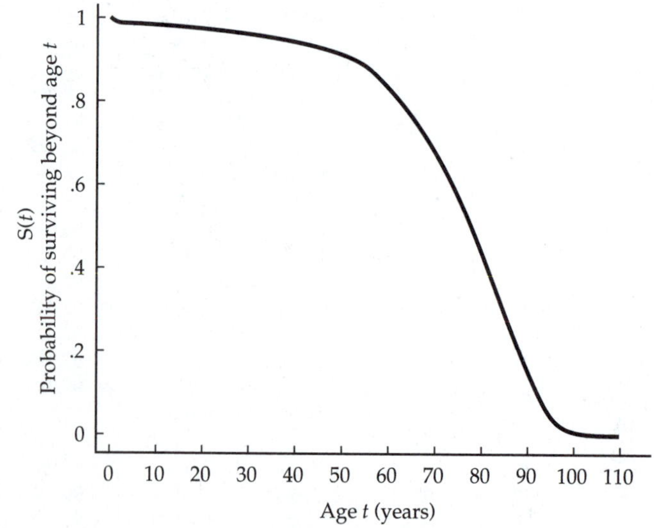

Figure 21.2. Survival curve for the U.S. population, 1979–1981

Table 21.2 Interval from primary AIDS diagnosis until death for a sample of 12 hemophiliac patients less than 41 years of age at HIV seroconversion

Patient Number	Survival (months)
1	2
2	3
3	6
4	6
5	7
6	10
7	15
8	15
9	16
10	27
11	30
12	32

Using the life table procedure described above, we could summarize the failure time data as in Table 21.3. Since one out of 12 individuals dies after two months, the proportion of individuals dying in the interval 2–3 is estimated as

$$\begin{aligned} {}_1q_2 &= \frac{1}{12} \\ &= 0.0833. \end{aligned}$$

One of the remaining 11 patients dies after three months; consequently

$$\begin{aligned} {}_1q_3 &= \frac{1}{11} \\ &= 0.0909. \end{aligned}$$

Similarly, since two of the remaining ten patients die after six months,

$$\begin{aligned} {}_1q_6 &= \frac{2}{10} \\ &= 0.2000. \end{aligned}$$

In time intervals not containing a death, such as 0–1 and 1–2, the estimated proportion of individuals dying is equal to 0.

Table 21.3 Life table method of estimating S(t) for a sample of 12 hemophiliac patients less than 41 years of age at HIV seroconversion

$[t, t+n)$	${}_nq_t$	l_t	S(t)
0–1	0.0000	12	1.0000
1–2	0.0000	12	1.0000
2–3	0.0833	11	0.9167
3–4	0.0909	10	0.8333
4–5	0.0000	10	0.8333
5–6	0.0000	10	0.8333
6–7	0.2000	8	0.6667
7–8	0.1250	7	0.5833
8–9	0.0000	7	0.5833
9–10	0.0000	7	0.5833
10–11	0.1429	6	0.5000
11–12	0.0000	6	0.5000
12–13	0.0000	6	0.5000
13–14	0.0000	6	0.5000
14–15	0.0000	6	0.5000
15–16	0.3333	4	0.3333
16–17	0.2500	3	0.2500
17–18	0.0000	3	0.2500
18–19	0.0000	3	0.2500
19–20	0.0000	3	0.2500
20–21	0.0000	3	0.2500
21–22	0.0000	3	0.2500
22–23	0.0000	3	0.2500
23–24	0.0000	3	0.2500
24–25	0.0000	3	0.2500
25–26	0.0000	3	0.2500
26–27	0.0000	3	0.2500
27–28	0.3333	2	0.1667
28–29	0.0000	2	0.1667
29–30	0.0000	2	0.1667
30–31	0.5000	1	0.0833
31–32	0.0000	1	0.0833
32–33	1.0000	0	0.0000

In the fourth column of Table 21.3, the survival function is calculated as

$$\begin{aligned} \mathrm{S}(t) &= \frac{l_t}{l_0} \\ &= \frac{l_t}{12} \end{aligned}$$

where l_t does not include any individuals who fail at time t. Since no patients die during the first two intervals,

$$\begin{aligned} S(0) &= S(1) \\ &= \frac{12}{12} \\ &= 1.0000. \end{aligned}$$

One patient dies after two months, therefore

$$\begin{aligned} S(2) &= \frac{11}{12} \\ &= 0.9167. \end{aligned}$$

After 32 months, all the patients in the study have died; as a result,

$$\begin{aligned} S(32) &= \frac{0}{12} \\ &= 0.0000. \end{aligned}$$

At this point, there are no individuals who have not yet failed.

21.2 The Product-Limit Method

In the calculation of the life table, the survival function $S(t)$ changes only when a death occurs. The *product-limit method,* also called the *Kaplan-Meier method,* of estimating a survival function is a nonparametric technique that uses the exact survival times for each individual instead of grouping them into intervals. Using the product-limit method, we would summarize the information in Table 21.2 as shown in Table 21.4. Not all of the one-month intervals are included; only the times at which a death occurs are important.

Note that the cumulative probability of survival $S(t)$ may be calculated using successive applications of the multiplicative rule. For example, let A be the event that an individual is alive prior to time 2 and B the event that he or she survives at time 2. Therefore, the event that the patient survives longer than two months may be represented by $A \cap B$. Recall that the multiplicative rule of probability states that

$$P(A \cap B) = P(A)\,P(B \mid A).$$

The probability that an individual survives longer than two months is $S(2)$, and the probability that he or she is alive before this time is $S(0)$. In addition, the

Table 21.4 Product-limit method of estimating S(t) for a sample of 12 hemophiliac patients less than 41 years of age at HIV seroconversion

Time	q_t	$1-q_t$	l_t	S(t)
0	0.0000	1.0000	12	1.0000
2	0.0833	0.9167	11	0.9167
3	0.0909	0.9091	10	0.8333
6	0.2000	0.8000	8	0.6667
7	0.1250	0.8750	7	0.5833
10	0.1429	0.8571	6	0.5000
15	0.3333	0.6667	4	0.3333
16	0.2500	0.7500	3	0.2500
27	0.3333	0.6667	2	0.1667
30	0.5000	0.5000	1	0.0833
32	1.0000	0.0000	0	0.0000

probability that a patient survives at time 2 given that he or she was alive up until this point is $1 - q_2$. As a result,

$$\begin{aligned} S(2) &= S(0)[1-q_2] \\ &= (1.0000)(1-0.0833) \\ &= (1.0000)(0.9167) \\ &= 0.9167. \end{aligned}$$

Similarly, the probability of living longer than three months is the probability of being alive just before time 3 multiplied by the probability of not failing at time 3, or

$$\begin{aligned} S(3) &= S(2)[1-q_3] \\ &= (0.9167)(1-0.0909) \\ &= (0.9167)(0.9091) \\ &= 0.8334. \end{aligned}$$

The survival curve corresponding to the data in Table 21.4 is plotted in Figure 21.3.

Since S(t) is calculated using the data in a sample of observations, it is merely an estimate of the true population survival function. To quantify the sampling variability involved, we can calculate the standard error of S(t) and construct a pair of confidence bands around the estimated survival curve [3].

The product-limit method can be adjusted to account for the partial information about survival times that is available from censored observations. Suppose that in the data for the 12 hemophiliac AIDS patients, individuals 2 and 6 had not yet died. Instead, they were still alive after 3 and 10 months of follow-up, respectively. A censored observation is commonly denoted by a plus (+) sign, as in Table 21.5. For these data, the product-limit estimate of the survival function

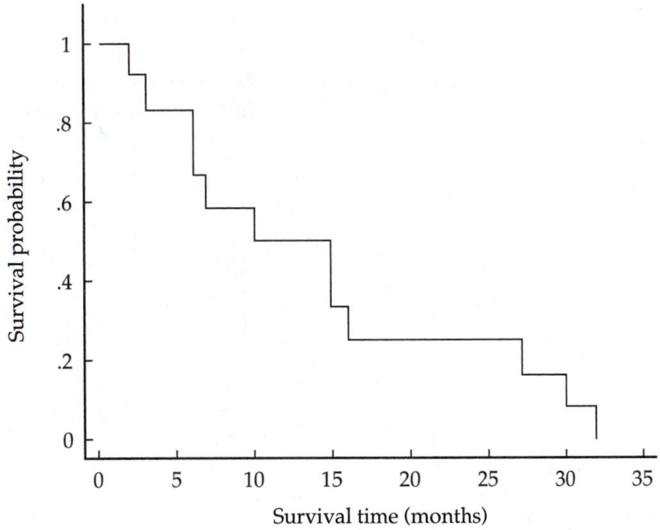

Figure 21.3. Survival curve for a sample of 12 hemophiliac patients less than 41 years of age at HIV seroconversion

Table 21.5 Interval from primary AIDS diagnosis until death for a sample of 12 hemophiliac patients less than 41 years of age at HIV seroconversion, censored observations included

Patient Number	Survival (months)
1	2
2	3+
3	6
4	6
5	7
6	10+
7	15
8	15
9	16
10	27
11	30
12	32

is calculated in Table 21.6. The corresponding curve is plotted in Figure 21.4; each small x on the graph denotes a censored survival time. Observe that $S(t)$ does not change from its previous value if the observation at time t is censored; however, this observation is not used to calculate the probability of failure in any subsequent interval. For instance, at time 6 one individual out of 12 has already

Table 21.6 Product-limit method of estimating S(t) for a sample of 12 hemophiliac patients less than 41 years of age at HIV seroconversion, censored observations included

Time	q_t	$1 - q_t$	S(t)
0	0.0000	1.0000	1.0000
2	0.0833	0.9167	0.9167
3	0.0000	1.0000	0.9167
6	0.2000	0.8000	0.7333
7	0.1250	0.8750	0.6417
10	0.0000	1.0000	0.6417
15	0.3333	0.6667	0.4278
16	0.2500	0.7500	0.3208
27	0.3333	0.6667	0.2139
30	0.5000	0.5000	0.1069
32	1.0000	0.0000	0.0000

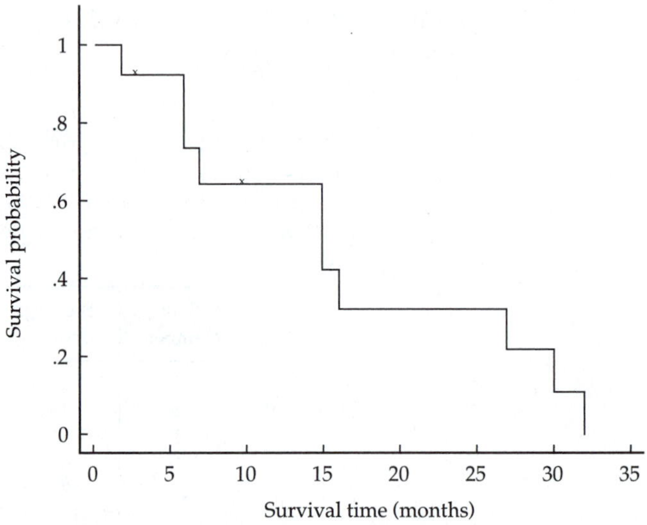

Figure 21.4. Survival curve for a sample of 12 hemophiliac patients less than 41 years of age at HIV seroconversion, censored observations included

died and another has been censored. Therefore, ten individuals remain at risk, and

$$\begin{aligned} q_6 &= \frac{2}{10} \\ &= 0.2000. \end{aligned}$$

21.3 The Log-Rank Test

Instead of characterizing the survival times in a single set of data, we often want to compare the distributions of times in two different populations. In this case, our goal is to determine whether survival differs systematically in the two groups. Recall the data for the 12 hemophiliacs–all 40 years of age or younger at the time of HIV seroconversion–that were presented in Table 21.2. We might wish to compare this distribution of times from primary AIDS diagnosis until death to the distribution for another group of hemophiliacs who were all over the age of 40 at the time of seroconversion. Survival times for both groups are listed in Table 21.7, and the corresponding curves are plotted in Figure 21.5. Patients undergoing HIV seroconversion when less than 40 years of age are designated Group 1 and are represented by the upper curve; patients over 40 at the time of HIV seroconversion are designated Group 2 and are represented by the lower curve. It appears that individuals who are younger at HIV seroconversion survive longer than those who are older.

The null hypothesis that the two distributions of failure times are identical may be tested using a nonparametric technique known as the *log-rank test*. The theory behind the log-rank test centers on constructing a 2×2 contingency table of group status versus survival for each time t at which a death occurs. When t is equal to one month, for example, no patients in Group 1 have died but four in Group 2 have. Therefore, the appropriate 2×2 table for $t = 1$ is as follows:

	Failure		
Status	**Yes**	**No**	**Total**
Group 1	0	12	12
Group 2	4	5	9
Total	4	17	21

After the sequence of tables has been constructed, the data are accumulated using the Mantel-Haenszel test statistic discussed in Chapter 16. This statistic compares the observed number of failures at each time to the expected number and, under the null hypothesis, has a chi-square distribution with 1 degree of freedom. The test of the hypothesis

$$H_0 : S_1(t) = S_2(t)$$

for the two groups of hemophiliacs in Table 21.7 generates a p-value of 0.025. The groups are significantly different at the $\alpha = 0.05$ level; we conclude that hemophiliacs experiencing seroconversion at a younger age tend to live longer after primary AIDS diagnosis than do individuals undergoing seroconversion at a later age.

As another example, consider a clinical trial that compares two different treatment regimens for moderate risk breast cancer [4]. Plots of the two survival

Table 21.7 Interval from primary AIDS diagnosis until death for a sample of 21 hemophiliac patients, stratified by age at HIV seroconversion

Age at SC ≤ 40 Years		Age at SC > 40 Years	
Patient Number	**Survival (months)**	**Patient Number**	**Survival (months)**
1	2	1	1
2	3	2	1
3	6	3	1
4	6	4	1
5	7	5	2
6	10	6	3
7	15	7	3
8	15	8	9
9	16	9	22
10	27		
11	30		
12	32		

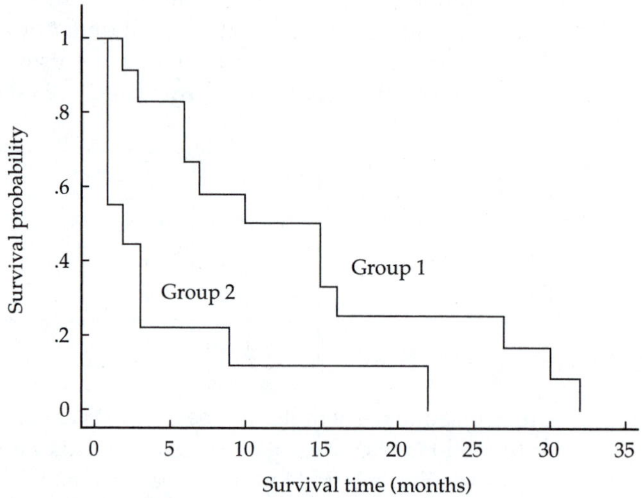

Figure 21.5. Survival curves for a sample of 21 hemophiliac patients, stratified by age at HIV seroconversion

curves are displayed in Figure 21.6. The small 1s on the graph indicate censored survival times for patients receiving the first treatment; similarly, the 2s are censored observations for individuals receiving the second treatment. There does not appear to be any difference in survival for patients in the two groups; there is a great deal of overlap in the curves. Also, a log-rank test of the null hypothesis

$$H_0 : S_1(t) = S_2(t)$$

results in a p-value of 0.883.

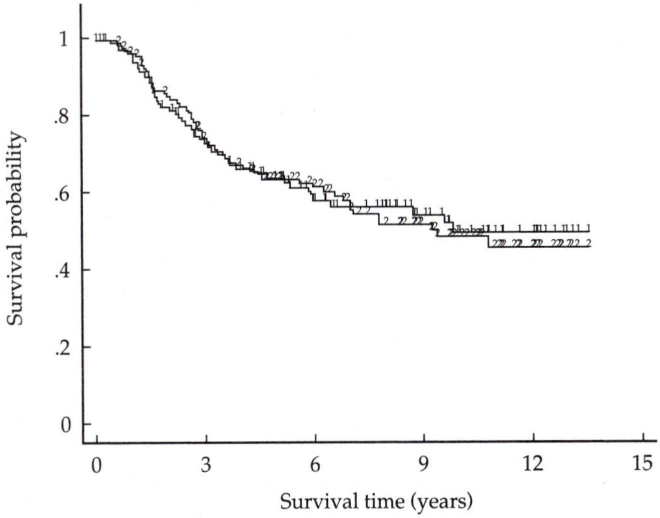

Figure 21.6. Survival curves for moderate risk breast cancer patients in two treatment groups

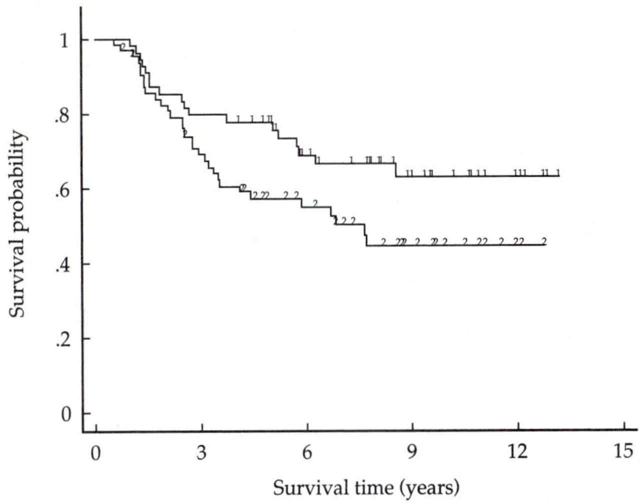

Figure 21.7. Survival curves for premenopausal, moderate risk breast cancer patients in two treatment groups

When the individuals enrolled in the trial are separated into two distinct subpopulations–premenopausal women versus postmenopausal women–then treatment group does appear to have a substantial effect on survival. As shown in Figure 21.7, treatment 1 seems to improve survival for premenopausal women. In this case, the log-rank test yields a p-value of 0.052. In contrast, Figure 21.8 suggests that treatment 2 is more effective for postmenopausal women. The p-value of the appropriate log-rank test is 0.086. Since the treatment effects are

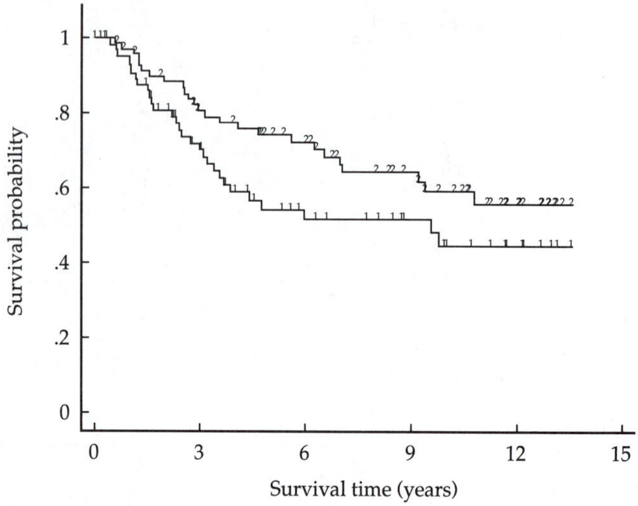

Figure 21.8. Survival curves for postmenopausal, moderate risk breast cancer patients in two treatment groups

going in opposite directions in the two different subpopulations, they tend to cancel each other out when the groups are combined. This serves as a reminder that care should be taken not to ignore important confounding variables.

21.4 Further Applications

Suppose that we are interested in studying patients with systemic cancer who subsequently develop a brain metastasis; our ultimate goal is to prolong their lives by controlling the disease. A sample of 23 such patients, all of whom were treated with radiotherapy, were followed from the first day of their treatment until recurrence of the original tumor. Recurrence is defined as the reappearance of a metastasis in exactly the same site, or, in the case of patients whose tumor never completely disappeared, enlargement of the original lesion. Times to recurrence for the 23 patients are presented in Table 21.8 [5]. What are we able to infer about the reappearance of brain metastases based on the information in this sample?

We could begin our analysis by summarizing the recurrence time data using the classic life table method. For intervals of length two weeks, we must first determine the proportion of patients who experience a recurrence in each interval. The results are presented in Table 21.9. Since four out of 23 individuals have a recurrence of the original brain metastasis at least two weeks but not more than four weeks after the start of their treatment, the proportion of individuals failing in the interval 2–4 is estimated as

$$
\begin{aligned}
{}_2q_4 &= \frac{4}{23} \\
&= 0.1739.
\end{aligned}
$$

Similarly, three of the remaining 19 patients have a recurrence between four and six weeks after treatment; therefore,

$$
\begin{aligned}
{}_4q_6 &= \frac{3}{19} \\
&= 0.1579.
\end{aligned}
$$

Table 21.8 Time to recurrence of brain metastasis for a sample of 23 patients treated with radiotherapy

Patient Number	Recurrence (weeks)	Patient Number	Recurrence (weeks)
1	2	13	14
2	2	14	14
3	2	15	18
4	3	16	19
5	4	17	20
6	5	18	22
7	5	19	22
8	6	20	31
9	7	21	33
10	8	22	39
11	9	23	195
12	10		

Table 21.9 Life table method of estimating S(t) for a sample of 23 patients treated with radiotherapy

$[t, t+n)$	${}_nq_t$	l_t	S(t)
0–2	0.0000	23	1.0000
2–4	0.1739	19	0.8261
4–6	0.1579	16	0.6957
6–8	0.1250	14	0.6087
8–10	0.1429	12	0.5217
10–12	0.0833	11	0.4783
12–14	0.0000	11	0.4783
14–16	0.1818	9	0.3913
16–18	0.0000	9	0.3913
18–20	0.2222	7	0.3043
20–22	0.1429	6	0.2609
22–24	0.3333	4	0.1739
24–26	0.0000	4	0.1739
26–28	0.0000	4	0.1739
28–30	0.0000	4	0.1739
30–32	0.2500	3	0.1304
32–34	0.3333	2	0.0870
34–36	0.0000	2	0.0870
36–38	0.0000	2	0.0870
38–40	0.5000	1	0.0435
40+	1.0000	0	0.0000

In time intervals not containing a failure, the estimated proportion of individuals experiencing a recurrence is equal to 0.

The survival function, or the proportion of individuals who have not yet failed, is defined as

$$\mathrm{S}(t) = \frac{l_t}{23}$$

where l_t does not include any individuals who experience a recurrence at time t. Since no patients fail in the first interval,

$$\begin{aligned}\mathrm{S}(0) &= \frac{23}{23} \\ &= 1.0000.\end{aligned}$$

Four patients have a recurrence between two and four weeks after the start of treatment, therefore

$$\begin{aligned}\mathrm{S}(2) &= \frac{19}{23} \\ &= 0.8261.\end{aligned}$$

Table 21.10 Product-limit method of estimating S(t) for a sample of 23 patients treated with radiotherapy

Time	q_t	$1 - q_t$	l_t	S(t)
0	0.0000	1.0000	23	1.0000
2	0.1304	0.8696	20	0.8696
3	0.0500	0.9500	19	0.8261
4	0.0526	0.9474	18	0.7826
5	0.1111	0.8889	16	0.6957
6	0.0625	0.9375	15	0.6522
7	0.0667	0.9333	14	0.6087
8	0.0714	0.9286	13	0.5652
9	0.0769	0.9231	12	0.5217
10	0.0833	0.9167	11	0.4783
14	0.1818	0.8182	9	0.3913
18	0.1111	0.8889	8	0.3478
19	0.1250	0.8750	7	0.3043
20	0.1429	0.8571	6	0.2609
22	0.3333	0.6667	4	0.1739
31	0.2500	0.7500	3	0.1304
33	0.3333	0.6667	2	0.0870
39	0.5000	0.5000	1	0.0435
195	1.0000	0.0000	0	0.0000

By the last interval in the table—the only one that is not of length two weeks—every patient in the study has experienced a recurrence; consequently,

$$\begin{aligned} S(40) &= \frac{0}{23} \\ &= 0.0000. \end{aligned}$$

Since we are dealing with a relatively small group of patients, we might prefer to summarize the data using the product-limit method. The product-limit method of estimating a survival function S(t) is a nonparametric technique that uses the exact recurrence times for each individual instead of grouping them into intervals. In this case, therefore, the three patients who experience a tumor recurrence two weeks after their initial treatment would not be grouped with the individual who fails at three weeks. For the data in Table 21.8, the product-limit estimate of S(t) is displayed in Table 21.10. Note that only the times at which a recurrence takes place are included.

The cumulative probability of recurrence S(t) may be calculated using successive applications of the multiplicative rule. The probability that an individual does not experience a recurrence within two weeks after the initial treatment, for instance, may be expressed as the probability that the patient has not had a recurrence prior to two weeks multiplied by the probability that he or she does not fail at time 2. Therefore,

$$\begin{aligned} S(2) &= S(0)[1 - q_2] \\ &= (1.0000)(1 - 0.1304) \\ &= (1.0000)(0.8696) \\ &= 0.8696. \end{aligned}$$

Similarly, the probability that a recurrence takes place more than three weeks after the beginning of treatment is the probability of no recurrence prior to time 3 multiplied by the probability of not failing at that time, or

$$\begin{aligned} S(3) &= S(2)[1 - q_3] \\ &= (0.8696)(1 - 0.0500) \\ &= (0.8696)(0.9500) \\ &= 0.8261. \end{aligned}$$

The survival curve corresponding to the data in Table 21.10 is plotted in Figure 21.9.

The product-limit method can be modified to take into account the information about recurrence times that is available from censored observations. In Table 21.11, censored observations are denoted by a plus (+) sign. These patients either died before experiencing a recurrence of their original brain metastasis or remained tumor-free at the end of the follow-up period. The product-limit estimate of the survival function is calculated in Table 21.12, and the corresponding survival curve is plotted in Figure 21.10. Note that if the largest observation in a sample is censored, then the curve does not come down to the horizontal axis.

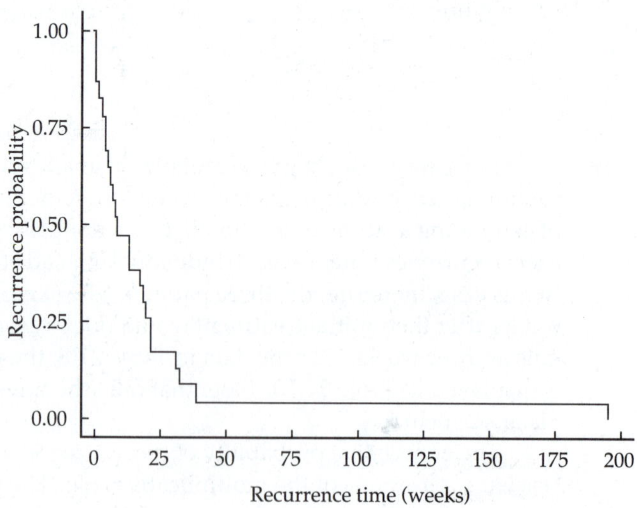

Figure 21.9. Survival curve for a sample of 23 patients treated with radiotherapy

Table 21.11 Time to recurrence of brain metastasis for a sample of 23 patients treated with radiotherapy, censored observations included

Patient Number	Recurrence (weeks)	Patient Number	Recurrence (weeks)
1	2+	13	14
2	2+	14	14+
3	2+	15	18+
4	3	16	19+
5	4	17	20
6	5	18	22
7	5+	19	22+
8	6	20	31+
9	7	21	33
10	8	22	39
11	9+	23	195+
12	10		

Table 21.12 Product-limit method of estimating $S(t)$ for a sample of 23 patients treated with radiotherapy, censored observations included

Time	q_t	$1 - q_t$	l_t	$S(t)$
0	0.0000	1.0000	23	1.0000
2	0.0000	1.0000	20	1.0000
3	0.0500	0.9500	19	0.9500
4	0.0526	0.9474	18	0.9000
5	0.0556	0.9444	16	0.8500
6	0.0625	0.9375	15	0.7969
7	0.0667	0.9333	14	0.7437
8	0.0714	0.9286	13	0.6906
9	0.0000	1.0000	12	0.6906
10	0.0833	0.9167	11	0.6331
14	0.0909	0.9091	9	0.5755
18	0.0000	1.0000	8	0.5755
19	0.0000	1.0000	7	0.5755
20	0.1429	0.8571	6	0.4933
22	0.1667	0.8333	4	0.4111
31	0.0000	1.0000	3	0.4111
33	0.3333	0.6667	2	0.2741
39	0.5000	0.5000	1	0.1370
195	0.0000	1.0000	0	0.1370

Also, the survival function $S(t)$ does not change if the observation at time t is censored. However, the observation is removed from subsequent intervals. At

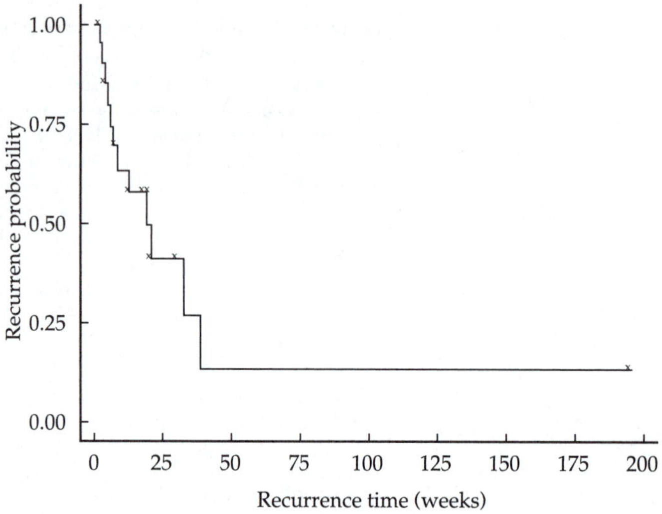

Figure 21.10. Survival curve for a sample of 23 patients treated with radiotherapy, censored observations included

time 3, for instance, three individuals out of 23 have been previously censored; therefore, 20 individuals remain at risk, and

$$q_3 = \frac{1}{20} = 0.0500.$$

Rather than work with a sample of survival times drawn from a single population, we often want to investigate the distributions of times in two different groups. For example, we might wish to compare the times to recurrence of brain metastasis for the patients treated with radiotherapy alone to the times for patients undergoing surgical removal of the tumor and subsequent radiotherapy. The data for the two groups are presented in Table 21.13; the corresponding survival curves are plotted in Figure 21.11. Patients treated with radiotherapy only are designated Group 1; those who undergo surgery as well are labeled Group 2. It appears that individuals treated with both surgery and postoperative radiotherapy have fewer recurrences of cancer in the brain and that the recurrences that do take place happen at a later time.

We can use the log-rank test to evaluate the null hypothesis that the distributions of recurrence times are identical in the two treatment groups. Although the calculations are somewhat complicated, they do not present a problem as long as a computer is available. The appropriate output from Stata is presented in Table 21.14. For each treatment group, the table contains both the observed and the predicted number of events—in this case, recurrences of brain tumor. It also

Table 21.13 Time to recurrence of brain metastasis for a sample of 48 patients, stratified by treatment

Radiotherapy Alone		Surgery/Radiotherapy	
Patient Number	Recurrence (weeks)	Patient Number	Recurrence (weeks)
1	2+	1	2+
2	2+	2	2+
3	2+	3	6+
4	3	4	6+
5	4	5	6+
6	5	6	10+
7	5+	7	14+
8	6	8	21+
9	7	9	23
10	8	10	29+
11	9+	11	32+
12	10	12	34+
13	14	13	34+
14	14+	14	37
15	18+	15	37+
16	19+	16	42+
17	20	17	51
18	22	18	57
19	22+	19	59
20	31+	20	63+
21	33	21	66+
22	39	22	71+
23	195+	23	71+
		24	73+
		25	85+

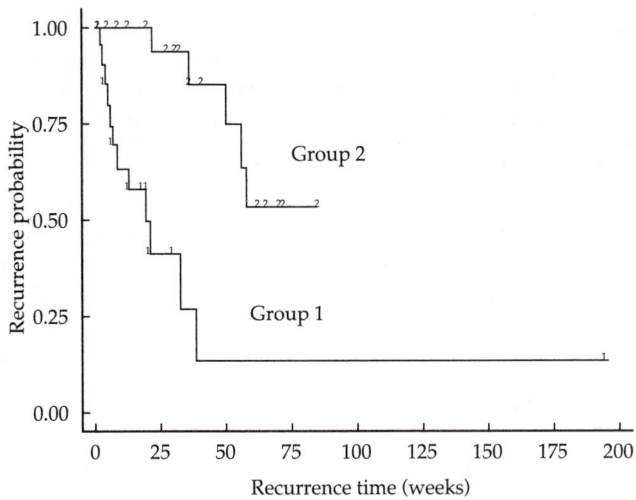

Figure 21.11. Survival curves for a sample of 48 patients, stratified by treatment

Table 21.14 Stata output displaying the log-rank test

Group	Events	Predicted
1	12	4.90
2	5	12.10
chi2(1)	=	14.44
Pr>chi2	=	0.0001

displays the test statistic and its corresponding p-value. Since $p \leq 0.0001$, we conclude that surgical removal of the brain metastasis followed by radiotherapy results in a longer time to recurrence of the original tumor than radiotherapy alone.

21.5 Review Exercises

1. Suppose that you are interested in examining the survival times of individuals who receive bone marrow transplants for nonneoplastic disease [6]. Presented below are the times in months for eight such patients. Assume that none of the observations is censored.

3.0 4.5 5.0 10.0 15.5 18.5 25.0 34.0

a. What is the median survival time for these patients?
b. For fixed intervals of length two weeks, use the life table method to estimate the survival function $S(t)$.
c. Is the life table cross-sectional or longitudinal?
d. Use the product-limit method to estimate $S(t)$.
e. Construct a survival curve for this sample of patients.

2. Displayed below are the survival times in months for ten AIDS patients suffering from concomitant esophageal candidiasis and cytomegalovirus [7]. Censored observations are denoted by a plus (+) sign.

Patient Number	Survival (months)
1	0.5+
2	1
3	1
4	1
5	2
6	5+
7	8+
8	9
9	10+
10	12+

a. How many deaths were observed in this sample of patients?
b. Use the product-limit method to estimate the survival function $S(t)$.
c. Construct a survival curve corresponding to these data.

3. A study was conducted to examine the effects of the drug ganciclovir on AIDS patients suffering from disseminated cytomegalovirus infection [8]. Two groups of patients were followed; 18 were treated with the drug, and 11 were not. The results of this study are contained in the data set `cyto`. Survival times are saved under the variable name `time`; indicators of censoring status–where 0 designates that an observation was censored and 1 that it was not–under the name `censor`; and indicators of treatment group under the name `group`.

a. What is the median survival time in each treatment group?
b. Generate survival curves for the two sets of patients.
c. Does it appear that the individuals in one group survive longer than those in the other?
d. Use the log-rank test to evaluate the null hypothesis that the distributions of survival times are identical in the two groups.
e. What do you conclude?

References

[1] Pearl, R., "Tobacco Smoking and Longevity," *Science*, Volume 87, March 4, 1938, 216–217.

[2] Ragni, M. V., and Kingsley, L. A., "Cumulative Risk for AIDS and Other HIV Outcomes in a Cohort of Hemophiliacs in Western Pennsylvania," *Journal of Acquired Immune Deficiency Syndromes*, Volume 3, July 1990, 708–713.

[3] Brown, B. W., "Estimation in Survival Analysis: Parametric Models, Product-Limit and Life Table Methods," *Statistics in Medical Research*, New York: Wiley, 1982.

[4] Shapiro, C. L., Henderson, I. C., Gelman, R. S., Harris, J. R., Canellos, G. P., and Frei, E., "A Randomized Trial of Cyclophosphamide, Methotrexate, and Fluorouracil Versus Methotrexate, Fluorouracil Adjuvant Chemotherapy in Moderate Risk Breast Cancer Patients," *Proceedings of the American Association for Cancer Research*, Volume 31, March 1990, 185.

[5] Patchell, R. A., Tibbs, P. A., Walsh, J. W., Dempsey, R. J., Maruyama, Y., Kryscio, R. J., Markesbery, W. R., MacDonald, J. S., and Young, B., "A Randomized Trial of Surgery in the Treatment of Single Metastases to the Brain," *The New England Journal of Medicine*, Volume 322, February 22, 1990, 494–500.

[6] Ash, R. C., Casper, J. T., Chitambar, C. R., Hansen, R., Bunin, N., Truitt, R. L., Lawton, C., Murray, K., Hunter, J., Baxter-Lowe, L. A., Gottschall, J. L., Oldham, K., Anderson, T., Camitta, B., and Menitove, J., "Successful Allogeneic Transplantation of T-Cell Depleted Bone Marrow from Closely HLA-Matched Unrelated Donors," *The New England Journal of Medicine*, Volume 322, February 22, 1990, 485–494.

[7] Laine, L., Bonacini, M., Sattler, F., Young, T., and Sherrod, A., "Cytomegalovirus and *Candida* Esophagitis in Patients with AIDS," *Journal of Acquired Immune Deficiency Syndromes*, Volume 5, June 1992, 605–609.
[8] Kotler, D. P., "Cytomegalovirus Colitis and Wasting," *Journal of Acquired Immune Deficiency Syndromes*, Volume 4, Supplement 1, 1991, S36–S41.

CHAPTER

22

Sampling Theory

When studying inference, we learn that one of the fundamental goals of statistics is to describe some characteristic of a population using the information in a sample of observations. In previous chapters, the underlying population was assumed to be infinite with mean μ and standard deviation σ. From this population, a random sample of size n was selected. The central limit theorem asserts that the distribution of the mean of the sample values is approximately normal with mean μ and standard deviation $\sigma/\sqrt{n}$. It is critical that the sample be representative of the population so that the conclusions drawn are valid. This chapter concerns itself with some of the important issues regarding sampling theory.

22.1 Sampling Schemes

Suppose that, instead of being infinite, our underlying population is finite and consists of a total of N elements. If N is large, it may still not be feasible–or even possible–to evaluate all elements of the population. Therefore, we would again like to make inference about a specified population characteristic using the information contained in a sample of the elements.

The individual elements in the population of interest are called *study units*, or *sampling units*; a study unit may be a person, a family, an object, or anything else that is the unit of analysis in a population. For example, suppose that we wish to determine the average amount of alcohol consumed each week by 15- to 17-year-olds living in the state of Massachusetts. In this case, the study units would be teenagers between the ages of 15 and 17 residing in Massachusetts at a particular time.

The ideal population that we would like to describe is known as the *target population*. In the preceding example, the target population is all 15- to 17-year-olds living in Massachusetts. In many situations, the entire target population is not accessible. If we are using school records to select our sample of teenagers, for instance, then individuals who do not attend high school or whose names are not on the list for some other reason would have no chance of being included. After we account for practical constraints, the group from which we are actually able to sample is known as the *study population*. A list of the elements in the study population is called a *sampling frame*. Note that a random sample, although representative of the study population from which it is selected, may not be representative of the target population. If the two groups differ in some way, then the selected sample may be *biased*.

22.1.1 Simple Random Sampling

The most elementary type of sample that can be drawn from the study population is a *simple random sample*. In simple random sampling, units are independently selected one at a time until the desired sample size is achieved. Since a given unit may be chosen only once, this strategy exemplifies *sampling without replacement*. Each study unit in the finite population has an equal chance of being included in the sample; the probability that a particular unit is chosen is equal to n/N where n is the size of the sample, and N is the size of the underlying population. The quantity n/N is known as the *sampling fraction* of the population.

When the underlying population of size N has mean μ and standard deviation σ, a finite version of the central limit theorem claims that the distribution of the sample mean has mean μ and standard deviation $\sqrt{1-(n/N)}(\sigma/\sqrt{n})$. The standard deviation of the sample mean for a finite population differs from that for an infinite population by a factor of $\sqrt{1-(n/N)}$. The square of this quantity, or $1-(n/N)$, is called the *finite population correction factor*. If n has some fixed value and N is very large, then n/N is close to 0. In this case, the finite population correction factor is approximately 1, and we return to the familiar situation in which the standard deviation of the sample mean is $\sigma/\sqrt{n}$. If the whole population is included in the sample, then n/N is equal to 1, and the standard deviation is 0. When the entire population is evaluated, there is no sampling variability in the mean.

One way to choose a simple random sample is to list and number each study unit, mix them up thoroughly, and then select units from this sampling frame until the required sample size is achieved. Another way is to use a computer or a table of random numbers to identify the units to be included in the sample. In either case, each unit should have an equal chance of being chosen. For example, to determine the average amount of alcohol consumed each week by 15- to 17-year-olds in Massachusetts, it would not be viable to interview every such teenager. Instead, a numbered list of these individuals could be compiled and one of the preceding methods used to select a sample of size n. The required information would then be obtained from the members of this group only, rather than the entire population.

22.1.2 Systematic Sampling

If a complete list of the N elements in a population is available, then *systematic sampling* may be used. Systematic sampling is similar to simple random sampling but is easier to apply. If a sample of size n is desired, then the sampling fraction for the population is n/N. This is equivalent to a sampling fraction of $1/(N/n)$, or 1 in N/n. Therefore, the initial sample unit is randomly selected from the first k units on the list where k is equal to N/n. If N/n is not an integer, then its value is rounded to the nearest whole number. Next, as we move down the list, every k^{th} consecutive unit is chosen. For example, to select a sample of n 15- to 17-year-olds living in Massachusetts, we would first randomly select a number between 1 and $k = N/n$. Suppose that we choose 4. We would then obtain information about the 4th person on the list, as well as persons $4 + k, 4 + 2k, 4 + 3k$, and so on.

Ideally, a sampling frame should be a complete list of all members of the target population. In reality, however, this is rarely the case. In some situations, it is impossible to devise a sampling frame for the population we wish to study. Suppose, for instance, that we are interested in the individuals who will use a particular health care service over the next year. Even when a sampling frame is not available, systematic random sampling may often be applied. We still wish to sample a fraction of 1 in N/n. If the population size N is unknown, then it must be estimated. In this case, the initial study unit is randomly selected from the first k units that become available. After this, each kth consecutive unit is chosen. Therefore, the sampling frame is compiled as the study progresses.

Unlike simple random sampling, systematic sampling requires the selection of only a single random number. Also, it distributes the sample evenly over the entire population list. Bias may arise if there is some type of periodic sequence to the list; such patterns are rare, however. If we are able to assume that the list is randomly ordered, we may treat the resulting sample as a simple random sample.

22.1.3 Stratified Sampling

Simple random sampling does not take into consideration any information that is known about the elements of a population and that might affect the characteristic of interest. Under this sampling scheme, it is also possible that a particular subgroup of the population would not be represented in a given sample. When selecting teenagers in Massachusetts, for instance, we may not choose any 17-year-old males. This could occur merely as the result of sampling variation. If we feel that it is important to include 17-year-old males, we could avoid this problem by selecting a *stratified random sample*. Our strata might consist of various combinations of sex and age. Using this method, we divide the population into H distinct subgroups or strata such that the hth stratum has size N_h. A separate simple random sample of size n_h is then selected from each stratum, resulting in a sampling fraction of n_h/N_h for the hth subgroup. This method ensures that each stratum is represented in the overall sample.

When stratified sampling is used, the population mean is estimated as a weighted average of the stratum-specific sample means. In addition, the variance is calculated as a weighted average of the within-strata variances. The variances between study units in different subgroups do not make a contribution. Therefore, we are able to make the overall variance—or standard deviation—smaller by choosing subgroups so that the study units within a particular stratum are as homogeneous as possible, decreasing the within-strata variances, whereas the units in distinct strata are as different as possible, increasing the between-strata variances. If the stratum-specific sample sizes are chosen properly, the estimated mean of a stratified sample has a smaller variance and is consequently more precise than the mean of a simple random sample.

22.1.4 Cluster Sampling

If study units form natural groups or if an adequate list of the entire population is difficult to compile, then *cluster sampling* might be considered. This involves selecting a random sample of groups or clusters and then looking at all study units within the chosen groups. In *two-stage sampling*, a random sample of clusters is selected and then, within each group, a random sample of study units. If our clusters were schools in Massachusetts, for instance, we could begin by selecting a sample of schools followed by a sample of 15- to 17-year-olds within each of the schools. Cluster sampling is usually more economical than other types of sampling; it saves both time and money. For samples of the same size, however, it does produce a larger variance than simple random sampling.

The choice of a sampling strategy depends on a number of factors, including the objectives of the study and the available resources [1]. The costs and benefits of the various methods should be weighed carefully. In practice, an investigator often combines two or more different strategies.

22.2 Sources of Bias

No matter what the sampling scheme, when we are choosing a sample, selection bias is not the only potential source of error. A second source is *nonresponse*. In situations in which the units of study are persons, there typically exist individuals who either cannot be reached or cannot or will not provide the information requested. Bias is present if these nonrespondents differ systematically from the individuals who do respond.

Consider the results of the following study in which a sample of 5,574 psychiatrists practicing in the United States were surveyed. A total of 1,442, or 26%, returned the questionnaire [2]. Of the 1,057 males responding to the question, 7.1% admitted to having sexual contact with one or more patients. Of the 257 females, 3.1% confessed to this practice. How might nonresponse affect these estimates?

Given that the Hippocratic oath expressly forbids sexual contact between physicians and their patients, it seems unlikely that there were claims of sexual contact that did not actually occur. It also seems probable that there were psychiatrists who did in fact have sexual contact with a patient and subsequently refused to return the survey. Therefore, it is more than likely that the percentages calculated from the survey data underestimate the true population proportions; by how much the proportions are underestimated we do not know.

A further potential source of bias in a sample survey results from the fact that a respondent may choose to lie rather than reveal something that is sensitive or incriminating; this might be true of the psychiatrists mentioned above, for example. Another situation in which an individual might not tell the truth is in a study investigating substance abuse patterns during pregnancy. In some states, a woman who confesses to using cocaine during her pregnancy runs the risk of having her child taken away from her; this risk might provide sufficient incentive to lie. One way to minimize this problem is to apply the technique of *randomized response*. By introducing an extra degree of uncertainty into the data, we are able to mask the responses of specific individuals while still making inference about the population as a whole. If it works, randomized response reduces the motivation to lie.

For example, suppose that the quantity we wish to estimate is the population prevalence of some characteristic, represented by π. A random sample of individuals from the population are questioned as to whether they possess this characteristic or not. Rather than being told to answer the question in a straightforward manner, a certain anonymous proportion of the respondents–represented by a where $0 < a < 1$–is instructed to answer "yes" under all circumstances. The remaining individuals are asked to tell the truth. Therefore, in a sample of size n, approximately na persons will always give an affirmative answer; the other $n(1 - a)$ will reply truthfully. Of these $n(1 - a)$, $n(1 - a)\pi$ will say "yes," and $n(1 - a)(1 - \pi)$ will say "no." If n^* is the total number of respondents answering "yes," then, on average,

$$n^* = na + n(1 - a)\pi.$$

If we subtract na from each side of the equation and divide by $n(1 - a)$, the population prevalence π may be estimated as

$$\hat{\pi} = \frac{(n^*/n) - a}{1 - a}.$$

For instance, a study conducted in New York City compared telephone responses obtained by direct questioning to those obtained through the use of randomized response. The study investigated the illicit use of four different drugs: cocaine, heroin, PCP, and LSD. Each individual questioned was asked to have three coins available; he or she was to toss the coins before being asked a question and to respond according to the outcome of the toss. The rules were a little more complicated than those described above. If all three coins were heads, the respondent was instructed to answer in the affirmative; if all three were tails, then he or she had to answer "no." If the coins were a mixture of both heads and tails, the respondent was asked to tell the truth. Therefore, the proportion of individuals always replying "yes" was

$$\begin{aligned} a_1 &= \left(\frac{1}{2}\right)\left(\frac{1}{2}\right)\left(\frac{1}{2}\right) \\ &= \frac{1}{8}. \end{aligned}$$

Similarly, the proportion always providing a negative response was

$$\begin{aligned} a_2 &= \left(\frac{1}{2}\right)\left(\frac{1}{2}\right)\left(\frac{1}{2}\right) \\ &= \frac{1}{8}. \end{aligned}$$

The remaining

$$\begin{aligned} 1 - \frac{1}{8} - \frac{1}{8} &= \frac{6}{8} \\ &= \frac{3}{4} \end{aligned}$$

were instructed to tell the truth. In a sample of size n, approximately $(3/4)n\pi$ of these would reply "yes," and $(3/4)n(1 - \pi)$ would reply "no." If n^* is the total number of individuals answering "yes," then

$$n^* = \frac{1}{8}n + \frac{3}{4}n\pi.$$

Consequently,

$$\hat{\pi} = \frac{8n^* - n}{6n}$$

would be our estimate of the proportion using a particular drug. For three of the four drugs in question, the proportions of individuals acknowledging use were higher when the answers were obtained by means of randomized response than they were when direct questioning was used; for cocaine the percentage increased from 11% to 21%, and for heroin it went from 3% to 10% [3]. This suggests that some individuals were not being completely truthful when they were questioned directly.

The primary advantage of randomized response is that it reduces the proportion of individuals providing untruthful answers. Although it is impossible to identify individual responses, aggregate information can still be obtained. However, since this technique introduces an extra source of uncertainty into the analysis, the estimator $\hat{\pi}$ has a larger variance than it would in the situation in which no masking device is used and everyone is assumed to answer the question honestly. What we lose in precision, however, we gain in accuracy.

22.3 Further Applications

In previous chapters we studied a sample of low birth weight infants born in two teaching hospitals in Boston, Massachusetts. To illustrate some of the practical issues of sampling, we now treat these children as if they comprise a finite population of size $N = 100$. We might wish to describe a particular characteristic of this population—their average gestational age, perhaps. The 100 measures of gestational age for the infants are displayed in Table 22.1 [4]; the true mean for the population is $\mu = 28.89$ weeks. Suppose that we do not know this and we also do not have the resources to obtain the necessary information for each child. Instead, we must estimate the mean by obtaining information about a representative fraction of the newborns. How would we proceed?

To select a simple random sample of size n, we choose study units independently from a list of the population elements—known as the *sampling frame*—until

Table 22.1 Measures of gestational age for a population of 100 low birth weight infants

ID	Age	ID	Age	ID	Age	ID	Age
1	29	26	28	[51]	23	76	31
2	31	27	29	52	27	77	30
3	33	[28]	28	53	28	78	27
4	31	29	29	54	27	79	25
5	30	30	30	55	27	80	25
[6]	25	31	31	56	26	81	26
7	27	32	30	57	25	82	29
8	29	33	31	58	23	83	29
9	28	34	29	59	26	84	34
[10]	29	35	27	60	24	85	30
[11]	26	[36]	27	61	29	86	29
12	30	37	27	62	29	87	33
13	29	38	32	63	27	88	30
14	29	39	31	64	30	89	29
15	29	40	28	65	30	[90]	24
16	29	41	30	66	32	91	33
17	29	42	29	67	33	92	25
18	33	43	28	68	27	[93]	32
19	33	44	31	69	31	94	31
20	29	45	27	70	26	95	31
21	28	46	25	71	27	96	31
[22]	30	47	30	72	27	97	29
23	27	[48]	28	73	35	98	32
24	33	49	28	74	28	99	33
25	32	50	25	75	30	100	28

we achieve the desired sample size. Suppose that we wish to draw a sample of size $n = 10$. One way to go about this would be to write the integers from 1 to 100 on slips of paper. After mixing them up thoroughly, we would select 10 different numbers. If we were working with a very large population then this method would be impractical. Instead, we could use a computer to generate the random numbers. In either case, each study unit has an equal chance of being selected. The probability that a particular unit is chosen is

$$\frac{n}{N} = \frac{10}{100} = 0.10.$$

The ratio n/N is the sampling fraction of the population.

Suppose that we follow this procedure for drawing a simple random sample and select the following set of numbers:

93 11 28 6 90 51 10 22 36 48

Returning to the population of low birth weight infants, we determine the gestational age of each of the newborns chosen; the appropriate values are marked in Table 22.1 and are listed below.

32 26 28 25 24 23 29 30 27 28

Note that the observations selected in a simple random sample need not be distributed evenly over the entire sampling frame. Using this random sample, we would estimate the population mean as

$$\bar{x} = \frac{32 + 26 + 28 + 25 + 24 + 23 + 29 + 30 + 27 + 28}{10} = 27.20 \text{ weeks.}$$

This value is a little smaller than the true population mean.

As an alternative to the procedure described above, we might prefer to apply the technique of systematic sampling. When a complete list of the N elements in a population is available, systematic sampling is easier to carry out; it requires the selection of only a single random number. As noted above, the desired sampling fraction for the population of low birth weight infants is 0.10, or 1 in 10. Therefore, we would begin by randomly selecting the initial study unit from the first 10 units on the list.

Suppose that we write the integers from 1 to 10 on slips of paper and randomly choose the number 5. In addition to identifying the gestational age of the 5th infant on the list, we would determine the gestational age of every 10th consecutive child—the 15th, the 25th, and so on. The appropriate data are displayed on the next page.

Number	Age
5	30
15	29
25	32
35	27
45	27
55	27
65	30
75	30
85	30
95	31

These observations are evenly distributed over the entire sampling frame. As long as the population list is randomly ordered–and we have no reason to believe that it is not–a systematic sample may be treated as a simple random sample. In this case, therefore, we would estimate the population mean as

$$\begin{aligned}\bar{x} &= \frac{30+29+32+27+27+27+30+30+30+31}{10} \\ &= 29.30 \text{ weeks.}\end{aligned}$$

This time, our estimate is slightly larger than the true population mean.

If we feel that it is important to include representative numbers of male and female infants in our sample–we might think that gender is associated with gestational age–then we could select a stratified random sample. To do this, we must first divide the population of low birth weight infants into two distinct subgroups of 44 boys and 56 girls. The 100 values of gestational age for the population, sorted by gender, are displayed in Table 22.2. Even though we are working with two separate subpopulations, we would still like to have an overall sampling fraction of 1/10. Therefore, we should select a simple random sample of size

$$\begin{aligned}44\left(\frac{1}{10}\right) &= 4.4 \\ &\approx 4\end{aligned}$$

from the group of males and a sample of size

$$\begin{aligned}56\left(\frac{1}{10}\right) &= 5.6 \\ &\approx 6\end{aligned}$$

from the group of females.

Using simple random sampling, we choose observations 2, 85, 61, and 54 for the males. For the females, we select 51, 14, 33, 25, 62, and74. These observations are marked in Table 22.2; the gestational ages of the infants are displayed below.

Males		Females	
Number	Age	Number	Age
2	31	51	23
85	30	14	29
61	29	33	31
55	27	25	32
		62	29
		74	28

Table 22.2 Measures of gestational age for a population of 100 low birth weight infants stratified by gender

Males				Females			
ID	Age	ID	Age	ID	Age	ID	Age
1	29	72	27	3	33	49	28
2	31	75	30	4	31	50	25
6	25	76	31	5	30	51	23
7	27	77	30	8	29	55	27
15	29	85	30	9	28	57	25
16	29	86	29	10	29	58	23
21	28	87	33	11	26	59	26
23	27	88	30	12	30	60	24
24	33	89	29	13	29	62	29
26	28	90	24	14	29	65	30
28	28	91	33	17	29	66	32
31	31	92	25	18	33	67	33
34	29	95	31	19	33	68	27
37	27	96	31	20	29	69	31
39	31	97	29	22	30	70	26
41	30	98	32	25	32	73	35
42	29			27	29	74	28
43	28			29	29	78	27
47	30			30	30	79	25
48	28			32	30	80	25
52	27			33	31	81	26
53	28			35	27	82	29
54	27			36	27	83	29
56	26			38	32	84	34
61	29			40	28	93	32
63	27			44	31	96	31
64	30			45	27	99	33
71	27			46	25	100	28

Using these data, we see that the stratum-specific sample means are

$$\bar{x}_{\text{males}} = \frac{31 + 30 + 29 + 27}{4} = 29.25 \text{ weeks}$$

and

$$\bar{x}_{\text{females}} = \frac{23 + 29 + 31 + 32 + 29 + 28}{6} = 28.67 \text{ weeks.}$$

The true population mean is estimated as a weighted average of these quantities; therefore,

$$\bar{x} = \frac{4(29.25) + 6(28.67)}{10} = 28.90 \text{ weeks.}$$

By chance, this value is nearly identical to the true population mean μ.

22.4 Review Exercises

1. The data set `lowbwt` contains information describing 100 low birth weight infants. Assume that these infants constitute a finite population. Their measures of systolic blood pressure are saved under the variable name `sbp`; the mean systolic blood pressure is $\mu = 47.08$ mm Hg. Suppose that we do not know the true population mean and wish to estimate it using a sample of 20 newborns.

   a. What is the sampling fraction of the population?
   b. Select a simple random sample and use it to estimate the true mean systolic blood pressure for this population of low birth weight infants.
   c. Draw a systematic sample from the same population and again estimate the mean systolic blood pressure.
   d. Suppose you believe that a diagnosis of toxemia in an expectant mother might affect the systolic blood pressure of her child. Divide the population of low birth weight infants into two groups: those whose mothers were diagnosed with toxemia and those whose mothers were not. Select a stratified random sample of size 20. Use these data to estimate the true population mean.
   e. What are the sampling fractions in each of the two strata?
   f. Could cluster sampling be applied in this problem?

2. A study was conducted to examine the effects of maternal marijuana and cocaine use on fetal growth. Drug exposure was assessed in two different ways: the mothers were questioned directly, and a urinalysis was performed [5].

a. Suppose that it is necessary to rely entirely on the information provided by the mothers. How might nonresponse affect the results of the study?
b. An alternative strategy might be to interview only those women who agree to be questioned. Do you feel that this would provide a representative sample of the underlying population of expectant mothers?

3. Each year, the U.S. Department of Agriculture uses the revenue collected from excise taxes to estimate the number of cigarettes consumed in this country. Over the 11-year period 1974 to 1985, however, repeated surveys of smoking practices can account for only about 72% of the total consumption [6].

a. How do you explain this discrepancy?
b. Which source are you more likely to believe?

4. Suppose that you are interested in conducting your own survey to estimate the proportion of psychiatrists who have had sexual contact with one or more patients. How would you carry out this study? Justify your method of data collection. Include a discussion of how you would attempt to minimize bias.

5. Return to the first exercise in Chapter 1, which asked you to design a study aimed at investigating an issue that you believe might influence the health of the world's people. Reread your original proposal and critique your study design. What would you do differently?

References

[1] Mendenhall, W., Ott, L., and Scheaffer, R. L., *Elementary Survey Sampling*, Belmont, Calif.: Wadsworth, 1971.

[2] Gatrell, N., Herman, J., Olarte, S., Feldstein, M., and Localio, R., "Psychiatrist–Patient Sexual Contact: Results of a National Survey, Prevalence," *American Journal of Psychiatry*, Volume 143, September 1986, 1126–1131.

[3] Weissman, A. N., Steer, R. A., and Lipton, D. S., "Estimating Illicit Drug Use Through Telephone Interviews and the Randomized Response Technique," *Drug and Alcohol Dependence*, Volume 18, 1986, 225–233.

[4] Leviton, A., Fenton, T., Kuban, K. C. K., and Pagano, M., "Labor and Delivery Characteristics and the Risk of Germinal Matrix Hemorrhage in Low Birth Weight Infants," *Journal of Child Neurology*, Volume 6, October 1991, 35–40.

[5] Zuckerman, B., Frank, D. A., Hingson, R., Amaro, H., Levenson, S. M., Kayne, H., Parker, S., Vinci, R., Aboagye, K., Fried, L. E., Cabral, H., Timperi, R.,

and Bauchner, H., "Effects of Maternal Marijuana and Cocaine Use on Fetal Growth," *The New England Journal of Medicine*, Volume 320, June 1989, 762–768.

[6] Hatziandreu, E. J., Pierce, J. P., Fiore, M. C., Grise, V., Novotny, T. E., and Davis, R. M., "The Reliability of Self-Reported Cigarette Consumption in the United States," *American Journal of Public Health*, Volume 79, August 1989, 1020–1023.

APPENDIX

A

Tables

Table A.1 Areas in one tail of the standard normal curve

z	0.00	0.01	0.02	0.03	0.04	0.05	0.06	0.07	0.08	0.09
0.0	0.500	0.496	0.492	0.488	0.484	0.480	0.476	0.472	0.468	0.464
0.1	0.460	0.456	0.452	0.448	0.444	0.440	0.436	0.433	0.429	0.425
0.2	0.421	0.417	0.413	0.409	0.405	0.401	0.397	0.394	0.390	0.386
0.3	0.382	0.378	0.374	0.371	0.367	0.363	0.359	0.356	0.352	0.348
0.4	0.345	0.341	0.337	0.334	0.330	0.326	0.323	0.319	0.316	0.312
0.5	0.309	0.305	0.302	0.298	0.295	0.291	0.288	0.284	0.281	0.278
0.6	0.274	0.271	0.268	0.264	0.261	0.258	0.255	0.251	0.248	0.245
0.7	0.242	0.239	0.236	0.233	0.230	0.227	0.224	0.221	0.218	0.215
0.8	0.212	0.209	0.206	0.203	0.200	0.198	0.195	0.192	0.189	0.187
0.9	0.184	0.181	0.179	0.176	0.174	0.171	0.169	0.166	0.164	0.161
1.0	0.159	0.156	0.154	0.152	0.149	0.147	0.145	0.142	0.140	0.138
1.1	0.136	0.133	0.131	0.129	0.127	0.125	0.123	0.121	0.119	0.117
1.2	0.115	0.113	0.111	0.109	0.107	0.106	0.104	0.102	0.100	0.099
1.3	0.097	0.095	0.093	0.092	0.090	0.089	0.087	0.085	0.084	0.082
1.4	0.081	0.079	0.078	0.076	0.075	0.074	0.072	0.071	0.069	0.068
1.5	0.067	0.066	0.064	0.063	0.062	0.061	0.059	0.058	0.057	0.056
1.6	0.055	0.054	0.053	0.052	0.051	0.049	0.048	0.047	0.046	0.046
1.7	0.045	0.044	0.043	0.042	0.041	0.040	0.039	0.038	0.038	0.037
1.8	0.036	0.035	0.034	0.034	0.033	0.032	0.031	0.031	0.030	0.029
1.9	0.029	0.028	0.027	0.027	0.026	0.026	0.025	0.024	0.024	0.023
2.0	0.023	0.022	0.022	0.021	0.021	0.020	0.020	0.019	0.019	0.018
2.1	0.018	0.017	0.017	0.017	0.016	0.016	0.015	0.015	0.015	0.014
2.2	0.014	0.014	0.013	0.013	0.013	0.012	0.012	0.012	0.011	0.011
2.3	0.011	0.010	0.010	0.010	0.010	0.009	0.009	0.009	0.009	0.008
2.4	0.008	0.008	0.008	0.008	0.007	0.007	0.007	0.007	0.007	0.006
2.5	0.006	0.006	0.006	0.006	0.006	0.005	0.005	0.005	0.005	0.005
2.6	0.005	0.005	0.004	0.004	0.004	0.004	0.004	0.004	0.004	0.004
2.7	0.003	0.003	0.003	0.003	0.003	0.003	0.003	0.003	0.003	0.003
2.8	0.003	0.002	0.002	0.002	0.002	0.002	0.002	0.002	0.002	0.002
2.9	0.002	0.002	0.002	0.002	0.002	0.002	0.002	0.001	0.001	0.001
3.0	0.001	0.001	0.001	0.001	0.001	0.001	0.001	0.001	0.001	0.001
3.1	0.001	0.001	0.001	0.001	0.001	0.001	0.001	0.001	0.001	0.001
3.2	0.001	0.001	0.001	0.001	0.001	0.001	0.001	0.001	0.001	0.001
3.3	0.000	0.000	0.000	0.000	0.000	0.000	0.000	0.000	0.000	0.000
3.4	0.000	0.000	0.000	0.000	0.000	0.000	0.000	0.000	0.000	0.000

Table A.2 Percentiles of the t distribution

	Area in One Tail					
df	**0.10**	**0.05**	**0.025**	**0.01**	**0.005**	**0.0005**
1	3.078	6.314	12.706	31.821	63.657	636.619
2	1.886	2.920	4.303	6.965	9.925	31.599
3	1.638	2.353	3.182	4.541	5.841	12.924
4	1.533	2.132	2.776	3.747	4.604	8.610
5	1.476	2.015	2.571	3.365	4.032	6.869
6	1.440	1.943	2.447	3.143	3.707	5.959
7	1.415	1.895	2.365	2.998	3.499	5.408
8	1.397	1.860	2.306	2.896	3.355	5.041
9	1.383	1.833	2.262	2.821	3.250	4.781
10	1.372	1.812	2.228	2.764	3.169	4.587
11	1.363	1.796	2.201	2.718	3.106	4.437
12	1.356	1.782	2.179	2.681	3.055	4.318
13	1.350	1.771	2.160	2.650	3.012	4.221
14	1.345	1.761	2.145	2.624	2.977	4.140
15	1.341	1.753	2.131	2.602	2.947	4.073
16	1.337	1.746	2.120	2.583	2.921	4.015
17	1.333	1.740	2.110	2.567	2.898	3.965
18	1.330	1.734	2.101	2.552	2.878	3.922
19	1.328	1.729	2.093	2.539	2.861	3.883
20	1.325	1.725	2.086	2.528	2.845	3.850
21	1.323	1.721	2.080	2.518	2.831	3.819
22	1.321	1.717	2.074	2.508	2.819	3.792
23	1.319	1.714	2.069	2.500	2.807	3.768
24	1.318	1.711	2.064	2.492	2.797	3.745
25	1.316	1.708	2.060	2.485	2.787	3.725
26	1.315	1.706	2.056	2.479	2.779	3.707
27	1.314	1.703	2.052	2.473	2.771	3.690
28	1.313	1.701	2.048	2.467	2.763	3.674
29	1.311	1.699	2.045	2.462	2.756	3.659
30	1.310	1.697	2.042	2.457	2.750	3.646
40	1.303	1.684	2.021	2.423	2.704	3.551
50	1.299	1.676	2.009	2.403	2.678	3.496
60	1.296	1.671	2.000	2.390	2.660	3.460
70	1.294	1.667	1.994	2.381	2.648	3.435
80	1.292	1.664	1.990	2.374	2.639	3.416
90	1.291	1.662	1.987	2.368	2.632	3.402
100	1.290	1.660	1.984	2.364	2.626	3.390
110	1.289	1.659	1.982	2.361	2.621	3.381
120	1.289	1.658	1.980	2.358	2.617	3.373
∞	1.282	1.645	1.960	2.327	2.576	3.291

Table A.3 Percentiles of the F distribution

Denominator df	Area in Upper Tail	Numerator Degrees of Freedom (df)										
		1	2	3	4	5	6	7	8	12	24	∞
2	0.100	8.53	9.00	9.16	9.24	9.29	9.33	9.35	9.37	9.41	9.45	9.49
	0.050	18.51	19.00	19.16	19.25	19.30	19.33	19.35	19.37	19.41	19.45	19.50
	0.025	38.51	39.00	39.17	39.25	39.30	39.33	39.36	39.37	39.41	39.46	39.50
	0.010	98.50	99.00	99.17	99.25	99.30	99.33	99.36	99.37	99.42	99.46	99.50
	0.005	198.5	199.0	199.2	199.3	199.3	199.3	199.4	199.4	199.4	199.5	199.5
	0.001	998.5	999.0	999.2	999.3	999.3	999.3	999.4	999.4	999.4	999.5	999.5
3	0.100	5.54	5.46	5.39	5.34	5.31	5.28	5.27	5.25	5.22	5.18	5.13
	0.050	10.13	9.55	9.28	9.12	9.01	8.94	8.89	8.85	8.74	8.64	8.53
	0.025	17.44	16.04	15.44	15.10	14.88	14.73	14.62	14.54	14.34	14.12	13.90
	0.010	34.12	30.82	29.46	28.71	28.24	27.91	27.67	27.49	27.05	26.60	26.13
	0.005	55.55	49.80	47.47	46.19	45.39	44.84	44.43	44.13	43.39	42.62	41.83
	0.001	167.0	148.5	141.1	137.1	134.6	132.9	131.6	130.6	128.3	125.9	123.5
4	0.100	4.54	4.32	4.19	4.11	4.05	4.01	3.98	3.95	3.90	3.83	3.76
	0.050	7.71	6.94	6.59	6.39	6.26	6.16	6.09	6.04	5.91	5.77	5.63
	0.025	12.22	10.65	9.98	9.60	9.36	9.20	9.07	8.98	8.75	8.51	8.26
	0.010	21.20	18.00	16.69	15.98	15.52	15.21	14.98	14.80	14.37	13.93	13.46
	0.005	31.33	26.28	24.26	23.15	22.46	21.97	21.62	21.35	20.70	20.03	19.32
	0.001	74.14	61.25	56.18	53.44	51.71	50.53	49.66	49.00	47.41	45.77	44.05
5	0.100	4.06	3.78	3.62	3.52	3.45	3.40	3.37	3.34	3.27	3.19	3.10
	0.050	6.61	5.79	5.41	5.19	5.05	4.95	4.88	4.82	4.68	4.53	4.36
	0.025	10.01	8.43	7.76	7.39	7.15	6.98	6.85	6.76	6.52	6.28	6.02
	0.010	16.26	13.27	12.06	11.39	10.97	10.67	10.46	10.29	9.89	9.47	9.02
	0.005	22.78	18.31	16.53	15.56	14.94	14.51	14.20	13.96	13.38	12.78	12.14
	0.001	47.18	37.12	33.20	31.09	29.75	28.83	28.16	27.65	26.42	25.13	23.79
6	0.100	3.78	3.46	3.29	3.18	3.11	3.05	3.01	2.98	2.90	2.82	2.72
	0.050	5.99	5.14	4.76	4.53	4.39	4.28	4.21	4.15	4.00	3.84	3.67
	0.025	8.81	7.26	6.60	6.23	5.99	5.82	5.70	5.60	5.37	5.12	4.85
	0.010	13.75	10.92	9.78	9.15	8.75	8.47	8.26	8.10	7.72	7.31	6.88
	0.005	18.63	14.54	12.92	12.03	11.46	11.07	10.79	10.57	10.03	9.47	8.88
	0.001	35.51	27.00	23.70	21.92	20.80	20.03	19.46	19.03	17.99	16.90	15.75

(continued)

Table A.3 (continued)

7	0.100	3.59	3.26	3.07	2.96	2.88	2.83	2.78	2.75	2.67	2.58	2.47
	0.050	5.59	4.74	4.35	4.12	3.97	3.87	3.79	3.73	3.57	3.41	3.23
	0.025	8.07	6.54	5.89	5.52	5.29	5.12	4.99	4.90	4.67	4.41	4.14
	0.010	12.25	9.55	8.45	7.85	7.46	7.19	6.99	6.84	6.47	6.07	5.65
	0.005	16.24	12.40	10.88	10.05	9.52	9.16	8.89	8.68	8.18	7.64	7.08
	0.001	29.25	21.69	18.77	17.20	16.21	15.52	15.02	14.63	13.71	12.73	11.70
8	0.100	3.46	3.11	2.92	2.81	2.73	2.67	2.62	2.59	2.50	2.40	2.29
	0.050	5.32	4.46	4.07	3.84	3.69	3.58	3.50	3.44	3.28	3.12	2.93
	0.025	7.57	6.06	5.42	5.05	4.82	4.65	4.53	4.43	4.20	3.95	3.67
	0.010	11.26	8.65	7.59	7.01	6.63	6.37	6.18	6.03	5.67	5.28	4.86
	0.005	14.69	11.04	9.60	8.81	8.30	7.95	7.69	7.50	7.01	6.50	5.95
	0.001	25.41	18.49	15.83	14.39	13.48	12.86	12.40	12.05	11.19	10.30	9.33
9	0.100	3.36	3.01	2.81	2.69	2.61	2.55	2.51	2.47	2.38	2.28	2.16
	0.050	5.12	4.26	3.86	3.63	3.48	3.37	3.29	3.23	3.07	2.90	2.71
	0.025	7.21	5.71	5.08	4.72	4.48	4.32	4.20	4.10	3.87	3.61	3.33
	0.010	10.56	8.02	6.99	6.42	6.06	5.80	5.61	5.47	5.11	4.73	4.31
	0.005	13.61	10.11	8.72	7.96	7.47	7.13	6.88	6.69	6.23	5.73	5.19
	0.001	22.86	16.39	13.90	12.56	11.71	11.13	10.70	10.37	9.57	8.72	7.81
10	0.100	3.29	2.92	2.73	2.61	2.52	2.46	2.41	2.38	2.28	2.18	2.06
	0.050	4.96	4.10	3.71	3.48	3.33	3.22	3.14	3.07	2.91	2.74	2.54
	0.025	6.94	5.46	4.83	4.47	4.24	4.07	3.95	3.85	3.62	3.37	3.08
	0.010	10.04	7.56	6.55	5.99	5.64	5.39	5.20	5.06	4.71	4.33	3.91
	0.005	12.83	9.43	8.08	7.34	6.87	6.54	6.30	6.12	5.66	5.17	4.64
	0.001	21.04	14.91	12.55	11.28	10.48	9.93	9.52	9.20	8.45	7.64	6.76
12	0.100	3.18	2.81	2.61	2.48	2.39	2.33	2.28	2.24	2.15	2.04	1.90
	0.050	4.75	3.89	3.49	3.26	3.11	3.00	2.91	2.85	2.69	2.51	2.30
	0.025	6.55	5.10	4.47	4.12	3.89	3.73	3.61	3.51	3.28	3.02	2.72
	0.010	9.33	6.93	5.95	5.41	5.06	4.82	4.64	4.50	4.16	3.78	3.36
	0.005	11.75	8.51	7.23	6.52	6.07	5.76	5.52	5.35	4.91	4.43	3.90
	0.001	18.64	12.97	10.80	9.63	8.89	8.38	8.00	7.71	7.00	6.25	5.42

(continued)

Table A.3 (continued)

Denominator df	Area in Upper Tail	Numerator Degrees of Freedom (df) 1	2	3	4	5	6	7	8	12	24	∞
14	0.100	3.10	2.73	2.52	2.39	2.31	2.24	2.19	2.15	2.05	1.94	1.80
	0.050	4.60	3.74	3.34	3.11	2.96	2.85	2.76	2.70	2.53	2.35	2.13
	0.025	6.30	4.86	4.24	3.89	3.66	3.50	3.38	3.29	3.05	2.79	2.49
	0.010	8.86	6.51	5.56	5.04	4.69	4.46	4.28	4.14	3.80	3.43	3.00
	0.005	11.06	7.92	6.68	6.00	5.56	5.26	5.03	4.86	4.43	3.96	3.44
	0.001	17.14	11.78	9.73	8.62	7.92	7.44	7.08	6.80	6.13	5.41	4.60
16	0.100	3.05	2.67	2.46	2.33	2.24	2.18	2.13	2.09	1.99	1.87	1.72
	0.050	4.49	3.63	3.24	3.01	2.85	2.74	2.66	2.59	2.42	2.24	2.01
	0.025	6.12	4.69	4.08	3.73	3.50	3.34	3.22	3.12	2.89	2.63	2.32
	0.010	8.53	6.23	5.29	4.77	4.44	4.20	4.03	3.89	3.55	3.18	2.75
	0.005	10.58	7.51	6.30	5.64	5.21	4.91	4.69	4.52	4.10	3.64	3.11
	0.001	16.12	10.97	9.01	7.94	7.27	6.80	6.46	6.19	5.55	4.85	4.06
18	0.100	3.01	2.62	2.42	2.29	2.20	2.13	2.08	2.04	1.93	1.81	1.66
	0.050	4.41	3.55	3.16	2.93	2.77	2.66	2.58	2.51	2.34	2.15	1.92
	0.025	5.98	4.56	3.95	3.61	3.38	3.22	3.10	3.01	2.77	2.50	2.19
	0.010	8.29	6.01	5.09	4.58	4.25	4.01	3.84	3.71	3.37	3.00	2.57
	0.005	10.22	7.21	6.03	5.37	4.96	4.66	4.44	4.28	3.86	3.40	2.87
	0.001	15.38	10.39	8.49	7.46	6.81	6.35	6.02	5.76	5.13	4.45	3.67
20	0.100	2.97	2.59	2.38	2.25	2.16	2.09	2.04	2.00	1.89	1.77	1.61
	0.050	4.35	3.49	3.10	2.87	2.71	2.60	2.51	2.45	2.28	2.08	1.84
	0.025	5.87	4.46	3.86	3.51	3.29	3.13	3.01	2.91	2.68	2.41	2.09
	0.010	8.10	5.85	4.94	4.43	4.10	3.87	3.70	3.56	3.23	2.86	2.42
	0.005	9.94	6.99	5.82	5.17	4.76	4.47	4.26	4.09	3.68	3.22	2.69
	0.001	14.82	9.95	8.10	7.10	6.46	6.02	5.69	5.44	4.82	4.15	3.38
30	0.100	2.88	2.49	2.28	2.14	2.05	1.98	1.93	1.88	1.77	1.64	1.46
	0.050	4.17	3.32	2.92	2.69	2.53	2.42	2.33	2.27	2.09	1.89	1.62
	0.025	5.57	4.18	3.59	3.25	3.03	2.87	2.75	2.65	2.41	2.14	1.79
	0.010	7.56	5.39	4.51	4.02	3.70	3.47	3.30	3.17	2.84	2.47	2.01
	0.005	9.18	6.35	5.24	4.62	4.23	3.95	3.74	3.58	3.18	2.73	2.18
	0.001	13.29	8.77	7.05	6.12	5.53	5.12	4.82	4.58	4.00	3.36	2.59

(continued)

Table A.3 (continued)

40	0.100	2.84	2.44	2.23	2.09	2.00	1.93	1.87	1.83	1.71	1.57	1.38
	0.050	4.08	3.23	2.84	2.61	2.45	2.34	2.25	2.18	2.00	1.79	1.51
	0.025	5.42	4.05	3.46	3.13	2.90	2.74	2.62	2.53	2.29	2.01	1.64
	0.010	7.31	5.18	4.31	3.83	3.51	3.29	3.12	2.99	2.66	2.29	1.80
	0.005	8.83	6.07	4.98	4.37	3.99	3.71	3.51	3.35	2.95	2.50	1.93
	0.001	12.61	8.25	6.59	5.70	5.13	4.73	4.44	4.21	3.64	3.01	2.23
60	0.100	2.79	2.39	2.18	2.04	1.95	1.87	1.82	1.77	1.66	1.51	1.29
	0.050	4.00	3.15	2.76	2.53	2.37	2.25	2.17	2.10	1.92	1.70	1.39
	0.025	5.29	3.93	3.34	3.01	2.79	2.63	2.51	2.41	2.17	1.88	1.48
	0.010	7.08	4.98	4.13	3.65	3.34	3.12	2.95	2.82	2.50	2.12	1.60
	0.005	8.49	5.79	4.73	4.14	3.76	3.49	3.29	3.13	2.74	2.29	1.69
	0.001	11.97	7.77	6.17	5.31	4.76	4.37	4.09	3.86	3.32	2.69	1.89
80	0.100	2.77	2.37	2.15	2.02	1.92	1.85	1.79	1.75	1.63	1.48	1.24
	0.050	3.96	3.11	2.72	2.49	2.33	2.21	2.13	2.06	1.88	1.65	1.32
	0.025	5.22	3.86	3.28	2.95	2.73	2.57	2.45	2.35	2.11	1.82	1.40
	0.010	6.96	4.88	4.04	3.56	3.26	3.04	2.87	2.74	2.42	2.03	1.49
	0.005	8.33	5.67	4.61	4.03	3.65	3.39	3.19	3.03	2.64	2.19	1.56
	0.001	11.67	7.54	5.97	5.12	4.58	4.20	3.92	3.70	3.16	2.54	1.72
100	0.100	2.76	2.36	2.14	2.00	1.91	1.83	1.78	1.73	1.61	1.46	1.21
	0.050	3.94	3.09	2.70	2.46	2.31	2.19	2.10	2.03	1.85	1.63	1.28
	0.025	5.18	3.83	3.25	2.92	2.70	2.54	2.42	2.32	2.08	1.78	1.35
	0.010	6.90	4.82	3.98	3.51	3.21	2.99	2.82	2.69	2.37	1.98	1.43
	0.005	8.24	5.59	4.54	3.96	3.59	3.33	3.13	2.97	2.58	2.13	1.49
	0.001	11.50	7.41	5.86	5.02	4.48	4.11	3.83	3.61	3.07	2.46	1.62
120	0.100	2.75	2.35	2.13	1.99	1.90	1.82	1.77	1.72	1.60	1.45	1.19
	0.050	3.92	3.07	2.68	2.45	2.29	2.18	2.09	2.02	1.83	1.61	1.25
	0.025	5.15	3.80	3.23	2.89	2.67	2.52	2.39	2.30	2.05	1.76	1.31
	0.010	6.85	4.79	3.95	3.48	3.17	2.96	2.79	2.66	2.34	1.95	1.38
	0.005	8.18	5.54	4.50	3.92	3.55	3.28	3.09	2.93	2.54	2.09	1.43
	0.001	11.38	7.32	5.78	4.95	4.42	4.04	3.77	3.55	3.02	2.40	1.54
∞	0.100	2.71	2.30	2.08	1.94	1.85	1.77	1.72	1.67	1.55	1.38	1.00
	0.050	3.84	3.00	2.60	2.37	2.21	2.10	2.01	1.94	1.75	1.52	1.00
	0.025	5.02	3.69	3.12	2.79	2.57	2.41	2.29	2.19	1.94	1.64	1.00
	0.010	6.63	4.61	3.78	3.32	3.02	2.80	2.64	2.51	2.18	1.79	1.00
	0.005	7.88	5.30	4.28	3.72	3.35	3.09	2.90	2.74	2.36	1.90	1.00
	0.001	10.83	6.91	5.42	4.62	4.10	3.74	3.47	3.27	2.74	2.13	1.00

Table A.4 Distribution function of T, the Wilcoxon signed-rank test

	Sample Size										
T_0	**2**	**3**	**4**	**5**	**6**	**7**	**8**	**9**	**10**	**11**	**12**
1	0.5000	0.2500	0.1250	0.0625	0.0313	0.0157	0.0079	0.0040	0.0020	0.0010	0.0005
2		0.3750	0.1875	0.0938	0.0469	0.0235	0.0118	0.0059	0.0030	0.0015	0.0008
3		0.6250	0.3125	0.1563	0.0782	0.0391	0.0196	0.0098	0.0049	0.0025	0.0013
4			0.4375	0.2188	0.1094	0.0547	0.0274	0.0137	0.0069	0.0035	0.0018
5			0.5625	0.3125	0.1563	0.0782	0.0391	0.0196	0.0098	0.0049	0.0025
6				0.4063	0.2188	0.1094	0.0547	0.0274	0.0137	0.0069	0.0035
7				0.5000	0.2813	0.1485	0.0743	0.0372	0.0186	0.0093	0.0047
8					0.3438	0.1875	0.0977	0.0489	0.0245	0.0123	0.0062
9					0.4219	0.2344	0.1250	0.0645	0.0323	0.0162	0.0081
10					0.5000	0.2891	0.1563	0.0821	0.0420	0.0210	0.0105
11						0.3438	0.1915	0.1016	0.0528	0.0269	0.0135
12						0.4063	0.2305	0.1250	0.0655	0.0337	0.0171
13						0.4688	0.2735	0.1504	0.0801	0.0416	0.0213
14						0.5313	0.3204	0.1797	0.0967	0.0508	0.0262
15							0.3711	0.2129	0.1163	0.0616	0.0320
16							0.4219	0.2481	0.1377	0.0738	0.0386
17							0.4727	0.2852	0.1612	0.0875	0.0462
18							0.5274	0.3262	0.1875	0.1031	0.0550
19								0.3672	0.2159	0.1202	0.0647
20								0.4102	0.2461	0.1392	0.0757

(continued)

Table A.4 (continued)

	Sample size										
T_0	2	3	4	5	6	7	8	9	10	11	12
21								0.4551	0.2784	0.1602	0.0882
22								0.5000	0.3125	0.1827	0.1019
23									0.3477	0.2066	0.1167
24									0.3848	0.2325	0.1331
25									0.4229	0.2598	0.1507
26									0.4610	0.2886	0.1697
27									0.5000	0.3189	0.1902
28										0.3501	0.2120
29										0.3824	0.2349
30										0.4156	0.2593
31										0.4493	0.2847
32										0.4830	0.3111
33										0.5171	0.3387
34											0.3667
35											0.3956
36											0.4251
37											0.4549
38											0.4849
39											0.5152

Table A.5 Distribution functions of W, the Wilcoxon rank sum test

	$n_2 = 3$		
W_0	$n_1 = 1$	**2**	**3**
1	0.25		
2	0.50		
3		0.10	
4		0.20	
5		0.40	
6		0.60	0.05
7			0.10
8			0.20
9			0.35
10			0.50

	$n_2 = 4$			
W_0	$n_1 = 1$	**2**	**3**	**4**
1	0.20			
2	0.40			
3	0.60	0.0667		
4		0.1333		
5		0.2667		
6		0.4000	0.0286	
7		0.6000	0.0571	
8			0.1143	
9			0.2000	
10			0.3143	0.0143
11			0.4286	0.0286
12			0.5714	0.0571
13				0.1000
14				0.1714
15				0.2429
16				0.3429
17				0.4429
18				0.5571

(continued)

Table A.5 (continued)

	$n_2 = 5$				
W_0	$n_1 = 1$	**2**	**3**	**4**	**5**
1	0.1667				
2	0.3333				
3	0.5000	0.0476			
4		0.0952			
5		0.1905			
6		0.2857	0.0179		
7		0.4286	0.0357		
8		0.5714	0.0714		
9			0.1250		
10			0.1964	0.0079	
11			0.2857	0.0159	
12			0.3929	0.0317	
13			0.5000	0.0556	
14				0.0952	
15				0.1429	0.0040
16				0.2063	0.0079
17				0.2778	0.0159
18				0.3651	0.0278
19				0.4524	0.0476
20				0.5476	0.0754
21					0.1111
22					0.1548
23					0.2103
24					0.2738
25					0.3452
26					0.4206
27					0.5000

(continued)

Table A.5 (continued)

	$n_2 = 6$					
W_0	$n_1 = 1$	**2**	**3**	**4**	**5**	**6**
1	0.1429					
2	0.2857					
3	0.4286	0.0357				
4	0.5714	0.0714				
5		0.1429				
6		0.2143	0.0119			
7		0.3214	0.0238			
8		0.4286	0.0476			
9		0.5714	0.0833			
10			0.1310	0.0048		
11			0.1905	0.0095		
12			0.2738	0.0190		
13			0.3571	0.0333		
14			0.4524	0.0571		
15			0.5476	0.0857	0.0022	
16				0.1286	0.0043	
17				0.1762	0.0087	
18				0.2381	0.0152	
19				0.3048	0.0260	
20				0.3810	0.0411	
21				0.4571	0.0628	0.0011
22				0.5429	0.0887	0.0022
23					0.1234	0.0043
24					0.1645	0.0076
25					0.2143	0.0130
26					0.2684	0.0206
27					0.3312	0.0325
28					0.3961	0.0465
29					0.4654	0.0660
30					0.5346	0.0898
31						0.1201
32						0.1548
33						0.1970
34						0.2424
35						0.2944
36						0.3496
37						0.4091
38						0.4686
39						0.5314

(continued)

Table A.5 (continued)

	$n_2 = 7$						
W_0	$n_1 = 1$	2	3	4	5	6	7
1	0.125						
2	0.250						
3	0.375	0.0278					
4	0.500	0.0556					
5		0.1111					
6		0.1667	0.0083				
7		0.2500	0.0167				
8		0.3333	0.0333				
9		0.4444	0.0583				
10		0.5556	0.0917	0.0030			
11			0.1333	0.0061			
12			0.1917	0.0121			
13			0.2583	0.0212			
14			0.3333	0.0364			
15			0.4167	0.0545	0.0013		
16			0.5000	0.0818	0.0025		
17				0.1152	0.0051		
18				0.1576	0.0088		
19				0.2061	0.0152		
20				0.2636	0.0240		
21				0.3242	0.0366	0.0006	
22				0.3939	0.0530	0.0012	
23				0.4636	0.0745	0.0023	
24				0.5364	0.1010	0.0041	
25					0.1338	0.0070	
26					0.1717	0.0111	
27					0.2159	0.0175	
28					0.2652	0.0256	0.0003
29					0.3194	0.0367	0.0006
30					0.3775	0.0507	0.0012
31					0.4381	0.0688	0.0020
32					0.5000	0.0903	0.0035
33						0.1171	0.0055
34						0.1474	0.0087
35						0.1830	0.0131
36						0.2226	0.0189
37						0.2669	0.0265
38						0.3141	0.0364
39						0.3654	0.0487
40						0.4178	0.0641
41						0.4726	0.0825
42						0.5274	0.1043
43							0.1297
44							0.1588
45							0.1914
46							0.2279
47							0.2675
48							0.3100
49							0.3552
50							0.4024
51							0.4508
52							0.5000

(continued)

Table A.5 (continued)

	$n_2 = 8$							
W_0	$n_1 = 1$	2	3	4	5	6	7	8
1	0.1111							
2	0.2222							
3	0.3333	0.0222						
4	0.4444	0.0444						
5	0.5556	0.0889						
6		0.1333	0.0061					
7		0.2000	0.0121					
8		0.2667	0.0242					
9		0.3556	0.0424					
10		0.4444	0.0667	0.0020				
11		0.5556	0.0970	0.0040				
12			0.1394	0.0081				
13			0.1879	0.0141				
14			0.2485	0.0242				
15			0.3152	0.0364	0.0008			
16			0.3879	0.0545	0.0016			
17			0.4606	0.0768	0.0031			
18			0.5394	0.1071	0.0054			
19				0.1414	0.0093			
20				0.1838	0.0148			
21				0.2303	0.0225	0.0003		
22				0.2848	0.0326	0.0007		
23				0.3414	0.0466	0.0013		
24				0.4040	0.0637	0.0023		
25				0.4667	0.0855	0.0040		
26				0.5333	0.1111	0.0063		
27					0.1422	0.0100		
28					0.1772	0.0147	0.0002	
29					0.2176	0.0213	0.0003	
30					0.2618	0.0296	0.0006	
31					0.3108	0.0406	0.0011	
32					0.3621	0.0539	0.0019	
33					0.4165	0.0709	0.0030	
34					0.4716	0.0906	0.0047	

(continued)

Table A.5 (continued)

	$n_2 = 8$ (continued)							
W_0	$n_1 = 1$	2	3	4	5	6	7	8
35					0.5284	0.1142	0.0070	
36						0.1412	0.0103	0.0001
37						0.1725	0.0145	0.0002
38						0.2068	0.0200	0.0003
39						0.2454	0.0270	0.0005
40						0.2864	0.0361	0.0009
41						0.3310	0.0469	0.0015
42						0.3773	0.0603	0.0023
43						0.4259	0.0760	0.0035
44						0.4749	0.0946	0.0052
45						0.5251	0.1159	0.0074
46							0.1405	0.0103
47							0.1678	0.0141
48							0.1984	0.0190
49							0.2317	0.0249
50							0.2679	0.0325
51							0.3063	0.0415
52							0.3472	0.0524
53							0.3894	0.0652
54							0.4333	0.0803
55							0.4775	0.0974
56							0.5225	0.1172
57								0.1393
58								0.1641
59								0.1911
60								0.2209
61								0.2527
62								0.2869
63								0.3227
64								0.3605
65								0.3992
66								0.4392
67								0.4796
68								0.5204

(continued)

Table A.5 (continued)

	$n_2 = 9$								
W_0	$n_1 = 1$	**2**	**3**	**4**	**5**	**6**	**7**	**8**	**9**
1	0.1000								
2	0.2000								
3	0.3000	0.0182							
4	0.4000	0.0364							
5	0.5000	0.0727							
6		0.1091	0.0045						
7		0.1636	0.0091						
8		0.2182	0.0182						
9		0.2909	0.0318						
10		0.3636	0.0500	0.0014					
11		0.4545	0.0727	0.0028					
12		0.5455	0.1045	0.0056					
13			0.1409	0.0098					
14			0.1864	0.0168					
15			0.2409	0.0252	0.0005				
16			0.3000	0.0378	0.0010				
17			0.3636	0.0531	0.0020				
18			0.4318	0.0741	0.0035				
19			0.5000	0.0993	0.0060				
20				0.1301	0.0095				
21				0.1650	0.0145	0.0002			
22				0.2070	0.0210	0.0004			
23				0.2517	0.0300	0.0008			
24				0.3021	0.0415	0.0014			
25				0.3552	0.0559	0.0024			
26				0.4126	0.0734	0.0038			
27				0.4699	0.0949	0.0060			
28				0.5301	0.1199	0.0088	0.0001		
29					0.1489	0.0128	0.0002		
30					0.1818	0.0180	0.0003		
31					0.2188	0.0248	0.0006		
32					0.2592	0.0332	0.0010		
33					0.3032	0.0440	0.0017		
34					0.3497	0.0567	0.0026		
35					0.3986	0.0723	0.0039		
36					0.4491	0.0905	0.0058	0.0000	
37					0.5000	0.1119	0.0082	0.0001	
38						0.1361	0.0115	0.0002	
39						0.1638	0.0156	0.0003	
40						0.1942	0.0209	0.0005	
41						0.2280	0.0274	0.0008	
42						0.2643	0.0356	0.0012	

(continued)

Table A.5 (continued)

W_0	$n_2 = 9$ (continued)								
	$n_1 = 1$	2	3	4	5	6	7	8	9
43						0.3035	0.0454	0.0019	
44						0.3445	0.0571	0.0028	
45						0.3878	0.0708	0.0039	0.0000
46						0.4320	0.0869	0.0056	0.0000
47						0.4773	0.1052	0.0076	0.0001
48						0.5227	0.1261	0.0103	0.0001
49							0.1496	0.0137	0.0002
50							0.1755	0.0180	0.0004
51							0.2039	0.0232	0.0006
52							0.2349	0.0296	0.0009
53							0.2680	0.0372	0.0014
54							0.3032	0.0464	0.0020
55							0.3403	0.0570	0.0028
56							0.3788	0.0694	0.0039
57							0.4185	0.0836	0.0053
58							0.4591	0.0998	0.0071
59							0.5000	0.1179	0.0094
60								0.1383	0.0122
61								0.1606	0.0157
62								0.1852	0.0200
63								0.2117	0.0252
64								0.2404	0.0313
65								0.2707	0.0385
66								0.3029	0.0470
67								0.3365	0.0567
68								0.3715	0.0680
69								0.4074	0.0807
70								0.4442	0.0951
71								0.4813	0.1112
72								0.5187	0.1290
73									0.1487
74									0.1701
75									0.1933
76									0.2181
77									0.2447
78									0.2729
79									0.3024
80									0.3332
81									0.3652
82									0.3981
83									0.4317
84									0.4657
85									0.5000

(continued)

Table A.5 (continued)

	$n_2 = 10$									
W_0	$n_1 = 1$	2	3	4	5	6	7	8	9	10
1	0.0909									
2	0.1818									
3	0.2727	0.0152								
4	0.3636	0.0303								
5	0.4545	0.0606								
6	0.5455	0.0909	0.0035							
7		0.1364	0.0070							
8		0.1818	0.0140							
9		0.2424	0.0245							
10		0.3030	0.0385	0.0010						
11		0.3788	0.0559	0.0020						
12		0.4545	0.0804	0.0040						
13		0.5455	0.1084	0.0070						
14			0.1434	0.0120						
15			0.1853	0.0180	0.0003					
16			0.2343	0.0270	0.0007					
17			0.2867	0.0380	0.0013					
18			0.3462	0.0529	0.0023					
19			0.4056	0.0709	0.0040					
20			0.4685	0.0939	0.0063					
21			0.5315	0.1199	0.0097	0.0001				
22				0.1518	0.0140	0.0002				
23				0.1868	0.0200	0.0005				
24				0.2268	0.0276	0.0009				
25				0.2697	0.0376	0.0015				
26				0.3177	0.0496	0.0024				
27				0.3666	0.0646	0.0037				
28				0.4196	0.0823	0.0055	0.0001			
29				0.4725	0.1032	0.0080	0.0001			
30				0.5275	0.1272	0.0112	0.0002			
31					0.1548	0.0156	0.0004			
32					0.1855	0.0210	0.0006			
33					0.2198	0.0280	0.0010			
34					0.2567	0.0363	0.0015			
35					0.2970	0.0467	0.0023			
36					0.3393	0.0589	0.0034	0.0000		
37					0.3839	0.0736	0.0048	0.0000		
38					0.4296	0.0903	0.0068	0.0001		
39					0.4765	0.1099	0.0093	0.0002		
40					0.5235	0.1317	0.0125	0.0003		
41						0.1566	0.0165	0.0004		
42						0.1838	0.0215	0.0007		
43						0.2139	0.0277	0.0010		
44						0.2461	0.0351	0.0015		
45						0.2811	0.0439	0.0022	0.0000	
46						0.3177	0.0544	0.0031	0.0000	
47						0.3564	0.0665	0.0043	0.0000	
48						0.3962	0.0806	0.0058	0.0001	
49						0.4374	0.0966	0.0078	0.0001	
50						0.4789	0.1148	0.0103	0.0002	
51						0.5211	0.1349	0.0133	0.0003	
52							0.1574	0.0171	0.0005	
53							0.1819	0.0217	0.0007	

(continued)

Table A.5 (continued)

	$n_2 = 10$ **(continued)**									
W_0	$n_1 = 1$	**2**	**3**	**4**	**5**	**6**	**7**	**8**	**9**	**10**
54							0.2087	0.0273	0.0011	
55							0.2374	0.0338	0.0015	0.0000
56							0.2681	0.0416	0.0021	0.0000
57							0.3004	0.0506	0.0028	0.0000
58							0.3345	0.0610	0.0038	0.0000
59							0.3698	0.0729	0.0051	0.0001
60							0.4063	0.0864	0.0066	0.0001
61							0.4434	0.1015	0.0086	0.0002
62							0.4811	0.1185	0.0110	0.0002
63							0.5189	0.1371	0.0140	0.0004
64								0.1577	0.0175	0.0005
65								0.1800	0.0217	0.0008
66								0.2041	0.0267	0.0010
67								0.2299	0.0326	0.0014
68								0.2574	0.0394	0.0019
69								0.2863	0.0474	0.0026
70								0.3167	0.0564	0.0034
71								0.3482	0.0667	0.0045
72								0.3809	0.0782	0.0057
73								0.4143	0.0912	0.0073
74								0.4484	0.1055	0.0093
75								0.4827	0.1214	0.0116
76								0.5173	0.1388	0.0144
77									0.1577	0.0177
78									0.1781	0.0216
79									0.2001	0.0262
80									0.2235	0.0315
81									0.2483	0.0376
82									0.2745	0.0446
83									0.3019	0.0526
84									0.3304	0.0615
85									0.3598	0.0716
86									0.3901	0.0827
87									0.4211	0.0952
88									0.4524	0.1088
89									0.4841	0.1237
90									0.5159	0.1399
91										0.1575
92										0.1763
93										0.1965
94										0.2179
95										0.2406
96										0.2644
97										0.2894
98										0.3153
99										0.3421
100										0.3697
101										0.3980
102										0.4267
103										0.4559
104										0.4853
105										0.5147

Table A.6 Percentiles of the chi-square distribution

	Area in Upper Tail				
df	**0.100**	**0.050**	**0.025**	**0.010**	**0.001**
1	2.71	3.84	5.02	6.63	10.83
2	4.61	5.99	7.38	9.21	13.82
3	6.25	7.81	9.35	11.34	16.27
4	7.78	9.49	11.14	13.28	18.47
5	9.24	11.07	12.83	15.09	20.52
6	10.64	12.59	14.45	16.81	22.46
7	12.02	14.07	16.01	18.48	24.32
8	13.36	15.51	17.53	20.09	26.12
9	14.68	16.92	19.02	21.67	27.88
10	15.99	18.31	20.48	23.21	29.59
11	17.28	19.68	21.92	24.72	31.26
12	18.55	21.03	23.34	26.22	32.91
13	19.81	22.36	24.74	27.69	34.53
14	21.06	23.68	26.12	29.14	36.12
15	22.31	25.00	27.49	30.58	37.70
16	23.54	26.30	28.85	32.00	39.25
17	24.77	27.59	30.19	33.41	40.79
18	25.99	28.87	31.53	34.81	42.31
19	27.20	30.14	32.85	36.19	43.82
20	28.41	31.41	34.17	37.57	45.31
21	29.62	32.67	35.48	38.93	46.80
22	30.81	33.92	36.78	40.29	48.27
23	32.01	35.17	38.08	41.64	49.73
24	33.20	36.42	39.36	42.98	51.18
25	34.38	37.65	40.65	44.31	52.62

APPENDIX

B

Data Sets

serzinc.raw

Table B.1 The variable `zinc`

50	70	74	77	81	85	88	93	97	103	109
51	70	74	77	81	85	88	93	97	103	110
53	70	74	77	81	85	88	93	97	103	110
55	70	74	77	81	85	88	93	97	103	111
56	70	74	77	82	85	88	93	97	103	111
58	70	74	77	82	85	88	93	97	103	111
60	70	74	77	82	85	89	93	97	104	111
60	70	74	78	82	85	89	93	98	104	112
60	71	74	78	82	85	89	94	98	104	112
61	71	74	78	82	85	89	94	98	104	112
61	71	74	78	82	85	89	94	98	104	113
61	71	75	78	82	86	89	94	98	104	113
61	71	75	78	82	86	89	94	98	104	113
62	71	75	78	82	86	89	94	98	105	114
62	71	75	78	82	86	89	94	98	105	114
62	71	75	78	82	86	90	94	98	105	114
62	71	75	78	82	86	90	94	99	105	114
63	71	75	78	83	86	90	94	99	105	115
63	71	75	78	83	86	90	94	99	105	115
63	72	75	79	83	86	90	95	99	105	115
64	72	75	79	83	86	90	95	99	105	116
64	72	75	79	83	86	90	95	99	106	116
64	72	75	79	83	86	90	95	100	106	116
64	72	75	79	83	86	91	95	100	106	116
64	72	75	79	83	87	91	95	100	106	117
65	72	75	80	83	87	91	95	100	106	117
65	72	76	80	83	87	91	95	100	106	117
65	72	76	80	83	87	91	95	101	107	118
66	72	76	80	84	87	91	95	101	107	118
66	72	76	80	84	87	91	95	101	107	119
66	72	76	80	84	87	91	96	101	107	119
67	72	76	80	84	87	91	96	101	107	121
67	73	76	80	84	87	91	96	101	107	123
67	73	76	80	84	87	92	96	101	107	124
67	73	76	80	84	88	92	96	102	108	125
67	73	76	80	84	88	92	96	102	108	128
68	73	76	81	84	88	92	96	102	108	131
68	73	76	81	84	88	92	96	102	108	135
68	73	76	81	84	88	92	96	102	108	142
68	73	77	81	84	88	92	96	102	108	147
69	73	77	81	84	88	92	97	102	108	151
70	74	77	81	84	88	93	97	102	109	153

Table B.2 The variables `nation` and `lowbwt`

Mozambique	20	Egypt	5	Mauritius	9
Afghanistan	20	Nicaragua	15	Chile	7
Angola	17	Myanmar	16	Yugoslavia	7
Mali	17	South Africa	12	Uruguay	8
Sierra Leone	17	El Salvador	15	Costa Rica	10
Malawi	20	Turkey	7	Jamaica	8
Niger	15	Zimbabwe	15	Kuwait	7
Central African Rep.	15	Iraq	9	Poland	8
Mauritania	11	Botswana	8	Bulgaria	6
Rwanda	17	Mongolia	10	Hungary	10
Burundi	9	Brazil	8	Portugal	5
Senegal	11	Ecuador	11	Cuba	8
Bangladesh	28	Viet Nam	18	Czechoslovakia	6
Madagascar	10	Papua New Guinea	25	Belgium	5
Tanzania	14	Dominican Rep.	16	Greece	6
Nigeria	20	Philippines	18	Israel	7
Bolivia	12	Tunisia	8	New Zealand	5
Pakistan	25	Iran	5	USA	7
Laos	39	Syria	11	Singapore	7
Benin	8	Paraguay	7	Ireland	4
Cameroon	13	Lebanon	10	Italy	7
Togo	20	Jordan	5	Austria	6
India	30	Oman	7	Denmark	6
Ghana	17	Mexico	15	Germany, Fed.	6
Cote d'Ivoire	14	Colombia	8	Norway	4
Haiti	17	Venezuela	9	Spain	1
Lesotho	11	China	9	United Kingdom	7
Zaire	13	Sri Lanka	28	Australia	6
Zambia	14	Thailand	12	Canada	6
Peru	9	USSR	6	France	5
Congo	16	Romania	6	Germany, Rep.	6
Kenya	15	Panama	8	Switzerland	5
Honduras	20	United Arab Emirates	7	Hong Kong	5
Algeria	9	Korea, Rep.	9	Finland	4
Indonesia	14	Albania	7	Sweden	4
Guatemala	14	Malaysia	10	Japan	5
Saudi Arabia	6				

Table B.3 The variables `sbp`, `duration`, and `time`

170	640	105	122	438	252	168	780	440
128	670	118	158	270	270	134	1,065	480
150	560	130	132	781	277	160	1,080	540
148	510	150	130	802	278	154	308	562
160	212	178	175	360	300	125	860	570
154	260	180	140	1,775	300	140	60	578
175	228	192	150	524	322	154	765	720
140	335	200	180	441	330	120	396	729
140	460	200	160	1,084	345	162	328	780
120	440	201	122	505	360	140	540	1,200
172	210	240	170	75	375	178	130	1,430
178	359	245	150	823	386			

Table B.4 The variables `bed80`, `bed86`, and `state`

4.7	4.2	1	5.0	4.7	35	7.4	7.2	18
4.4	4.0	3	4.6	4.0	37	6.0	5.9	20
3.8	3.5	5	5.9	5.7	39	3.6	3.4	22
4.5	4.3	7	3.6	4.4	41	7.3	7.7	24
4.8	4.6	9	3.1	2.9	43	5.5	5.1	26
4.5	4.2	11	3.1	2.7	45	3.9	3.4	28
4.4	4.0	13	3.1	3.0	47	5.1	4.4	30
5.7	5.0	15	3.6	3.1	49	5.5	5.3	32
5.7	5.1	17	3.1	2.6	51	5.3	5.2	34
5.5	6.5	19	3.9	3.3	2	4.8	4.5	36
5.8	5.0	21	4.4	4.3	4	4.7	3.8	38
3.6	3.3	23	3.5	3.1	6	3.7	3.3	40
4.1	3.7	25	4.2	3.9	8	4.2	3.5	42
4.2	3.5	27	4.7	4.4	10	3.6	3.1	44
4.6	4.3	29	5.1	4.5	12	4.2	3.7	46
4.5	4.4	31	4.9	4.5	14	3.5	3.1	48
5.1	5.0	33	5.7	5.2	16	2.7	2.4	50

Table B.5 The variables `age` and `center`

60	1	73	1	65	3
66	1	56	2	61	3
65	1	65	2	63	3
55	1	65	2	59	3
62	1	63	2	42	3
70	1	57	2	53	3
51	1	47	2	63	3
72	1	72	2	65	3
58	1	56	2	60	3
61	1	52	2	57	3
71	1	75	2	62	3
41	1	66	2	70	3
70	1	62	2	73	3
57	1	68	2	63	3
55	1	75	2	55	3
63	1	60	2	52	3
64	1	73	2	58	3
76	1	63	2	68	3
74	1	64	2	70	3
54	1	67	3	72	3
58	1	56	3	45	3

Table B.6 The variables `hiv` and `ivdu` (This data set consists of a large number of repeated observations; rather than display them all, we summarize the outcomes.)

prison.raw

61	pairs "positive yes" followed by
27	pairs "positive no" followed by
75	pairs "negative yes" followed by
312	pairs "negative no"

Table B.7 The variables fluoride and caries

1.9	236	.3	652
2.6	246	0	673
1.8	252	.2	703
1.2	258	.1	706
1.2	281	0	722
1.2	303	.2	733
1.3	323	.1	772
.9	343	0	810
.6	412	.1	823
.5	444	.1	1,037
.4	556		

Table B.8 The variables length and weight

41	1,360	32	880
40	1,490	39	1,130
38	1,490	38	1,140
38	1,180	39	1,350
38	1,200	37	950
32	680	39	1,220
33	620	38	980
38	1,060	42	1,480
30	1,320	39	1,250
34	830	38	1,250

Table B.9 The variables rate and year

.2419	1	.0904	7
.1732	2	.0792	8
.1361	3	.0701	9
.1108	4	.0891	10
.0996	5	.0799	11
.0952	6	.1084	12

Table B.10 The variables `sbp`, `sex`, `tox`, `grmhem`, `gestage`, and `apgar5`

43	Male	No	No	29	7	26	Female	No	No	23	8
51	Male	No	No	31	8	27	Male	No	No	27	9
42	Female	No	No	33	0	35	Male	No	No	28	8
39	Female	No	No	31	8	40	Male	No	No	27	7
48	Female	Yes	No	30	7	44	Female	No	No	27	6
31	Male	No	Yes	25	0	66	Male	No	No	26	8
31	Male	Yes	No	27	7	59	Female	No	No	25	3
40	Female	No	No	29	9	24	Female	No	No	23	7
57	Female	No	No	28	6	40	Female	No	Yes	26	3
64	Female	Yes	No	29	9	49	Female	No	No	24	5
46	Female	No	No	26	7	53	Male	Yes	No	29	9
47	Female	No	Yes	30	6	45	Female	No	No	29	9
63	Female	No	No	29	8	50	Male	No	Yes	27	8
56	Female	No	No	29	1	64	Male	No	No	30	7
49	Male	No	No	29	8	48	Female	No	No	30	6
87	Male	No	No	29	7	48	Female	No	Yes	32	4
46	Female	No	No	29	8	58	Female	Yes	No	33	7
66	Female	No	No	33	8	67	Female	No	No	27	8
42	Female	Yes	No	33	8	40	Female	No	Yes	31	8
52	Female	No	No	29	7	48	Female	No	No	26	8
51	Male	No	No	28	7	36	Male	No	No	27	5
47	Female	No	No	30	9	44	Male	No	No	27	6
54	Male	No	No	27	4	53	Female	Yes	No	35	9
64	Male	No	No	33	9	45	Female	Yes	No	28	6
37	Female	No	No	32	7	54	Male	No	No	30	8
36	Male	Yes	No	28	3	44	Male	Yes	No	31	2
45	Female	No	Yes	29	7	42	Male	No	No	30	5
39	Male	No	No	28	7	50	Female	No	No	27	0
29	Female	No	No	29	4	48	Female	No	No	25	5
61	Female	No	No	30	3	29	Female	No	Yes	25	5
53	Male	No	No	31	7	30	Female	No	Yes	26	2
64	Female	No	No	30	7	36	Female	No	No	29	0
35	Female	No	No	31	6	44	Female	No	No	29	0
34	Male	No	No	29	9	46	Female	Yes	No	34	9
62	Female	No	No	27	7	51	Male	Yes	No	30	4
59	Female	No	No	27	8	51	Male	No	No	29	5
36	Male	No	No	27	9	43	Male	Yes	No	33	7
47	Female	No	No	32	8	48	Male	No	No	30	5
45	Male	No	Yes	31	2	52	Male	No	No	29	8
62	Female	No	Yes	28	5	43	Male	No	No	24	6
75	Male	Yes	No	30	7	42	Male	Yes	No	33	8
44	Male	No	No	29	0	48	Male	No	Yes	25	5
39	Male	No	No	28	8	49	Female	Yes	No	32	8
48	Female	No	Yes	31	7	62	Male	Yes	No	31	7
43	Female	Yes	No	27	6	45	Male	No	No	31	9
19	Female	No	Yes	25	4	51	Female	Yes	Yes	31	6
63	Male	No	No	30	7	52	Male	No	No	29	8
42	Male	No	No	28	6	47	Male	Yes	No	32	5
44	Female	No	No	28	9	40	Female	Yes	No	33	8
25	Female	No	No	25	8	50	Female	No	No	28	7

Table B.11 The variables `time`, `censor`, and `group`

Cyto.raw

11	1	1
26	1	1
35	1	1
60	1	1
89	1	1
101	1	1
126	1	1
142	1	1
149	1	1
191	1	1
204	1	1
213	1	1
229	1	1
261	1	1
362	1	1
368	0	1
387	0	1
400	0	1
1	1	2
1	1	2
1	1	2
1	1	2
16	1	2
47	1	2
61	1	2
82	1	2
90	1	2
121	1	2
162	1	2

Table B.12 The variables `state`, `expadm`, `los`, and `salary`

Connecticut	3328	7.8	16957	Florida	2780	7.5	14840
Maine	2810	7.7	14317	Georgia	2156	6.7	12923
Massachusetts	4105	8.9	15696	Kentucky	1974	6.7	13504
New Hampshire	2487	7.1	13542	Louisiana	2412	6.3	14169
Rhode Island	3380	8.5	14947	Mississippi	1772	7.0	11928
Vermont	2478	8.2	13432	North Carolina	2196	7.5	13070
Delaware	2957	8.2	15717	South Carolina	2134	7.4	12973
Washington, D.C.	4612	8.7	18700	Tennessee	2176	7.2	12923
Maryland	3210	8.3	15213	Virginia	2575	7.9	14107
New York	2712	8.4	15573	West Virginia	2248	7.2	13731
New Jersey	3607	9.7	16657	Arizona	3101	6.5	15307
Pennsylvania	3194	8.5	15256	New Mexico	2626	6.7	14573
Illinois	3351	8.0	16872	Oklahoma	2425	6.5	13916
Indiana	2592	7.8	13946	Texas	2272	6.6	13847
Michigan	3351	8.0	16635	Colorado	2782	7.1	16618
Ohio	3007	8.1	15492	Idaho	2072	6.5	13211
Wisconsin	2724	8.3	14951	Montana	2152	8.3	14635
Iowa	2361	8.0	13579	Utah	2384	5.4	14192
Kansas	2600	7.8	14107	Wyoming	2009	5.5	14686
Minnesota	2730	9.4	14503	California	3886	6.5	18611
Missouri	2915	8.0	14950	Nevada	3500	6.3	17898
Nebraska	2448	8.4	13584	Oregon	2660	5.9	15578
North Dakota	2277	8.6	13559	Washington	2524	5.7	15617
South Dakota	2059	8.7	12508	Alaska	3633	5.9	23594
Alabama	2174	7.2	12834	Hawaii	2780	8.3	14758
Arkansas	1859	6.6	12737				

APPENDIX C

Solutions to Selected Exercises

Chapter 2

1a. The number of suicides is discrete.

1b. The concentration of lead is continuous.

2b. Histogram for the 23 values of per capita health care expenditure:

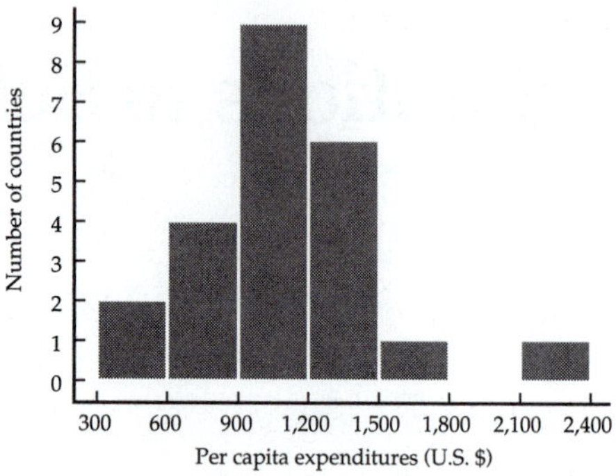

Health care expenditures, 1989

5a. The workers have lower blood lead levels in 1987.

5b. Cumulative relative frequencies of blood lead concentrations for the two groups of Canadian workers:

Blood Lead (μg/dl)	**1979 (%)**	**1987 (%)**
<20	11.5	37.8
20–29	23.6	52.5
30–39	37.5	65.6
40–49	52.9	80.9
50–59	69.4	91.4
60–69	82.2	98.2
70–79	90.6	99.6
≥80	100.0	100.0

Chapter 3

1a. i. The mean is 25.9 months.

ii. The median is 24 months.

iii. The modes are 12 months and 24 months.

iv. The range is 95.9 months.

v. The standard deviation is 27.4 months.

4a. The grouped mean of these data is 149.79 mm Hg; the grouped standard deviation is 19.32 mm Hg.

4b. The ungrouped mean is 150.26 mm Hg; the ungrouped standard deviation is 18.66 mm Hg.

Chapter 4

4a. The crude cancer mortality rate for 1940 is 120.2 per 100,000 population; the crude rate for 1986 is 194.7 per 100,000 population. The crude death rate is considerably higher in 1986 than it is in 1940. In fact, it increases 62% in 46 years.

4e. If we use the direct method of standardization, the age-adjusted cancer mortality rate for 1940 is 120.2 per 100,000 population, and the age-adjusted rate for 1986 is 133.9 per 100,000 population.

4j. If we use the indirect method of standardization, the age-adjusted cancer mortality rate for 1940 is 120.2 per 100,000 population, and the age-adjusted rate for 1986 is 137.0 per 100,000 population. The results obtained using the indirect method do correspond to those obtained when the direct method was applied; in both cases, the age-adjusted death rate for 1986 is somewhat higher than the age-adjusted rate for 1940. We do not expect the results to be exactly the same.

Chapter 5

1b. Columns (7) and (8) of the life table:

(7) Total Person-Years Outside Institutions from Age x On	(8) Average Remaining Years Outside of Institutions
7,005,116	70.1
5,532,705	56.5
4,562,720	47.3
2,671,609	28.9
982,595	13.6

4a. The relevant probabilities are listed below.

	1909–1911	1969–1971
P(0 to 15)	0.831	0.979
P(15 to 45)	0.834	0.967
P(45 to 75)	0.383	0.669

5a. The probability of surviving from birth until age 80 is 0.460.

5b. If an individual reaches his or her 50th birthday, the probability of surviving to age 80 is 0.514.

Chapter 6

2a. The probability that the woman was 24 years of age or younger is 0.4065.

2b. The probability that she was 40 or older is 0.0095.

2c. Given that she is under 30 years of age, the probability that she is not yet 20 is 0.1710.

4a. The probability that an individual who is 60 years of age will survive for ten years is 0.826.

4b. The probability that both the woman and her husband are alive at age 70 is 0.682.

4c. The probability that either the woman or her husband but not both is alive at age 70 is 0.287.

6a. The predictive value of a positive test result is 0.220.

7b. The probability that an individual has coronary artery disease given that he or she tests positive is 0.283.

Chapter 7

2a. The ten persons can be ordered in 3,628,800 different ways.

2b. Four individuals can be selected from the group of ten in 210 different ways.

2c. The probability that exactly three of the ten persons are left-handed is 0.0549.

5a. The probability that exactly one case of tetanus will be reported is 0.0500.

5b. The probability that at most two cases will be reported is 0.1736.

8a. Approximately 7.1% of the women have a diastolic blood pressure less than 60 mm Hg.

8b. About 13.1% have a diastolic blood pressure greater than 90 mm Hg.

10a. The probability of predicting coronary heart disease for a man who will never develop it is 0.159.

10b. The probability of failing to predict disease for a man who will develop it is 0.622.

Chapter 8

2a. Approximately 0.4% of the means of samples of size 20 lie above 35 mg/100 ml.

2b. Virtually none of the means—approximately 0.0%—lie below 22 mg/100 ml.

3a. Roughly 70.2% of the samples of size 15 will have a mean hemoglobin level between 13.0 and 13.6 g/100 ml.

Chapter 9

1a. A two-sided 95% confidence interval for μ_s is (122.7, 137.3).

1c. A two-sided 90% confidence interval for μ_d is (79.3, 88.7).

3a. A 95% confidence interval for the true mean length of stay is (7.6, 11.4).

3b. The length of the interval is 3.8 days.

Chapter 10

1a. The null hypothesis of the test is $H_0: \mu = 74.4$ mm Hg.

1b. The alternative hypothesis is $H_A: \mu \neq 74.4$ mm Hg.

1c. The test statistic is $z = 3.34$ and $p < 0.001$.

1d. We conclude that the mean diastolic blood pressure for the population of female diabetics between the ages of 30 and 34 years is not equal to 74.4 mm Hg; in fact, it is significantly higher.

4b. The probability of committing a type II error is 0.081.

4e. A sample of size 30 would be required.

Chapter 11

1b. The test statistic is $t = 3.35$ and $0.001 < p < 0.01$. We reject H_0 at the 0.05 level of significance.

1c. We conclude that the true difference in population means is not equal to 0; the mean LDL cholesterol level is lower for individuals adhering to the oat bran diet.

4a. The test statistic is $t = 0.86$, and the p-value is greater than 0.10. Therefore, we are unable to reject H_0 at the 0.05 level of significance. We conclude that the two populations of women may indeed have the same mean arterial blood pressure.

4b. A 90% confidence interval for the true difference in population means is
$(-1.93, 5.93)$.

Chapter 12

2a. The within-groups estimate of the variance is 154.88 $(\text{mm Hg})^2$.

2b. The between-groups estimate of the variance is 448.98 $(\text{mm Hg})^2$.

2c. The test statistic is $F = 2.90$, and $0.025 < p < 0.05$. We reject H_0 at the 0.05 level of significance.

2d. We conclude that the mean systolic blood pressures of the four groups are not identical; at least one of the means differs from another.

Chapter 13

1a. The test statistic is $z_{T^+} = 2.69$, and $p = 0.008$. Alternatively, $T^+ = 84$, and, from the table of critical values, $p < 0.01$. We reject H_0 and conclude that the median difference is not equal to 0. Resting energy expenditure is higher among patients with cystic fibrosis.

1b. The results are the same as those obtained when the sign test was used.

4a. The test statistic is $z_W = -3.31$, and $p < 0.001$. Therefore, we reject the null hypothesis at the 0.05 level of significance.

4b. It appears that adolescents with bulimia do require a lower daily caloric intake than healthy adolescents.

Chapter 14

3a. A 95% confidence interval for the population proportion p is (0.368, 0.743).

3d. The test statistic is $z = 2.52$, and $p = 0.0118$. Although it is extremely close, we are unable to reject H_0 at the 0.01 level of significance.

3e. Based on this sample, we are unable to conclude that the proportion of mothers who smoked during pregnancy for children born with an oral cleft differs from the proportion of mothers who smoked for children with other types of malformations.

Chapter 15

2a. The test statistic is $X^2 = 54.92$, and $p < 0.001$. Therefore, we reject the null hypothesis at the 0.05 level of significance.

2b. There does appear to be an association between driving while drinking and calendar year; a smaller proportion of college students were drinking while driving in 1987 than in 1983.

6a. The test statistic is $X^2 = 8.10$, and $0.001 < p < 0.01$. We reject H_0 at the 0.05 level of significance; these data suggest that the diagnosis is more likely to be "present" at the first screening and "absent" at the second than it is to be "absent" at the first screening and "present" at the second.

9a. The estimated relative odds of invasive cervical cancer are 1.52.

9c. The test statistic is $X^2 = 5.95$, and $0.01 < p < 0.05$. At the 0.05 level of significance, we reject H_0 and conclude that the odds of developing cervical cancer among smokers are greater than the odds among nonsmokers.

Chapter 16

1a. For women who have had at most one sexual partner, the estimated odds ratio is 2.70.

1b. For women who have had two or more sexual partners, the estimated odds ratio is 1.10.

1d. The test statistic is $X^2 = 3.76$, and $0.05 < p < 0.10$. We cannot reject H_0 at the 0.05 level of significance; therefore, we may proceed with the Mantel-Haenszel procedure.

1e. The estimate of the summary odds ratio is 1.26.

Chapter 17

1a. Two-way scatter plot of cholesterol level versus triglyceride level (see figure on p. 519).

1c. The Pearson correlation coefficient is $r = 0.6497$.

2c. The Pearson correlation coefficient is $r = -0.0737$.

2d. The duration of angina tends to decrease as time to angina increases.

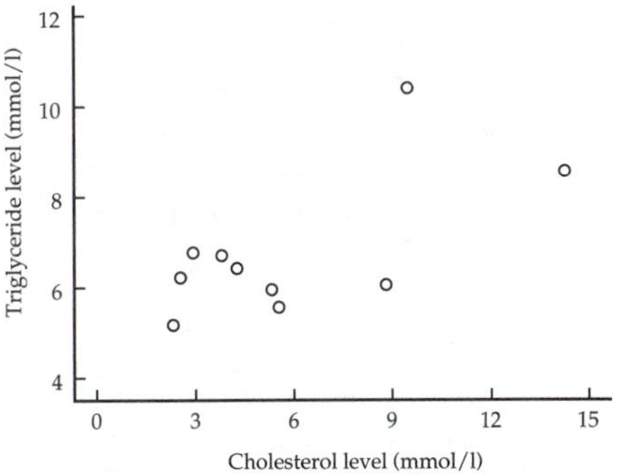

Triglyceride versus cholesterol levels

Chapter 18

1c. The least squares regression line is $\hat{y} = 10.552 + 1.264x$. The slope of the line implies that, for each one-week increase in gestational age, an infant's systolic blood pressure increases by 1.264 mm Hg on average. The y-intercept of 10.552 is the predicted systolic blood pressure for an infant whose gestational age is 0; note that this does not make sense in this case.

1d. The test statistic is $t = 2.898$, and $p = 0.005$. We reject H_0 and conclude that, for low birth weight infants, systolic blood pressure increases in magnitude as gestational age increases.

1e. For the population of infants whose gestational age is 31 weeks, the estimated mean systolic blood pressure is 49.75 mm Hg.

1f. A 95% confidence interval for the true mean value of systolic blood pressure is (40.11, 46.74).

2c. The least squares regression line is $\hat{y} = -760.52 + 51.24x$.

2e. The new least squares regression line is $\hat{y} = -1864.27 + 79.89x$.

Chapter 19

1a. The explanatory variables diet group and baseline cholesterol have a significant effect on serum cholesterol level.

2c. The least squares regression model is $\hat{y} = 9.803 + 1.185x_1 + 0.488x_2$. The estimated coefficient of x_2 suggests that, if gestational age remains constant while five-minute apgar score increases by one point, then an infant's systolic blood pressure increases by 0.488 mm Hg on average.

2f. The test statistic is $t = 1.057$, and $p = 0.293$. Therefore, we are unable to reject the null hypothesis that the population parameter is equal to 0.

3a. The male would tend to have the higher systolic blood pressure.

3c. There does not appear to be a different relationship between systolic blood pressure and gestational age for males as opposed to females; the coefficient of sex is not significantly different from 0.

Chapter 20

1a. At the 0.05 level of significance, the explanatory variables gender and HIV knowledge influence intravenous drug use in the past year.

2a. The probability decreases.

3a. The fitted model is $\ln[\hat{p}/(1-\hat{p})] = -0.304 - 0.250x_1$.

3b. If a child has a five-minute apgar score of 3, the estimated probability that this infant will experience a germinal matrix hemorrhage is 0.259.

3e. The estimated odds of suffering a germinal matrix hemorrhage are 0.232.

Chapter 21

1a. The median survival time is 12.75 months.

1c. The life table is longitudinal.

2c. Survival curve for ten AIDS patients suffering from concomitant esophageal candidiasis and cytomegalovirus:

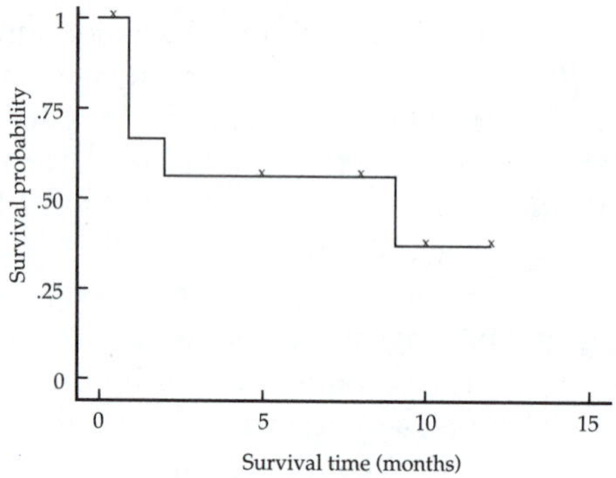

Data for 10 AIDS patients

Index